The Modeling Process in Geography

The Modeling Process in Geography

From Determinism to Complexity

Edited by

Yves Guermond

ISTE &(W)WILEY

First published in France in 2005 by Hermes Science/Lavoisier entitled: "Modélisations en géographie : déterminismes et complexités"
First published in Great Britain and the United States in 2008 by ISTE Ltd and John Wiley & Sons, Inc.

ISTE Ltd
27-37 St George's Road
London SW19 4EU
UK

John Wiley & Sons, Inc.
111 River Street
Hoboken, NJ 07030
USA

www.iste.co.uk

www.wiley.com

Library of Congress Cataloging-in-Publication Data

Modélisations en géographie. English.
  The modeling process in geography : from determinism to complexity / edited by Yves Guermond.
     p. cm.
  "First published in France in 2005 by Hermes Science/Lavoisier entitled: 'Modélisations en géographie : déterminismes et complexités'."
  Includes bibliographical references and index.
  ISBN 978-1-84821-087-5
  1. Regional planning. 2. Human geography. 3. Geographic information systems. I. Guermond, Yves. II. Title.
  HT391.M625513 2008
  307.1'2--dc22

                                                                                            2008035145

British Library Cataloguing-in-Publication Data
A CIP record for this book is available from the British Library
ISBN: 978-1-84821-087-5

Printed and bound in Great Britain by CPI Antony Rowe Ltd, Chippenham, Wiltshire.

FSC
Mixed Sources
Product group from well-managed forests and other controlled sources
Cert no. SGS-COC-2953
www.fsc.org
© 1996 Forest Stewardship Council

Table of Contents

Foreword

The Taste for Measuring and Modeling

This book gives me the opportunity to reflect upon the reasons behind my complex relationship with mathematics, as well as the frustrations revealed by an insufficient deepening of the relationship between, on the one hand, the human and "interdisciplinary" geography that I practice, and, on the other hand, a question that remains fundamental: which models for which science?

The evidence of a taste…

Before analyzing the reasons that make me think that my use of the scientific practice referred to as "modeling" – of which I would have liked so much to be a part[1] – has been insufficient, I must recall a few steps along my career that testify to this interest in quantification, that is, the concept of models and modeling.

First come the taste and need for measurements in order to identify facts and geographical processes, and to test the hypotheses used to understand them. Despite Hubert Béguin's reply of "But how do you measure it!", when I enthusiastically told about the content of my talk for the *VIIth European Colloquium on Theoretical and Quantitative Geography* in September 1991 in Hasseludden (Sweden), from the beginning, and guided by Ernest Labrousse, I have known that one way or another

Written by Nicole MATHIEU, Emeritus Research Director, CNRS, Paris.
1 In particular when Alain Pavé started, at the beginning of the 1990s, a special program in the Programme Environment of the CNRS, "Method Models, Theories", whose results (along with others) led to a conference in 1996 *Tendances nouvelles en modélisation pour l'environnement*, Paris, CNRS, Actes des journées du Programme Environnement, Vie et Sociétés.

"everything can be measured" and must be for an assertion to be credible. Measuring is also at the center of politics, as attested by the responsibility given by Napoleon to the two scientists Chaptal and Montalivet, for collating the agricultural statistics of France.

The *Tableaux de l'agriculture française* with its previously unpublished maps (1966, 1968), the attempt to measure the degrees of urbanization at the scale of the *départements* (1973), the *arrondissements* (1971) and of a sample of undefined *districts* between rural and urban (1974), are proof of the early use of mathematics in building clever indicators revealing the unequal spatial distribution of situations evaluated through the complex indicators of density, urban frame and the combined dynamics of the demographic components of rural *districts*. During those years, following what I had been taught by Ernest Labrousse and then Pierre Coutin (2001), the link between mathematics and politics became obvious as an ordinary experience of social sciences researchers. This is how, attempting to translate Pierre Coutin's prospective vision regarding the ways in which to modernize French agriculture while respecting the local and regional farming communities, Jean-Claude Bontron and myself have used standard deviation in order to theoretically calculate the "technically necessary agricultural population" in France, for each *département*, and to suggest to the Commissariat Général du Plan the objective of reducing the active agricultural population in equal proportion, calculated in relation to the level of overpopulation reached in each *département*.

You could say, as my colleague Denise Pumain did at the end of the 1970s, that I am "of the pioneer pre-quantitative generation", because "it is not by using measurements, as cleverly as this may be done, that geography is *theoretical and quantitative*, but that it can rather be identified by what is defined by the term spatial analysis!"

Why was my way of using mathematics not theoretical? Was it because of an overly "applied" approach, as was said then? Yet, in my research on low-density areas, and still with the complicity of J.C. Bontron and Lucette Vélard, whose skills in statistical processing, multivariate analyses, and the ascending hierarchical classification method never ceased to grow, our aim was to test hypotheses concerning the functioning of these areas as a spatial system and to build a theory of the dynamics of the "reverse side" of urbanization processes! While these studies underlined measured spatial discontinuities[2] (and were not ideological, as in *La France du vide*), at the same time as they highlighted (as pioneers, and going against the prevailing analysis techniques of the day) the fact that this level of organization

---

2 See Map "Zones des faibles densités et écarts de densité avec les régions voisines" in Bontron, J.C., Mathieu, N., 1977. *La France des Faibles Densités, Délimitation Problèmes Typologies*, Paris, ACEAR/Segesa, p. 32.

and spatial structuring was not dependent on demographic evolution[3], why was it that they not enter the canons of the theoretical and quantitative geography being built then.

Was it a matter of the cultural relationship with mathematics? Whatever the importance I granted to the dimension of data analysis in examining causes and effects hierarchically, it is true that I never used it exclusively. As was the case for the generation of which I was part, I had to distance myself with mathematical reasoning (the concept of modeling) and its top-down application to social and spatial facts. Above all, I had to confront what quantitative analysis proved in terms of what could be called the level of experiences, as suggested to me by P. Coutin, referring to Leplay, that is the experience of the complex object that are local monographs, field studies as models of a relationship system between populations and territories, between societies and living environments. This was probably the weakness in the eyes of Theoquant geographers looking for a science and purified spatial laws in the field.

Furthermore, in the 1980s, I sometimes used the concept of a model in a sense that diverted it from the mathematical or physical model. Among the various meanings of this term, I found it efficient and relevant to use the terms "prototype", "object to imitate" or "exemplary". Thus, the various situations of rural development politics at the local level that I had been observing from the 1970s until the end of the 1980s (rural planning schemes, national, then regional *pays* contracts), always complicated to analyze, appeared as being part of either one of two "models", one based more on centrality and spatial equity, and the other "local", i.e. giving free rein to the specific social dynamics of a territory. In this, it was both a model for analysis (for the researcher) and a model for action (for the politician). Once again, this multiple usage of the term and the incongruity of a quantified translation of this type of model led me away from the hard core of theoretical and quantitative geography.

Modeling as a necessity…

However, I was never discouraged, and the issue of method, the necessity of models and modeling to the study of complex objects has been a recurring motif of my research in the 1990s, when it took a clear turn towards environmental concerns. What was maybe only an "opinion", or rather a "certainty", then became the awareness of a necessity. The decision I made to research "complex objects" that

---

3 Hence the notice taken, as early as 1975, of the reversal of the century-old tendency to exodus and depopulation in rural districts, as well as highlighting the importance of non-agricultural activities and jobs, and of the new living practices.

"cannot be decomposed and made simple without being modified, and their nature transformed, by the reductions used" is a decision that necessarily entails interdisciplinarity. The social issue of the environment reactivates the paradigm of society/nature relationships and requires the modeling of interactions between social systems, natural systems and technico-political systems within the complex object that is an environmental problem. In this case, the modeling can only be local, which, translated into geographical language, means that the identification of the relationships of the complex system is only valuable in their strict co-localization. Thus, the two epistemological requirements of geography linked to environmental issues: the revival of what I have called "inner interdisciplinarity", meaning 1) a work articulated between physical geography and human geography; 2) the modeling *in situ* of processes with distinct natures and times specific to this type of object. Hence also the importance of tools such as the geo-referencing with a constant grid and GIS, or the imperative, for all disciplines, to work on the same microsite.

From the Observatory for ecological, economic and social changes in Causse/Cévennes which Marcel Jollivet was in charge of and in which I was in charge of coordinating the teams for Causse Méjan, to the Méjan Observatory that followed that first PIREN program, and then during the PEVS program "Co-evolutions of the dynamics of the natural environment and the society of Méjan cattle breeders: the bush progression" coordinated by Marianne Cohen, I have never ceased to assert, as did the whole group of border crossers (we must keep in mind that Jean-Marie Legay was the leader of this group on the natural sciences side) the absolute necessity of using all the methods and tools of modeling and GIS to study these crossover issues between social and natural sciences, while advocating internal interdisciplinarity in geography, that is the re-articulation of the systemic knowledge possessed by physical geographers with that of spatial analysis geographers that was, at the time, more widely used in human geography. It was obvious and I was certain that, whatever method was used to build the models, be it a deductive method (a theory → a model → a situation) as used by mathematicians, physicists, biophysicists, or even chemists and some geographers, or an ascending method (a situation → a model → a theory) for which agronomists and physicians know the difficulties linked to the constraining hypotheses imposed by the situation, and which I preferred due to my attachment to the field, it was truly the back and forth movement between model and field, "this to-and-fro between model and experimentation", which is the core of the method used to highlight the functioning of a complex geographic object at the boundary of physical and social systems.

Reasons for dissatisfaction and incompletion...

However, there then crept into my research practice, subtly but inevitably, a dissociation between what I was expecting others to do, in particular young researchers I oversee or those who are part of research collectives that I am a part of, and what I would do myself, thereby leading me further and further away from mathematical and modeling skills. In other words, while I am convinced that, in order to be heuristic, the geography that studies the urban environment, risk management, territory sustainability – be it sustainable cities or neighborhoods, or agricultural systems or rural and periurban territories – must be both model-dependent and multidisciplinary, I myself tend more and more to position myself as an observer of what is brought into environmental research by modeling without immersing myself in the new modeling tools that keep invading that field (fractals, MAS and cellular automata etc.). While I recommend this methodological orientation and support those who apply it (who can be found in C. Soulard and W. Hucy's work), while I even try through them to introduce with all its force the idea that spatial analysis methods are an aspect of "workshop site" programs that cannot be ignored and the aim of which is the cognitive and continuing observation of "eco-sociosystems", I take a critical stance regarding some works in "spatial modeling" that, and I will come back to this, seem to me to be not only simplistic but antithetical to the complex objects they claim to be studying.

Out of respect for the way I am being welcomed, through this book, into the community of spatial analysis, I must decipher the undercurrents of this attitude bordering on schizophrenia. Thus I must first answer the question: is it a strictly personal issue, of a judgment cast on the way some people use modeling, or an awareness of the difficulties in "bringing together volunteers from all disciplines", in particular those "good at math and modeling" in order to accomplish my own research ambitions?

Let us review these hypotheses one after the other. There is indeed, at the point of origin of this lack of enthusiasm in going from "pre-quantitative" to quantitative and model-based skills a matter of personal and theoretical perspective. Well trained in mathematics in high school, surrounded during my first research years by mathematicians and philosophers who reflected on the relationship between politics and sciences, mathematics and models[4], I have come to think that mathematics does

---

4 I am simply referring to my acquaintance in the 1960s-1970s with Louis Althusser and Alain Badiou at the École Normale Supérieure, they themselves being friends with Maurice Mathieu who was then a mathematician at the Collège de France in Perrin's team. I am also referring to conversations with the mathematicians at the ENS, including Adrien Douady who was connected to the Bourbaki school (formalist, metamathematician, structuralist), but also Benzécri and Françoise Badiou.

not consist of taking reality as a starting point, since mathematicians (whose intuitive gift for formalization is often detected in their early childhood) make discoveries in abstraction or rather in a realm of reasoning that do not go beyond mathematics as a discipline. Nurtured on many anecdotes about the career of Sophie Germain, Poincaré, etc. that all showed how mathematical discoveries are ill-adapted to life in academic society and would rather be fitted with social isolation, it seemed to me hard, even impossible, to reconcile my taste for the social and current aspects of the world in which I was living (which had made me choose to join the CNRS as a geographer rather than a historian) with a deepening sense of the heuristic virtues of mathematics applied to geography. More than that, whenever, led by a then poorly defined intuition of the importance of multidiciplinarity in solving complex issues, I tried to engage the attention of my mathematician and/or philosopher friends, I was immediately faced with a negative judgment of my attempt. The arguments used against it were quite similar: either the critique was aimed at the conceptual perversity of modeling that I have referred to before, or based on harsh judgments of my first attempts at applying mathematics to social sciences, considered as simplistic and, lacking conceptualization, as unconsciously serving the dominant ideology. If, I was told, quantum physics has made progress thanks to mathematics and has helped mathematics evolved, it is because the level of conceptualization was maximum. By choosing to apply mathematics to my research[5], I was running the risk of weakening my theoretical capacity and my results through a mediocrity of mathematical foundation. In other words, it was better for me to deepen my hypotheses and build a system allowing a stronger conceptualization, rather than depend on already existing models and modelings (for instance regarding the processes of dissemination and polarization), which would modify my research goal, and maybe even put it under the influence of the then dominant ideology.

In short, from a personal perspective, doubt took hold: was I capable of being heavily involved, both in mathematics and geography, until I found the mathematical expression fitting each of my research objectives, which were oriented more and more towards the study of complex objects? This doubt was reinforced when I read Edgar Morin, who did not have to use mathematics in order to "introduce us to complex thought", and this at the time when the "mathematics expert" Le Moigne joined him in his "theory of the general system" as a "theory of modeling".

---

5 I wanted to try and build a typology of farms in which I could integrate temporal processes (dynamics of the family and the reproduction of the farm) and spatial processes (layout of crop parcels, proximity and contiguity, etc.) See MATHIEU N., 1972, "Typologie dynamique d'exploitations agricoles des plateaux de Haute-Saône", in *Approche géographique des exploitations agricoles, Cahier no.1*, Paris, April, pp. 9-24 (Équipe rurale du LA de Géographie Humaine).

What can be said, then, about my relationship to others, geographers or those close to geography, who deliberately committed themselves to a path I was not willing to follow? A retrospective piece of internal enquiry has led me to distinguish three attitudes towards them that may be linked to the way I evaluate differences in the epistemological, even the ethical scope of these practices. It is true that generally I had a positive prejudice regarding all those who embarked on the adventure of models and modeling. Being curious about all the accomplishments, about *progress in geography*, I have always made a point of participating in the Dupont Géopoint group and in the European conferences on theoretical and quantitative geography. However, and this I admit is the first position that I took, I am somewhat wary of those who, compulsive and eager to be regarded as the most effective in spatial analysis, seem to forget the meaning of the research objects whose systemic functioning they claim to analyze. To try out a new method *in se* and *per se* is more important for them than the cognitive goal which seems to me to be the core, the ultimate value of research. I do not need to dwell on the texts that have led to my theoretical wariness of this usage of models and modeling. It may be enough to refer to my outrage when I read that the best example of urban growth following the fractal model was the town of Nouakchott! Nouakchott, the city of all poverties, but also of all the craftiness of informal economy, exploited in the quest for survival! How could anyone call this growth, what was no more than the extension of a spatial form emptied of its social and human content? How, when the research was supposed to be theoretical and fundamental, could anyone thus simplify the city to the variable of developed sites and to a demographic dynamic? Was the craving for a "mathematically expressed" result and a rigorous proof antagonistic to the effort to think complexly, to bring to light the intricacies of elements and processes that form urban spatial systems? Of course, this is an extreme example that does not represent all the attempts at modeling which are more concerned about the social dimension of geography, and also more concerned about the relationship between physical processes and natural processes than such a simulation model allows. Yet, it is representative of a tendency to use a method for its own sake without insisting upon the results yielded being repositioned within the broader conceptualization of the research field and the discipline.

Although they are in a very different position from those mentioned above, certain well-used studies in the scientific milieu concerned with the environment and more specifically on the management of renewable resources also make me circumspect. Here is the second reason inhibiting my personal involvement in the use of modeling tools. From reading the journal *Natures Sciences Sociétés*, I cannot help but notice the current craze for "modeling as an accompanying tool" corollary to the valorization of "action research" (or "development research"), corollary also to praising the virtues of spatial modeling as a decision-making aid. New computer tools such as GIS, MAS, cellular automata, etc. that is to say artificial intelligence applied to localized (geographical, territorialized) complex situations, are at the core of this type of modeling. This research trend is being used more and more in the big applied research institutions such as INRA[6], CIRAD[7] or CEMAGREF[8] and suggests models with joint "resource/exploitation" dynamics, between field and theory, that are supposed to both produce knowledge about complex systems and facilitate the dialog between users and the learning of collective decision making concerning the management of ecosystems and renewable resources. We may wonder if in this shift from systemic analysis to systemic modeling, and then the building of expert systems using computing modeling tools, there might be some confusion between what is called a mathematical model, which is supposed to be extremely reliable in its own realm of application, and mathematico-computing models that are supposed to simulate various dynamic behaviors (some of which cannot be expressed mathematically) in scripts that impact the spatial system. However, this is not really the issue since, as I have already mentioned, the conceptual clarification that comes from going back and forth between a situation (or an experience of reality) and the model built to explain it is in itself positive. What raises questions is the risk taken by these researchers, even when they do try to follow deontological principles of respect for participants who do not have scientific expertise, of missing out on certain scientific knowledge without which the "decisions" made by the actors can in no way be understood. Who are the "actors"? What does it mean to make a decision? What is the meaning of territory in the simulation? What do "landscape dynamics" mean to the researcher in the simulation model and in the mind of the people to whom it is presented, and from whom a decision, or even a consensus is expected? In other words, once again, the risk of simplifying complexity to the point of misrepresenting the object regarding which a decision must be made, is important. Is obtaining a consensus with the use of scripts simulating consequences not a way of making use, as being blind, of those who are supposed to make decisions and about whom very little is being said? Once again, skills are considered as most important and weaken the awareness of being in a position of power. How could I not choose to be careful when confronting experiments already considered as models to help make decisions, and which I think are premature and insufficiently thought out in relation to the social stake they raise?

Thus, it is a matter of science partners and trust in a collective of researchers intent on studying a complex object even if, as is the case for the sustainable city, the study of the object depends on "social demand", or even the well-being desired by its inhabitants. Here is the third reason for my relative neglect of spatial analysis and modeling. In order to overcome the criticism I just referred to, the only tenable

---

6 Institut National de Recherche Agronomique: National Institute for Agronomical Research.
7 Centre de coopération Internationale en Recherche Agronomique: Center for International Cooperation in Agronomical Research.
8 Institut de recherche pour l'ingénierie de l'agriculture et de l'environnement: public agricultural and environmental research institute.

position, on the theoretical and practical levels, is to be certain I am part of a multidisciplinary team aware of the skills of everyone, and its complementarities. As I have written earlier, interdisciplinarity is a practice in which, step-by-step, a conceptual approach and a multidisciplinary research plan is built around a complex object, the study of which is of equal interest to all scientific partners. For us geographers, it is often an issue involving interactions between natural systems and social systems (for instance, flooding risks due to erosive run-off, or the management of biodiversity in an urban environment, etc.). This type of complex issue requires a broadened multidisciplinarity, at least between physical geographers and human geographers, which is still an exceptional occurrence. Modeling no doubt has a place, but not exclusively, as must be the case for all disciplines involved, and above all, under the condition that it is introduced when the problem is very clearly expressed and the need to model is clearly identified, and also when, as mentioned before, there is a to-and-fro between the model (modeling) and the experience (field work), the latter being defined as "any organized way of acquiring information that includes, in the perspective of an expressed goal, a confrontation of reality".

However this internal interdisciplinarity in geography aiming to build a common approach to spatial analysis, social geography and physical geography, is still a utopia. Not that I underestimate the results obtained in the Causse Méjean Observatory! Not that I deny the forward strides of the MTG group, in particular around Daniel Delahaye, in articulating physical issues and social dimension! But the interdisciplinary practice in geography, as I have tried to define it, is still a minority, and its results are still too meager. Each research group tries to innovate within its own activity, with its scientific capital, without trying to move *in terra incognita*, or beyond its recognized horizons. As I did when I was part of MTG and tried in vain to build a research program bringing together Patrice Langlois and Marianne Cohen, I still regret that a more vigorous work is not being implemented between our two laboratories in order to think together about the place of modeling in the advances of our research.

The acceptance of a conceptual modeling based on the statement of an interaction system

While at the onset of this reflection I pointed to the incongruity of my being one of the authors of this book, I find myself able to conclude a conciliating approach that would reinvigorate the dialog between the "modeling and graphic processing" and "social dynamics and recomposition of space" laboratories, precisely on the subject of *modeling in geography*. My suggestion is a mutual recognition of the importance of conceptualizing the issue to be studied before using models and modeling. Indeed, I think that in the research school to which I belong, the enunciating of hypotheses regarding the interaction between elements whose mutual connection is not obvious, in particular interactions between social and spatial practices and natural elements, should be taken into account. This logical statement of the relationships between natural systems and social systems relies on the identification and the construction of concepts that can open mediation in these relationships (for instance, the practice/representation duo, or the concept of mode of inhabitance). The logical statement also has a temporal value and must articulate the various temporalities of nature and society. These properties are found in the "heuristic research model on sustainable development" suggested by Monique Barrué-Pastor: examining all the terms of the relationship; discussing notions down to the definition of useful concepts; building a hierarchy of concepts and relationships, etc. Enunciating a relationship system seems to me to be a scientific result, but that is not really recognized by the specialists of spatial analysis and modeling because it is a conceptual approach that cannot be immediately translated into measuring methods that do not immediately call for a certain already tested model. Would it not be a worthy intellectual adventure to bring together means of thought that, in the end, leaves a large place to the conceptualization of a complex system?

References

[ANT 05] ANTONA M., *et al.* 2005, "La modélisation comme outil d'accompagnement", *Natures Sciences Sociétés*, forthcoming.

[BAD 69] BADIOU A., 1969, *Le concept de modèle*, Paris, Maspéro.

[BAL 93] BALLEY C., COHEN J., LENORMAND P., MATHIEU N., 1993, "Réseaux d'acteurs et territoires en milieu rural: test d'un pré-modèle de l'ancrage territorial de l'emploi" in *Propositions méthodologiques pour une prospective sur les espaces ruraux français, sous-groupe "Emploi et nouvelles ressources"*, Nanterre, AGRAL (Observatoire AGRER), pp. 136-148, (Contrat DATAR/CNRS no. 50 1555, 158p.).

[BAR 04] BARRUE-PASTOR M., 2004, "La construction d'un modèle heuristique de recherche sur le développement durable. Le système "Acteurs-Territoire multicritère" de la région des Lacs" in *Forêts et développement durable au Chili. Indianité Mapuche et mondialisation*, Monique BARRUE-PASTOR (ed.), Presses Universitaires du Mirail, pp. 21-48.

[BON 66] BONTRON J.C., MATHIEU N., 1966, "La population agricole, évolution au cours du Vème Plan" *Paysans*, no. 60, June–July, pp. 23-32.

[COH 69] COHEN M., MATHIEU N., ALEXANDRE F., 1997, "Modeling interactions between biophysical and social systems: the example of dynamics of vegetation in Causse Mejan", *10th European Colloquium on Theoretical and Quantitative Geography*, 1997.

[COH 03] COHEN M., 2003, *La brousse et le berger. Une approche interdisciplinaire de l'embroussaillement*, Paris, CNRS Editions, p.366.

[COS 98] COSTANZA R., RUTH M., 1998, "Using dynamic modeling to scope environmental problems and build consensus", *Environmental Management*, 22, pp. 183-195.

[EQU 79] Équipe Analyse des Espaces Ruraux, 1979, *La sécheresse de 1976 dans le canton de Luzy*, Laboratoire de Géographie Humaine, CNRS, 105 p., maps and appendices.

[GAI 04] GAILLARD D., 2004, Gestion concertée du ruissellement-érosif dans les espaces agricoles des plateaux de grande culture. Analyse spatiale, approche socio-économique et mise en place d'une dynamique collective d'aménagement. Exemple de la Seine-Maritime, Thesis, University of Rouen, Volume 1, 1-364 p. Volume 2, appendices 365-508.

[LEG 97] LEGAY J.-M., 1997, *L'expérience et le modèle. Un discours sur la méthode*, Paris, INRA Editions, p.111.

[LEG 04] LEGAY J.-M., SCHMID, A.-F., 2004, *Philosophie de l'interdisciplinarité. Correspondance (1999-2004) sur la recherche scientifique, la modélisation et les objets complexes*, Paris, Editions PETRA, p.300.

[LEM 77] LE MOIGNE, J.-L., 1977, *La Théorie du système général, théorie de la modélisation*, Paris, P.U.F.

[MAT 70] MATHIEU N., *et al*, 1970, *Étude pour le classement d'un échantillon de districts (mesure du degré d'urbanisation)*, 194 dossiers + note méthodologique et de synthèse p.15 + tableaux de valorisation et de classement, Paris, SEGESA, June.

[MAT 88] MATHIEU N., *et al*, 1988, "La géographie et la mesure de l'homme", *Esprit*, December, pp. 87-105 (round table organized by Y. GUERMOND).

[MAT 92] MATHIEU N., 1992, "Un outil d'observation de l'interaction entre systèmes naturels et systèmes sociaux: l'observatoire du Méjan en Lozère", in *Trente communications au 27ᵉ Congrès International de Géographie*, Washington, INTERGEO/AFDG.

[MAT 92] MATHIEU N., 1992, "Géographie et interdisciplinarité, rapport naturel ou rapport interdit ?" in *Sciences de la nature, sciences de la société, les passeurs de frontiers*, M. JOLLIVET (ed.), Paris, CNRS Editions, pp. 129-154.

[MAT 97] MATHIEU N., 1997, "French geography status in interdisciplinary research" in *Modeling Space and Networks, Progress in Theoretical and Quantitative Geography*, Einar HOLM (ed.), Gerum Kulturgeography, Umea Universitet, pp. 325-342.

[MAT 97] MATHIEU N., 1997, "Interdisciplinarité interne, interdisciplinarité externe, quel intérêt heuristique pour la géographie: Réflexion à partir d'une confrontation de pratiques", *Actes du Colloque Géographie Interface, Représentation Interdisciplinarité*, Institut Universitaire Kurt BÖSCH, Sion (Switzerland).

[MAT 99] MATHIEU N., "L'expérience et le modèle, un discours sur la méthode, compte rendu de l'ouvrage de Jean Marie LEGAY", *Economie rurale*, 251, May–June 1999, p.61-62.

[MAT 01] MATHIEU N., "Pierre Coutin et la formation des élites. Une certaine idée du rôle des sciences socials", *Les Etudes Sociales*, 134, 2001, pp. 23-34.

[MAT 71] MATHIEU N., BONTRON J.C., 1971, *Les zones à faible densité de peuplement*, Note created for Intergroupe CNAT/Espace rural, 4 p. mimeo + 1 map: Types d'espaces ruraux selon la densité de population et l'encadrement urbain par arrondissement.

[MAT 73] MATHIEU N., BONTRON J.C., 1973, "Les transformations de l'espace rural. Problèmes de méthode", *Études rurales*, no. 58-59, January–June, pp. 137-159.

[MAT 96] MATHIEU N., COHEN M., FRIEDBERG C., LARDON S., OSTY P.L., 1996, "Approches pour la modélisation des interactions entre dynamiques de la végétation, dynamiques sociales et techniques: confrontation des énoncés logiques et des méthodes: l'embroussaillement sur trois sites du Causse Méjan", in *Tendances nouvelles en modélisation pour l'environnement*, Paris, CNRS, Actes des journées du Programme Environnement, Vie et Sociétés, session A, pp. 37-42.

[MAT 96] MATHIEU N., COHEN M., 1996, "Approches pour la modélisation des interactions entre dynamiques de systèmes naturels et de systèmes sociaux: l'embroussaillement du Causse Méjan (France)", 28th *International Geographical Congress, Abstract Book*, The Hague, p. 289.

[MAT 85] MATHIEU N., DUBOSCQ P., 1985, *Voyage en France par les pays de faible densité*, Toulouse, CNRS Editions, 179 p.

[MAT 85] MATHIEU N. and Équipe Analyse des Espaces Ruraux, 1985, "Un nouveau modèle d'analyse des transformations en cours: la diversification/spécialisation de l'espace rural français", *Économie Rurale*, no. 166, pp. 38-44 (abstract in English).

[MAT 86] MATHIEU N., MENGIN J. DALLA ROSA G. and AVENTUR D., 1986, "Nature et signification de la diversité des modèles de développement rural", *Min. de la Recherche et de la Technologie*.

[MAT 86] MATHIEU N., MENGIN J., 1986, *La diversité des modèles de développement rural: histoire, nature et signification*. Paris, FORS, Ministère de la Recherche et de la Technologie, p. 49.

[MAT 88] MATHIEU N., MENGIN J., 1988. "Les politiques de développement rural: unité ou diversité", in JOLLIVET M., (ed.), *Pour une agriculture diversifiée*, Paris, Ed. L'Harmattan, 335 p., pp. 268-282.

[MOR 01] MORIN, E., 1977-2001, *La Méthode*, Paris, Le Seuil.

[ROB 82] ROBIC M.-C., MATHIEU N. (en collaboration), 1982, "Accident climatique et fonctionnement de la société agricole, la sécheresse de 1976 chez les éleveurs d'un canton de la Nièvre", *Espace géographique*, 2, pp. 111-123.

[ROB 01] ROBIC M.-C., MATHIEU N., "Géographie et durabilité: redéployer une expérience et mobiliser de nouveaux savoir-faire", in *Le développement durable, de l'utopie au concept: de nouveaux chantiers pour la recherche*, Marcel Jollivet, Editeur scientifique Paris; Amsterdam; New York: Elsevier, 2001, pp. 167-19.

Preface

In a book published in 2007, Lena Sanders [SAN 07] revealed the great variety of choices made by geographers in the field of spatial analysis modeling. Our aim is not to produce a new inventory, but to propose general reflections about the realizations and perspectives of modeling research, both in the field of theoretical geography and in the field of applied geography in town and country planning. The tools are widely available, and are continuously improving, for spatial analysis as well as for geographic information systems. The MTG research group (models and graphic processing in geography) was created in 1986, with the ambitious target of keeping "close control of the new technical tools, with a permanent link to social demand, and to discover all the opportunities of interface between science and technology". These 20 years of collective research have now given us an opportunity to propose this "reflection". The chapters below are the work of researchers currently working in the laboratory, as well as former members of the initial team, who are now working in other universities.

The first two chapters situate our research program: what does a modeling process mean, and what is the specificity of this process in the field of human and social sciences? The path covered since the early realizations of spatial analysis is a basis from which new research has developed, mainly in terms of simulation techniques, thanks to recent computing developments.

In Chapters 3 to 8, we see how these models are confronted with the reality of what geographers are being asked to do in the field of land planning and management: cultural policy, territorial forecasting, socio-spatial segregation, inequity of regional dynamics, polarization, enclosing. Geography is, by definition, engaged in a process of understanding the relationship between society and space, but these confrontations with material work must not occlude the importance of a permanent evolving theory.

Towards that aim, the final chapters make it clear that some distance is necessary in responding to social demand, to enhance a new reflection on the fundamental concepts structuring the discipline, as well as the weaving of new links with the present level of science and technology. This distance is the only means of progressing towards the new horizons of a theoretical geography allowing numerical experimentation, or, in other words, an "artificial geography". However, this research is only valuable if it prevents a retreat into previously tested methods. By keeping a concern for a constant reference to socially suitable themes, this reflection must allow methodological transfers towards the social agents. This to-and-fro gives its value to theoretical geography and prevents the interpretation models of the social life from staying set.

Bibliography

[SAN 07] SANDERS L., *Models in Spatial Analysis*, ISTE, 2007.

Acknowledgements

I wish to thank the members of the "translation team" who helped me in finalizing the English version: Sandrine Baudry, Lyla Bradley, Lee Campbell, Bobby Hiltz, Shivani Khosla, Donnacha O'Ceallaigh, Aruna Popuri, Philana Rustin, James Taylor and Pierrick Tranouez.

Yves Guermond

Chapter 1

The Place of Both the Model
and Modeling in HSS

The aim of this chapter is to present a few points of view on the concept of the model or on the modeling process. In Human and Social Sciences (HSS), modeling can cause some specific problems because of the immersion of the human researcher in his object of study, which is equally human. Our goal is to show the specificity of modeling in HSS, and the conditions of its utilization. The rigor with which the modeler will demonstrate the conditions of use of his own tool will allow the precision of the field of its utility in HSS.

It is helpful to specify the definition of the model and modeling utilization because of the different assertions in common sense, but also in HSS. The same definition in the same discipline can hide paradigms, methodologies and different issues, diverging or contradictory. The same theoretical posture in two different disciplines can lead to the use of two different words.

We will thus start from the definition that common sense gives to the word "model". This is the object from the beginning. Modeling being used most often to mathematically formalize a reality, we will explore the notion of a model in mathematics. Modeling's different utilities and issues in social sciences will thus be examined before putting them in perspective with mathematical language.

Chapter written by Patrice LANGLOIS and Daniel REGUER.

1.1. Models and modeling: definitions

The term "modeling" means both the activity required to produce a model as well as the result of this activity. From this distinction, the concept of modeling is larger than that of the model as it corresponds to the human activity producing a finished model, while the model is an object (concrete or abstract), voluntarily drawn from the activity. The model does not appear all of a sudden at the end of the modeling activity, it is progressively formed like a vase from the hands of a potter. It establishes itself in an activity, without identifying itself with it, it existed before (during the conception phase), it exists during the utilization phase, it exists even after its rejection, or in the will to create a better model, one which surpasses the first.

First we discuss definitions of the word "model".

Among the many definitions in the Encyclopedia Britannica, we will retain two:

1. on the one hand, a "model" is a "formalized structure to realize a set of phenomena, which between them possess certain links". In the mathematical model, this is the case defined as a mathematical representation of a physical, economical and human phenomenon...;

2. on the other hand, a model is a "schematic representation of a process, of a sound approach".

These two definitions are on different levels; however they still possess certain connections.

The first definition is associated with the relation in the middle of a structure. It implies two notions: that of totality and that of interdependence between elements which is not the result of accidental accumulations. Thus, in this definition, the use of models would consist of "taking the totalizing attitude in any case", as with what Sartre says about structuralism [SAR 60]. The catchphrase would be: "We don't know if what we say is true, but we know that it makes sense." This definition also returns to the system's notion addressed in Chapter 11. In this category (the structure-model) a mathematical sense is given to the term "model".

In the second definition, we can use the example of the geographical map. This is also the case for a Conceptual Data Model (CDM) in the framework of the elaboration of a database. However, we must acknowledge that the schematic-model is not a long way from the first definition, in as much that "a schematic representation" can very well be a graphical representation of the formalized structure returning back to the first definition. Frequently we associate a verbal formalization (like in mathematics, physics, chemistry, geography, etc.); a graphical formalization (like a picture associated with a graph; a diagram of phases associated with a differential system; a molecular schema of a chemical formula; and a map associated with the values of a structure-model). The schematic-model is thus a representation of the structure-model. In summary, it is a model of a model. It is possible that a schematic-model is not associated with a more formalized structure, like the water cycle schema or a "choreme" [BRU 86]. It then corresponds to a more empirical measure, which can be a stage in the modeling activity becoming (emerging towards) a formalized model.

Is the forming of verified but not yet explained observations already a model? The catchphrase for this radical empiricism would be "we don't know if what we are saying makes sense, but we know it's true".

We think that there is a gradation in the models and that it is impossible to fix absolute criteria of "modelicity". In fact, a model is always preceded and followed by a complex scientific procedure, since the reflection on the choice of data, and after on the tools (physical, institutional or methodological) allowing the collection, the observation, the organization, the structure, the digitizing capability, until the final formatting of the model's data. Also with respect to the downstream of the modeling, we must define some forms of selection and observation from the model's results. We must translate the results in the framework of theoretical interpretation. All of these stages also contain modeling forms. The execution of a map necessitates different sources of data: a census report on the population that gives databases, the remote sensing that gives images after complex processes of satellite pictures are already forms of abstraction of reality, which we can qualify as models. The map which results is in itself a model resulting from the former. This map, numerically structured under a GIS form, can lead to a mathematical model, which can then generate several results. These results will themselves be formatted to be interpreted in the frame of a theoretical corpus, this translation phase is also a form of modeling, as the same results of a model can produce very different theoretical interpretations. Thus, we can see that the model does not have to be extracted from the general scientific approach.

Even though it is not our goal to bring a general and unifying semantic clarification, it would seem useful in the pursuit of our study to formulate four positions concerning modeling:

– establishing the norm, stating the pros and cons. We are not concerned herein with "modeling morale";

– explanatory, which consists of finding a general law outside of the object;

– comprehensive, which consists of understanding motivations that have a meaning for each person;

– interpretative, which is the will to give a significance by putting a field of representation (signified) in relation with another (signifying).

Let us note here that the explanatory and comprehensive procedures are complementary, but the modeling in a comprehensive perspective seems much more delicate.

In a contemporary economic dictionary (Mokhtar Lakehal, *Dictionary of Contemporary Economy*, Ed. Wuibert, November 2002), five pages are dedicated to the word "model". In fact, it has very little to do with presenting a definition of the concept, while this seems obvious. For the economic dictionary, it has to do with presenting different models, with which their authors sometimes associated their names (Walras' equilibrium model, the Keynesian model, the Marxist model, Makowitz' model, etc.). This is evidence of the importance of the modeling practice in this discipline, which has for that matter won many Nobel prizes awarded for the development of these models.

After having noted the Italian origin of the word model (figure destined to be reproduced), the Robert *Dictionary of Sociology*, in a chapter written by Pierre Ansart [ANS 99] distinguishes two assertions on the concept of the word "model". The first one, relative to social practices, would be a "reality that we force ourselves to reproduce" (here again?). The second one, relating to methodology, would be a "constructed representation, more or less abstract, of a social reality". One is the reality as an object of reproduction; the other is a representation of reality.

The first sense thus returns to reproduction, but the model is the reality, it is the object of reproduction. It could consist, in the common sense, of the artist's model. Meanwhile even Miro, who is not even known for the figurative character in his work, used models, which he did not even reproduce. "In my paintings, each form, each color is taken from a fragment of reality." In this sense, a reproduction practice would not be associated with the use of models, but they would be a source of inspiration. Miro added that a moving object, like a jack-in-the-box surprisingly springing from its box, could serve as a model for him. Thus, the painter's model would not be an object of reproduction. It would only be supporting the imagination, maybe even a suggestion of dynamics.

In the second definition given by the Robert Dictionary of Sociology, the model "doesn't reproduce reality, it simulates it". We notice that if the modeling is an instrument, a technique "that enables us to think and interpret reality", we can apply a technical definition to "simulation" which is none other than a "method.... that consists of replacing a phenomenon... by a more simple model, but which has an analogous behavior". In this definition, the model is a simulation, always approximate, of reality. In this case, the model sets its heart on coming closer to it, to the best of its ability, without pretending to return all of its complexity. The choice of the components of reality, integrated into the model, results in the construction of the research agenda, of its theoretical frame. The components thus selected are seen as fundamental for the purpose of the study. Therefore, the model can pretend to return all of the components, but not the wholeness of reality. The parts of reality that are beyond the object of study are voluntarily excluded from these components. The aspects of reality that are not linked between themselves by relations leave the scope of the parameters of the model, even if these aspects of reality are a part of the object of study. A provisional use is not more stated for simulation than for modeling.

After having defined various forms of the model's concept, we will study precisely the model in mathematics before putting in perspective its use in human sciences.

1.2. The mathematical concept of a model

There are at least two mathematical definitions of the term model: the first one is situated in the framework of model theory, and the second one in the interface between mathematics and the other sciences.

1.2.1. *The semantic conception*

In the framework of model theory, the notion of a model is used in a rather particular manner, since the term is used as something that allows us to give a "meaning" to a theoretical discussion by end-to-end correspondence between the model and the formal theory. A model is thus a sort of reference example, of the fulfillment of the theory, allowing the justification of the theory by an external significance. However, this also gives the model a theoretical framework, allowing us to rigorously formalize it. Moreover, the same theory possibly having various models in different contexts, their comprehension reinforces them mutually and they can be studied in the framework of a formalized theory with a great economy of thought, in so far as the same (theoretical) thinking scheme is used in different contexts. If likewise, all of the model's elements and properties correspond to the theory's symbols and formulae, the model, in this theory, is then known as complete. We then see the convergence interest between syntactic and semantic aspects and the importance of the theorems of completeness or of incompleteness. Thus, Gödel enunciated the incompleteness of arithmetic by proving that there exists at least one property of arithmetic that cannot be demonstrated nor refuted starting from the axioms. This result ruined Hilbert's plan to constitute a totally formalized and coherent foundation of mathematics, and disproved the Vienna Circle's formalist theses.

1.2.2. *The empirical concept*

The second aspect of the mathematical model's concept, which we could call *empirical*, or simply a *mathematical model*, is much more widespread, as it largely overlaps the frame of pure mathematics and is seen in all sciences. A *mathematical model* is a representation by a formulation or a mathematical formalization of a portion of reality (whether static or dynamic).

The thinking scheme is contrary to the preceding one, in so far as in the first case, the model is a fulfillment which gives significance to a theory, whereas in this case, it is an operation of abstraction that allows us, by simplifying it, to give an *explanation* of reality... Furthermore, the link between the model's mathematical formulation and the reality to which it refers itself, is not mathematically formalized as before, from where its denomination of *empirical* stems. We can tell that the meaning of empirical conception used here is very large, whereas the notion of *simulation* is much stronger, as it holds a will to reproduce reality, to imitate it in certain dynamics, consequently in time. Thus, the model's notion cannot be confused with that of simulation, especially when it is applied to human behaviors.

We are necessarily in an interdisciplinary situation here, where we correspond a certain mathematical formulation to a concrete reality. What we call *concrete reality* is quite relative, this only means that we are referring to a non-mathematical area, such as actual objects or phenomena, but this can be non-material, such as information (ideas, texts, images, observations, measures, etc.) and this can even be a part of the psychic universe, such as mental representations, fantasies, desires, etc. as could be used in psychology, psychoanalysis or sociology. Let us think about the considerable development of cognitive sciences that have produced models for multiple applications like neuronal networks, self-adapting systems, etc. Another example, in the very different context of lacanian psychoanalysis, is the torus as a topological surface modeling the neurosis; the subject's desire and pleasure are modeled by the projective plan, illustrated by the *Cross-Cap* (a figure obtained by the suture of a hemisphere and a Moebius strip). By contrast, in social sciences, only the observable externalized concrete realities can be studied.

Similarly, what we call *mathematical formulation*, may also be very diversified, going from the simple number (the number of sheep in the flock) to the statistical chart (a population census), then to formulae and equations (Newton's law of gravity), or a mathematical structure having certain properties (vector space of the representation of variables from a statistical table, in an principal components analysis), and then going all the way to the formalized theory's enunciations (the quantum theory of fields).

1.2.3. *Links between the mathematical model and its object*

The link between concrete reality and mathematical formulation cannot be in itself mathematized because the so-called "concrete reality" should also be a mathematical formalization. We would then fall back on the former semantic conception. The link is then built up empirically. The shepherd who brings back his sheep every night decides to model his herd using an integer. Putting the correspondence between the herd and the mathematical object "integer" depends only on its observational capacities, bringing his counting technique into play. He would then use mathematics to compare both this evening's and last night's numbers; the results give him indications that he should interpret in terms of reality, by using all of his experience as a shepherd: if the two numbers are equal, he can interpret this by saying that there has been no change in his herd. But he can also wonder if this result does not hide an equal number of losses and births, making him reflect upon the appropriateness of his model, relatively to the knowledge and the mastering that he seeks of his herd... He will perhaps consider the set theory, or develop a much more complex specific theory, to better model his herd. His science progresses this way, as does science in general ... by confronting theory with reality, going back to it and making it evolve.

Thus we must consider the two arrows of correspondence: the one that makes it possible to pass from concrete to abstract, which is the activity of *modeling*, then of *observation-measure and information of the model*, and the other, from the abstract to the concrete, which corresponds to the activities of *interpretation* of the results and *validation* of the model. These activities include almost the entire scientific procedure and it would be vain to give it a definition here. We often call it the modeling context. Nonetheless, this makes it obvious that the existence of a mathematical model is not a guarantee of scientificity, "truth", or of the control of reality that this entails, since all of that depends on the quality of the modeling context. Stated in a caricatured manner, a solely mathematical formula has no significance if we do not give the components of this formula the precise correspondence that it symbolizes together with reality. When this demand is carried out, it can then obtain the status of model.

1.3. Is there a specificity of HSS?

As we have just stated, mathematics cannot take the place of scientific truth independently of the problems to which they are supposed to answer and which are firstly the result of human activity and social constructs. Thus, they impose perpetually, not a doubt, which is a posture of retreat, but the critical verification, the explanation of the procedure, and the reinforcement of the coherence. This procedure addresses not the purely mathematical aspect, which by its essence is the most verifiable, but the empirical aspect that links the purely mathematical discussion to the reality that it is supposed to describe or explain. It addresses the translation of hypotheses in mathematical terms and some conclusions in terms relative to the disciplinary problem and the interpretation that results from it. Thus, the risk is either to caricaturize these problems by an overly simple mathematical formalization or to delegate to "specialists" who are not in the field of HSS, resulting in an incomprehensible formalization. On the other hand, because of the plural character of the different models available to the researcher, their choice cannot escape the ideological, political and economic challenges and the theoretical postures that run throughout the social sciences.

The HSS phenomena are "multi-determined". The instability, inherent to the complex systems, offers, at certain times, degrees of freedom between these determinisms with the possibility for mankind not only to change its behavior but also to influence these determinisms. A phenomenon in HSS is thus defined as "a succession of choices to make in situations of tension balance joined by portions of determinist trajectories".

Already, the phenomenon of the living individual (that also produces social issues as much as it is produced by them) bears the unique capacity of auto-reproduction, not identical reproduction but with the possibility of mutation, which is generating a Darwinist evolution, by growing complexity. This goes against the rest of the physical world, ruled by the second law of thermodynamics, which stipulates that the universe, globally, always tends towards more disorder. Thus, life seems to be the bearer of a "project", that of self-perpetuation. For that, it must be able to adapt to environmental changes. The chance of mutation plays a constructive role in as far as it permits only the choice of those forms of life capable of surviving, therefore producing an evolution towards forms of life that are more and more complex. However, the cultural dimension of mankind cannot be solely explained by biology.

The mechanism of evolution of life that is based on the diversity of its production and on the selection that only conserves the most adapted productions is a wasting mechanism of time and energy. By conscience, mankind, individually or collectively, has the possibility to construct an ideal, to make projects, which are projections in the future. More generally, the dynamics of evolved living do not depend only on its past and present condition, but become dependent on the future, because life is capable of anticipating and representing possible futures that are more or less close, and directing its actions in accordance with this representation, such as changing a trajectory in order to avoid an obstacle, or making choices in order to be closer to a reachable objective and to learn efficient behaviors. Moreover, mankind is not only "dependent on the future", but above all, the future is dependent on it, on its conscience to act on its environment and on the historicity of society. Thus, modeling in HSS becomes much more complex if it must take into account this new type of mechanism.

The distinction between, on the one hand, natural and life sciences and, on the other hand, HSS is nevertheless no more satisfying than that between natural sciences and cultural or spiritual sciences that Weber describes. It has a binary and rather Manichean character. It separates the world in half, the world of mankind and the rest. This classification presents a practical character. However, when we define mankind as an animal who thinks, things start to become complicated, since we cannot rigorously state the complete negation of thought in animals. Mankind is not the only one who thinks and all animal thought is not reducible to reflex. That is to say, the application of a model, for example, the behavior in accordance with external variables to the object, will be eternally capable of being reproduced in natural sciences. That will not be inexorably so when the object is mankind, individually and collectively, with the exception of the interference of chance in extreme situations of equilibrium. Mankind has in fact the capacity to think and thus to act upon itself, and indeed upon its environment, in order to obtain control, even marginally, over its future. If we hold a difference between the types of modeling according to the capacity of thinking of the modeled object, then the criterion of classification is no longer mankind, but the faculty of thinking. This is the reason why we will bear in mind a second classification, which is no longer around the central object that is mankind, but the thought, or at least the conscience, which is to say an internal representation of the environment, in which the "subject" evolves:

– Material sciences, where the objects solely obey laws, deterministic or not, and have no conscience either of themselves or of their environment, and are only reactive.

– Life sciences, where the objects have no real autonomous thought, but where, for example, each cell possesses a representation of the whole, by genetic code, allowing it to know its place, its role and its future in the living organism in which it finds itself. Moreover, living multi-cellular beings possess a representation of their own identity and of their environment, which can be extremely limited, or relatively evolved (inter-active objects).

– "Human life" sciences (of the social animal, capable of abstract thinking of itself or its environment). They do not have the agent as their only object, but also the subject or the actor upon its environment, capable of projects and invention. Thus, in this environment, there is observation and the conditions of observation, upon which man has the capacity to act (teleological object).

On the other hand, for ethical reasons, the HSS cannot be conceived as experimental sciences. The notion of experience to be scientifically validated means that the population object of that experiment must not be informed of its existence. This absence of information is nothing other than a manipulation. As soon as people are informed of the experiment, they act differently due to the experimental situation, which makes it invalid. This ethical and scientific difficulty forces the researcher to have recourse to other investigation methods, such as numerical simulation. To provoke a "calculated phenomenon" on "numerical objects" is a representation that may be very far from reality. We cannot confuse simulation and experience. This more frequent recourse to computerized simulation is a meeting point with other sciences, without substitution of paradigms or methodologies belonging to HSS.

The HSS have something in particular, and that is to look at mankind with a perspective of mankind. "To undertake as a research project a group of phenomena pre-defined by some external characteristics" [DUR 77] consists of adopting with regards to the study of mankind, the same social position (external) as with the objects of nature, whether in the order of the living or of the material. The researcher in HSS is therefore always more or less concerned directly by his object, being himself within society. Considering the phenomena only by their external character is only a part of the research. Thus, for Weber, "the knowledge of causality *laws* cannot be the *goal*, but only the means of research" [WEB 04].

The construction of the hypotheses is a matter of thought, representations, perception, opinions and values that mankind produces at a time in the history of humanity, at a place on Earth (and now outside it), in the social position that characterizes it. "What is stated as the knowledge object… is largely imposed by the instance that holds the cultural codes" [VAL 96]. Although we can establish laws of probability, we cannot exactly predict a particular random event that would be submitted to that law of probability. We can simulate the dice game but we cannot know the number that will fall at the next throw. Thus, how is it possible to model the portion of human activity that would result not only from a random event, but also from the will to escape it? Therefore, it seems to us that all human activity may be modeled, but only partially.

The HSS phenomena involves a system with a great number of elements of behavior, all different in a more or less direct interdependence according to social groups. Each system element is already itself a greatly complex system with a smaller or larger number of levels of freedom. The difficulty of the mathematical formalization of that field is a result of that. Social sciences should use more non-accountable mathematics for the benefit of more adaptable mathematics, translating the social dynamics more than the quantitative comparison, the solidarity more than the total, the complex organizations more than the holistic simplifications or some generalizations of the methodological individualism. Many mathematical and theoretical studies agree on that, such as the system and complexity theory, morphological theories, topology, fuzzy set theory, the theory of possibilities, the theories of languages, of knowledge-based systems, of distributed artificial intelligence, of cellular automata and multi-agent systems, and many more. These works should be revisited by HSS researchers to make them evolve and to invent others always moving in the direction of their research agenda.

Thus, we must avoid reducing the reflection on the notion of the model as much as a modeling practice to the exportation of physical or natural sciences methods to HSS. If this exportation was productive at the beginning, it seems now that HSS must find their own tools of thought, their own models, their own formalizations and theories through the problems, objects and phenomena that are specific to HSS.

1.4. Modeling: explain to understand?

The model is a type of explanation of the fact that it represents well enough and is simply a part of certain characteristics of reality or helps foresee its evolution within certain limitations, this being the reason that J. L. Lemoigne [LEM 77] established his objective of "modeling in order to understand".

Since A. Comte [COM 94], the positivist ambition of being completely able to explain the world by a combination of rational and mathematical laws must of course be tempered. Heisenberg's principle of uncertainty, Gödels's theorem of incompleteness, the various theorems of incalculability, of indecisiveness, the diversity of interpretations (the duality of the corpuscular and ondulatorial interpretation of the atom by Niels Bohr) have somewhat modified this vision towards a vision whereby science can only approach reality without ever understanding it or totally mastering it, or even being able to impartially observe it and unable to interact with it. We must now accept many interpretations of the same reality. We must accept not being able to always find simple explanations to complex phenomena. We must accept that phenomena may occur at random, according to a stochastic process if there exists a law of probability and at worst according to a chance with an unknown law. We must accept that chance can produce organization and that an organization may become chaotic. We must accept that there exists at the same time universal principles and singularities, individualities, and that there exists uniformity and diversity. Finally, we must accept that nature is both determinist and non-determinist, and that there exists phenomena that will never be explained.

A model *represents*, in the universe of knowledge, a portion of the universe of reality, but cannot be its equivalent. The universe is too vast in its scope, in its infinitesimal granularity, in the diversity of its objects and its phenomena, to hope to be able to wholly explain it in an exhaustive representation. Are there enough atoms in the universe to store all of this knowledge? Not unless its construction could be resumed in a few elementary laws.

Even if there were only a small number of laws enabling us to produce all of the workings of the universe, including the living, and that we could know them, this would in no way hinder the other disciplines from continuing to pose themselves pertinent problems at their level, because the knowledge of the elementary mechanisms of the production of reality does not produce the keys to the intelligibility of each level of complexity, permitting them to be efficient at this level. Each level of complexity produces its own coherence, and is only correctly understood and formalized at its own level. Chemistry is explained but is not well understood using only the physics of the particles; turbulence is not well understood based on the individual movement of each molecule; psychology is not well understood by biochemistry and society is not well understood by the psychology of the individual, etc.

If reductionism arrived at its ultimate goal this would be a great step forward in the intimate comprehension of the universe, but would not resolve, however, the scientific problems of each discipline and in particular those of HSS, from biology to sociology and even medicine, because the elementary mechanisms of reality are not generally better adapted to the possibilities of representation, verbalization and formalization of our brain.

Everything occurs as though, at each level of complexity, there existed certain properties or "laws of nature", intelligible and specific to this level, which allow us to efficiently approach it in order to understand it, formalize it, act upon it or make correct previsions. However, these laws do not exist in reality, they are only the result of the stepping back of the observer who erases the true constituting details, such as the painting by Magritte entitled "this is not a pipe". In effect, there are only color pigmentations side by side that make an illusion.

Thus, science does not have an objective reason to believe it could understand reality in its totality. Every bit of reality should be assumed not to be totally observable and *a fortiori* not totally intelligible, thus not totally able to be modeled. On the other hand, in model theory, we take a simpler position and on the contrary there may exist a complete semantic model of a formal theory that defines a perfect equivalence between the syntax level of the theory and the semantic level of the model. However, our position in principle concerning reality is reinforced by the theorems of incompleteness, since even basic mathematics such as arithmetic cannot be completely formalized. Thus, if the universe of the integers is already too complex to be completely formalized, we have strong presumptions to think that it is worse for "the reality".

The modeling must, though, content itself with a posture "between two chairs". Nothing is completely able to be modeled, but nothing is totally unable to be modeled. The problem with the "truth" of a mathematical model, defined as a perfect equivalence between the model and the reality that it represents, is thus a scientist and Aristotelian myth. It is a false problem, as if the reality or science could be resumed to a binary logic. Rather than knowing if a model is true or false, it is worthwhile defining its characteristics, its qualities of correspondence to reality. They must be researched not only through observation, but also in comparison with the objectives and the problems created for this modeling. This is expressed in terms of model category (qualitative, quantitative, probabilistic, etc. model), of precision, of the field of validity (in time and space). In 1998, Dahan, Dalmedico and Pestre criticized Sokai's theses [DAH 98] saying that "the question is not so much to say or assure oneself that reality exists, as saying how we apprehend that reality, saying how humans can judge the adequateness of their constructions and theories to this reality. That is the only important question, the only difficult question, the one that deserves consideration by us."

1.5. Bibliography

[AND 04] D. ANDLER, D. LASCAR, G. SABBAGH, *Théorie des modèles*, Encyclopaedia Universalis V9, 2004.

[ANS 99] P. ANSART, *Dictionnaire de sociologie*, Robert, 1999.

[BAR 01] J. BARZMAN, M. BROCARD, C. DUCRUET, A. FRÉMONT, D. REGUER, *Transports: Mutations des techniques, emploi et cultures professionnelles*, Les rapports du CIRTAI, UMR 6063 CNRS, Le Havre.

[BOU 90] R. BOUDON, F. BOURRICAUD, *Dictionnaire critique de la sociologie*, PUF, 1990.

[BRU 86] R. BRUNET, "Chorèmes et Modèles", *Revue Mappemonde*, no. 4, 1986.

[COM 94] A. COMTE, "Cours de philosophie positive", in *Œuvres* 1894, reprinted by Anthropos, Paris, 1968

[DAH 98] A. DAHAN DALMEDICO, D. PESTRE, "Comment parler des sciences aujourd'hui?", *Alliage*, no. 35-36, 1998.

[DUR 77] E. DURKHEIM, *Les règles de la méthode sociologique* (19th ed.), Paris, PUF.

[DAR 82] C. DARWIN, *De l'origine des espèces par voie de sélection naturelle ou des lois de transformation des êtres organisés*, Paris, Marpon et Flammarion (4th edition), trad. C. Royer.

[GUR 56] G. GURVITCH, "La crise de l'explication en sociologie", *Cahiers internationaux de sociologie*, Vol. XXI, PUF.

[LAB 70] H. LABORIT, *L'agressivité détournée*, Ed. 10/18.

[LEN 69] R. LENOBLE, *Histoire des idée de nature*, A. Michel.

[SAR 60] J.P. SARTRE, *Critique de la raison dialectique*, Gallimard.

[SPE 96] H. SPENSER, *Principes de sociologie*.

[VAL 96] B. VALADE, *Introduction aux sciences sociales*, PUF.

[WEB 04] M. WEBER, "l'Objectivité de la connaissance dans les sciences et la politique sociale", in *Essais sur la théorie de la science*, Plon, 1965 pp. 164.

Chapter 2

From Classic Models to Incremental Models

Mathematical models applied to urban planning underwent major developments in the 1960s. Since that time, new scientific and technological breakthroughs have led to deep-rooted changes in the domain of spatial modeling, as a reaction against the idea that reality could be reduced to a few fixed models, whereas, as Ilya Prigogine writes in his preface to [PUM 89], "the city is an open system, a permanent site for micro-events, some of which are amplified and others attenuated".

Progress in information technology brought about new perspectives, notably for urban simulation models, relying on the launch of Geographic Information Systems. The appearance of new software in the 1980s encouraged more and more simulation to be substituted by the analytical resolution of mathematical models. The model losing in generality and reproducibility what it gains in adaptation to specific local situations, we must not underestimate the risk of a certain disjunction between theorizing and modeling [PUM 96]. Traditional modeling appeared to be, in fact, better adapted to the reproduction of a past structure rather than to the prediction of an evolution, whereas in the case of the management of social objects, the challenge, as Jean-Louis Le Moigne [LEM 77] states, is to "pilot projects and not structures".

The introduction of prediction in models of complex socio-economic systems thus led to the development of paradigms of auto-organization, fractal geometry or a "deterministic chaos", but research stumbled upon the simulation of the emergence of a veritable novelty in the running of a system.

Chapter written by Yves GUERMOND.

2.1. The geographic "object"

The processes that enable interaction between elements of a system are the result of complex actions by an environment, which are expressed in a combination of spatial links. A system, according to the definition proposed by Patrice Langlois (see Chapter 11), "is supposed to describe a portion of reality unequivocally restricted between two levels of scale and knowledge: externally restricted by the environment which encompasses it and internally limited by its terminal objects, which we do not seek to explain (or understand) the workings of, but each of these nevertheless operating as a system". "The geographic object" is thus a confusing term, albeit a frequently used one because "the object" itself is, on another scale, a system: the city, the factory, the agricultural farm, the house are "objects" only at a given level of observation. Each magnitude can be considered as being indivisible with respect to a higher magnitude, and as being successively made up of an infinite number of lower indivisibles. It can thus also be stated, notes Jean-Pierre Cléro [CLE 03], "that a point does not have a dimension with respect to a straight line, but it can be considered as originating from an infinite number of indivisibles, in which case, it would have one dimension more than these, and the straight line, made up of an infinite number of points, would then have two dimensions. The infinitesimal calculation has made the number of dimensions relative".

In the geographic space, the difference of scale between the "objects" includes differences in operating and in the capacity for intervention in the modeling process. While households can be considered as actors in the development of the city, what about a localized group of households, a town, a city, a district? To what extent can we speak of the influence "of a town" or "of a district" on the neighboring districts? The spatial partition is, in fact, always a mental construction relying on subjective criteria. Administrative and political divisions are fluctuating and artificial, divisions based on demographic densities or on the landscape have fuzzy borders, and ethnic criteria are very unstable. It is possible, as Roger Brunet [BRU 90] writes, that the levels of expression of the geographic spaces are just a question of representation, but "these levels of emergence, organization and integration are related to masses and distances that make it possible to integrate enough numerous and coherent elements to create a mass, and this close enough to ensure intercommunication". It is from this idea of the emergence of "geographic spaces" at certain size levels that traditional spatial models have been developed.

2.2. Lessons from the "classic models"

Emulating Léon Walras's [WAL 00] "pure economy", traditional models intend to build a "pure geography" based on some general rules establishing what could be called "social physics". Needless to say, this approach was taken to pieces under the pretext that it was exclusively technical, if not technocratic in nature. The models by Lowry, Wilson and Forrester can be put down as milestones for the 1960s, 1970s and 1980s, before the introduction of the idea of auto-organization by Peter Allen and the Prigogine team.

Lowry's model [LOW 64] is the oldest (a very clear and detailed presentation can be found in [AKI 86]). Based on the mechanism of the "economic base", its objective was the spatial allocation of the residences of a population generated by the setting-up of new jobs. The model starts by allocating the "population to be distributed" (PD) among the residential areas located around the basic jobs placed at the starting stage of the model. This distribution is performed by a gravity model, according to the distance function $f(c_{ij})$, defined with reference to routinely observed work commuting. Every zone j receives a population according to the ratio between its distances from the employment locations and the sum of all the distances from the jobs of the different zones taken into account:

$P_j = PD ((\Sigma_i E_i f(c_{ij})) / (\Sigma_{ij} E_i f(c_{ij})))$, with:

PD: population to be distributed;

E: jobs placed at the starting stage of the model;

$f(c_{ij})$: transportation cost (in accordance with the distance) between the areas i and j.

The service jobs required for this population are subsequently calculated. These service jobs to be distributed (SD) are proportional to the population of the different zones, according to a "service rate" to be fixed for each region. The spatial distribution of this service jobs is carried out by a second gravity model, in accordance with the location of the populations already settled (Pj). The distance decay function taken into account (f') may be different from the function used for the migrations towards the basic jobs.

Constraints may be decided, for example a maximum density of inhabitants per habitable acre as well as a minimum number of service employment for each zone, according to the types of service. Due to its rigor and theoretical simplicity, we can talk of a "model of models", following the example of what could be said of Von Thünen's model, although with, in comparison, a complication (not a complexity) that absolutely requires recourse to computerization. While it is true that the calculations are iterative, they are considerable in number, in accordance with the number of zones to be taken into account.

The quantity of data required at the starting point of the model constituted a restraint to its use. Among this data, some may be estimated with a good empirical knowledge: the service rate, the constraints of maximum density or minimum number of jobs. Many, however, require rather in-depth preliminary studies: the transportation cost (or simply the distance) between the zones, and mainly the ratio between surfaces and jobs. The area available for the dwellings in each zone A^d_j necessitates being able to deduct from the total area A_j, the unusable area A^u_j as well as the area devoted to the basic employment A^b_j and the area devoted to the service employment A^s_j:

$$A^d_j = A_j - A^u_j - A^b_j - A^s_j$$

We should thus be able to convert the jobs into surfaces, that is, to evaluate two parameters: the average surface occupied by one basic employment, and the average surface occupied by one service employment (according to the types of service employment).

Wilson's model [WIL 74] resumes a gravity model proposed by Huff at the same time as Lowry's model [HUF 63] by analyzing it further. Applied to the distribution of the customers among various commercial centers (Huff) or to the distribution of journey-to-work patterns (Wilson), the basic principle is still the same: the number of persons affected by the journeys (T_{ij}) is proportional to the mass of the origin (O_i) and destination (D_j) poles, and is inversely proportional to the distance d_{ij} between these poles. The "most probable" configuration is the one that maximizes the entropy of the trip distribution pattern. Peter Haggett [HAG 77] gives a very simple example illustrating it on a 6 box table whose margins are known (Figure 2.1).

The maximization of the entropy of the table can be calculated by Shannon's formula, which had been applied to mathematical expectation of the quantity of information expected at the exit point of a channel. The entropy function to be maximized is as follows: $H(\{T_{ij}\}) = -\Sigma_i \Sigma_j t_{ij} \log t_{ij}$.

	D_1	D_2	D_3	
O_1				4
O_2				3
	2	3	2	

Only 8 configurations satisfy these constraints:

(1)	1	3	0	**(2)**	2	2	0	**(3)**	0	3	1	**(4)**	2	0	2
	1	0	2		0	1	2		2	0	1		0	3	0

(5)	2	1	1	**(6)**	0	2	2	**(7)**	1	2	1	**(8)**	1	1	2
	0	2	1		2	1	0		1	1	1		1	2	0

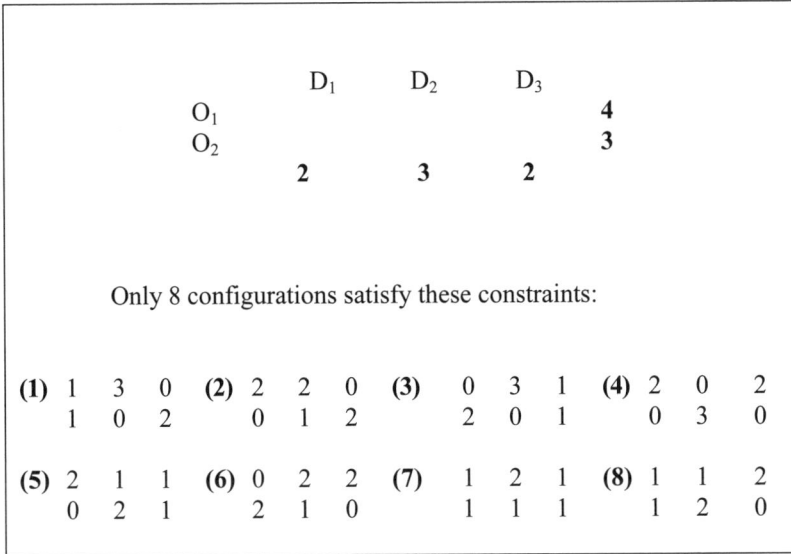

Figure 2.1. *Entropy of a table*

In the example given above, the entropy maximization is obtained by the 7th pattern (in which there is no zero box remaining):

$$H(\{T_{ij}\}) = -(1/7 \log 1/7) - (2/7 \log 2/7) - (1/7 \log 1/7) - (1/7 \log 1/7) - (1/7$$
$$\log 1/7) - (1/7 \log 1/7) = 0.759$$

The maximum possible entropy of this 6-box table would be:

$$H(\{T_{ij}\}) = \log 6 = 0.778$$

A.G. Wilson demonstrates that, under the constraints of O_i and D_j and of a transportation function cost to be determined, the most probable distribution (which maximizes the entropy) corresponds to a gravity model:

$$T_{ij} = O_i D_j A_i B_j e^{-\beta C_{ij}}$$

Wilson's equation consists of three elements:

– O_i and D_j are the emission and attraction indexes;

– A_i and B_j are weighting coefficients of the geographic areas ("Lagrange's multipliers"), which correspond to the characteristics of the different localities:

$$A_i = 1 / \Sigma_j \, B_j \, D_j \, e^{-\beta C_{ij}}$$

$$B_j = 1 / \Sigma_i \, O_i \, A_i \, e^{-\beta C_{ij}}$$

In order to calculate A_i, first B_j is fixed at 1, then B_j is calculated with the approximate value of A_i thus found, and the values are progressively improved by a series of iterations.

– $e^{-\beta C_{ij}}$ is a " generalized cost of transportation".

This "generalized cost" is related to a modification of the Euclidian distance by four elements:

– the layout of the network;

– the frequency of the means of transportation;

– the travel cost;

– the journey time.

These four elements are combined in a complex way, but always according to the distance. There is an abundant geographic literature on the distance decay function, which we will not discuss here, but it is evident that this reflection on the different "friction" of the distance related to the nature of the observed phenomena is an essential element of reflection on the geographic space. The migrations towards hypermarkets cannot be analyzed in the same way as the migrations towards corner shops, and journey-to-work trips of the employees of a small town have nothing in common with those of Parisians or New Yorkers.

The "geographic space" is the spatial framework in which the activity or the lifestyle of a social group is exercised. This "interaction field" I, as defined by Hägerstrand [HAG 67], is not the same as the "perceived space" of the sociologists, since it is not in the domain of cognitive psychology. The geographic space of the executives is not that of the cleaning staff and this depends not on their perception, but on the manner in which their network of relations is physically structured, independently of themselves. It is not "the lived space" either, since within the same field of interactions, for example that of the urban middle classes, several "lived spaces" coexist, differentiated only by the preferences and activities of each one.

The degree of interaction thus takes different shapes according to the distance. Those more frequently retained for investigation can be summed up by "Goux's typology" (Figure 2.2, from [TAY 75]).

The distance function may be a power function, of the form

$$f(d_{ij}) = d_{ij}{}^{a}$$

In the urban space, it is often an exponential function, of the form

$$f(d_{ij}) = e^{-\beta C_{ij}}$$

but numerous shapes of the role of the geographic space are possible.

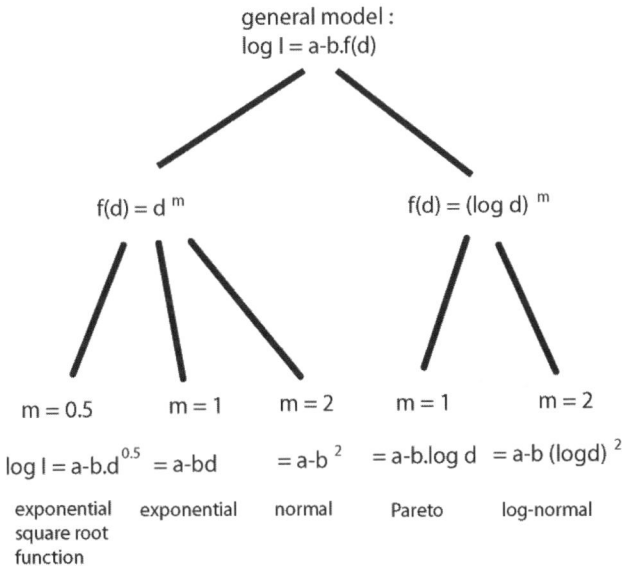

general model :
log I = a-b.f(d)

f(d) = d ^{m} f(d) = (log d) ^{m}

m = 0.5   m = 1   m = 2   m = 1   m = 2

log I = a-b.d$^{0.5}$ = a-bd   = a-b ^{2} = a-b.log d   = a-b (logd) ^{2}

exponential   exponential   normal   Pareto   log-normal
square root
function

Figure 2.2. *Interaction field according to distance*

There are a lot of various calibration programs based on the analysis of the differences between theoretical and actual flows observed on a sample panel or on a past situation [FOT 89]. A paper from M.A. Laurent and I. Thomas [LAU 97] for example, illustrates this issue by applying it to the residence of thieves with respect to the location of the thefts…

The analysis of classic models may arouse the feeling that they are purely descriptive and static. This is not completely true, since while it is certain that they seek to reproduce an observed situation, they also have an ability to put forward new solutions. They adapt themselves perfectly to the objectives of the "geographer-developer", the one who says "where the main equipment must be located", in this often criticized role of "advisor of the Prince". It is not a vain function, but it is obviously an encouragement for the technocratic tendency of geography: scientific research intervenes on a fundamental invariant, which may be considered as unbiased, and the consequences to be drawn on the action level subsequently belong to the domain of the "politics", eluding scientific analysis.

2.3. Introduction to dynamics and auto-organization

In a context of contestation of the deductive methods in the 1960s, the models of system dynamics represented, in the 1970s, another way of modeling. The inspiration came from J.W. Forrester's work [FOR 69], whose objective was to "study the problems of the ageing urban areas by using methods developed recently with the view of understanding the complex social systems". For him, "the interactions between social and economic activities are so complex that intuition alone cannot make it possible to conceive policies making it possible to anticipate a decline... There is thus no reference to urban literature in this book", since the author claims that "it originates really from a different corpus of knowledge", in which discussion holds an important place: "the most valuable source of information [comes] not from documents, but from persons with practical experience in urban affairs".

The elaboration of a graph is a pre-requisite for the construction of the model, in which the state variables ("stocks") are governed by the flows. The differential equation that takes them into account (d STOCK/d t) depends on the entries and the exits. In order to control these entry and exit gates, it is necessary to introduce some "auxiliary variables", obtained either by threshold functions (if X > Y then Z) or by tabulated functions.

The tabulated functions make it possible to express non-quantified relations, originating from an intuitive terrain knowledge, which is certainly an advantage of this type of models as compared to the strictly quantitative models. The other positive features of these models have often been emphasized: they make it possible to take into account simultaneously numerous related phenomena linked by non-linear relations of a different nature; they also make it possible to integrate time and to conduct multiple and fruitful simulation experiments. Inversely, this empiricism leads to the loss of any ability of generalization, and even of prediction. A famous application was developed in France by geographers in the AMORAL model "Analysis and rural modeling of the Alps" [CHA 84].

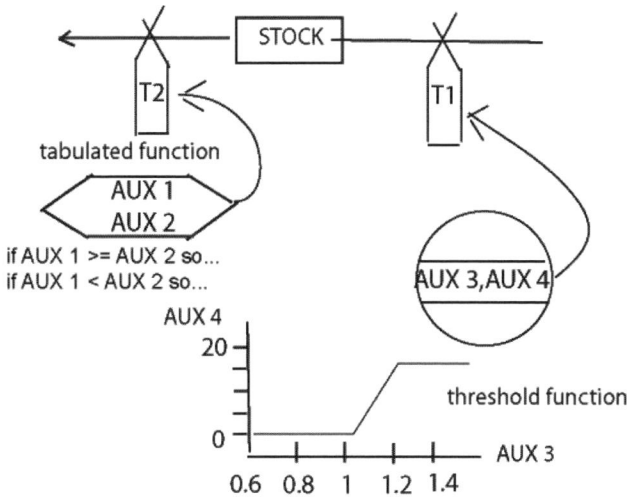

Figure 2.3. *Graph of a dynamic model*

The aim of the study on the Alps was to simulate demographic evolution of an alpine "country": the "stock" was thus the local population and the entry and exit gates are immigration and emigration. The auxiliary variables are elements that are supposed to play a role in this residential mobility. The effective immigration is a function of a "potential immigration", which itself is a function of the "country's image", as it is perceived from the outside. The figure given above expresses this tabulated function: the abscissa (the country's image) is calibrated from 0.6 to 1.4, with a step of 0.2, with the value 1 representing the initial image of the country at the starting point of the modeling process. The successive values of the potential immigration were placed on the ordinate, in accordance with this image variation. If the country's image is worse than the initial image, the potential immigration is presumed to be zero and it increases to a maximum of 15 (saturation threshold) when the image attains or surpasses 20% improvement.

The development of the program gradually generates a rather complicated structure as new explanatory elements are integrated: the country's image depends for example on the rate of settlement of secondary residents, but a threshold function can intervene... Beyond a certain rate of secondary residents, the internal relations of the village are modified, which can bring about the departure of certain initial inhabitants of the country, and thus have a bearing on emigration. It can be seen how this modeling approach can facilitate the debate with the concerned populations and lead to a certain formalization of arguments exchanged during a discussion, even when these arguments are not quantified.

The main deficiency in this type of model has been for a long time their non-geographic nature, as the spatial interactions have not been taken into account initially. Each regional evolution is analyzed separately and the influences of the neighboring regions, depending on the closeness of their proximity, could not be integrated.

As noted by Paul-Marie Boulanger [BOU 03], the "system-dynamics", after having witnessed a phenomenal success, "went through a long period in the doldrums following numerous criticisms leveled mainly by the economists about the models of the world that it inspired" (the World 2 and World 3 models by the *Club of Rome*). However, at present, adds the author, "it seems to have found a second youth with the question of durable development and the recognition of the environmental issues". Fairly simple simulation software has contributed to this diffusion: P.M. Boulanger notably cites the model (with the software STELLA) of the evolution of a coastal landscape bearing 2,479 interconnected spatial cells [MAX 94].

The recognition of the spatial interaction in the dynamic models led to the auto-organization model proposed by Peter Allen [ALL 81], based on Ilya Prigogine's work. This model was presented in detail by D. Pumain *et al.* [PUM 89]. Its great advantage, for geographic analysis, is that it links with one another the differential equations which refer to the evolution of each region. The management of the model needs to have an access, at the beginning, for every locality, to the distribution of industrial and tertiary jobs and that, among the tertiary employment, we could be in a position to distinguish between a "fundamental" employment and a regional and local employment. The distribution of the resident population can be limited, although it is a bit old-fashioned, to a distinction between the "blue-collar workers" and the "white-collar workers". However, as the authors emphasize, "the parameterization of the model is a complex stage, given the very high number of unknown parameters. Each of them has an influence on all the system variables in a more or less direct manner, through the close interrelations that exist between them. While the direct effects can be predicted and easily dominated, the indirect effects do not always act as predicted and make the calibration difficult". The application of the model makes it possible to perform simulations of the spatial evolution among a lot of different hypotheses, and for example, to choose the size, the optimal location and the date for setting up new amenities, in order to optimize their insertion in an urban network.

The main force behind the evolution comes from the difference, for each zone, between the real rate of occupation, and its "potential" (its ceiling value), this potential being itself constantly reviewed. Population growth is in fact limited in every region by the quantity of resources available: it should thus follow a logistical curve, that is, tending towards saturation.

The equation of this curve is of the form:

$$dY / dt = \varepsilon Y (1 - Y/P)$$

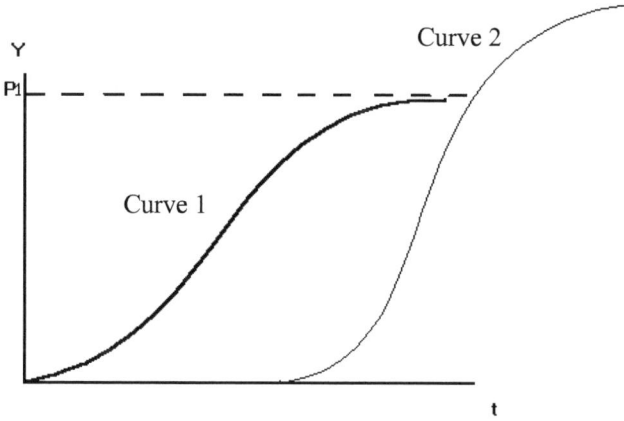

Figure 2.4. *Evolution over a long period of time*

The Y variation is proportional to the time t, according to a growth rate ε, but this growth is increasingly limited as the threshold P draws closer. Each ceiling thus attained is provisional, depending on the interference with the neighboring regions. The evolution curves over the long term show a succession of growths and ceilings (Figure 2.4).

The authors thus construct a simulation of evolving distribution of jobs and resident population in an agglomeration (Rouen). For the employment, a distinction is made between export activities ("basic" employment) – industrial (E1) or fundamental tertiary (E2) – and induced employment (local tertiary E3 and regional tertiary E4). For each of these employment categories E, the surface evolution which is allocated to it in each zone j, dS_j^E, is defined by a logistical equation:

$$\frac{\partial S_j^E}{\partial t} = \varepsilon^E S_j^E \cdot \left(1 - \frac{S_j^E}{D^E \cdot \dfrac{A_j}{\sum_i A_i}} \right)$$

In this formula, the "potential" of each zone for a given activity is determined by the proportion that the attractiveness A_j of the zone j represents with respect to all the attractivenesses of all the zones studied. This proportion is balanced by the "external demand" D^E in the type of employment considered. In the absence of data enabling estimation of this external demand, the simulations simply used, for retrospective studies, the total number of jobs actually attained at the end of a period in the activity in question, which makes it possible to calibrate the model for a historical study, but obviously does not resolve the issue for a predictive study.

What remains to be defined is the comparative "attractiveness" of each zone, which is defined by the number of jobs (balanced by accessibility) according to the available surface. It requires quite a complex calibration of a certain number of parameters, for which we must admit that existing geographical research leaves us with few resources, resulting in rough estimations. Let us cite, among these parameters, the propensity of an activity to agglomerate, the accessibility of each zone, the sensitivity of entrepreneurs to location, the surface required per employment in the analyzed activity, etc.

A similar analysis is carried out for the evolution of the active resident populations, by distinguishing between the "blue-collar" workers and "white-collar" workers. There is thus a system of six sets of simultaneous differential equations (for the four categories of employment and the two categories of population), that continuously interact with each other, particularly by competing for land use. This acknowledgement of spatial interaction entails the implementation of a relatively complex device.

2.4. From auto-organization to complexity

In the construction of an auto-organization model "the local interactions eventually produce a structure, which cannot be determined *a priori* by accountable equations…Validation of local interaction and growth mechanisms depends on the model's ability to bring about the emergence of plausible macro-geographical structures" [SAN 97] . Edgar Morin's words may be recalled here [MOR 82]: "The idea of permanent, anonymous, sovereign laws guiding everything in nature, is substituted by the idea of laws of interactions, which means depending on interactions between physical bodies which depend upon these laws".

The concept of auto-organization raises many questions. "In the real world", writes Denise Pumain [PUM 03], "novelty is produced, it actually emerges from what exists…but simulation of the emergence of a true novelty by means of models continues to be very difficult". The "auto-organization" of Peter Allen's model was mono-scalar, which restricted the possibilities of emergence apart from those that were fixed by the initial rules of the model. The researcher's dream would be to succeed in stimulating this emergence not by introducing events outside the model, but as an outcome of an interaction between the internal elements of the model, that is to say eventually by coupling simple mechanisms. According to J.P. Dupuy [DUP 99] auto-organization is impossible, since a computer program cannot go against its own rules…We can however assume that evolution may be brought about by elementary interactions, and that these interactions create a complexity that eventually generates new properties: the CNRS program "Complex systems in Social and Human Sciences" (2003) in France thus affirms that a "complex system possesses a holistic behavior which renders futile any attempt to analyze it by division into simpler sub-systems". In other words, according to the recognized formula, the whole is greater than the parts, since certain mechanisms inherent to complexity, can appear by chance, simply due to the interference of several deterministic processes. A vast literature has been developed on these "emergence" phenomena. However, as André Dauphiné notes [DAU 03], real geographical dynamics are more often uniform than chaotic, since space, he thinks, "is a constraint that is opposed to mechanisms that generate complexity". Space transforms itself slowly, which often favors adaptations, sometimes bifurcations, but rarely chaos.

A chaotic evolution can be produced by a "relatively simple" system of differential equations. A single non-linear equation is sufficient to cause unforeseeable system behavior, as demonstrated by Lorenz's system [DAU 03], which is why it is impossible to make long-term meteorological predictions. We can thus talk of a "deterministic chaos" (that is, not caused by intrusions from outside the system). Thus the objective of complexity theories, we could say, is to limit the hazard factor. It can be considered (although it is something of a caricature) that, in "traditional modeling", the temptation was to explain the presence of model residuals as a random occurrence. The chart below (Figure 2.5) attempts to express this desirable reduction in the part that randomness plays in the functioning of systems.

The objective of modeling is to carry out a stepwise increase in the search for determinisms, to which "division into sub-systems" can contribute effectively, contrary to the affirmation of the CNRS document cited above. This role of determinism is evidently limited when we are confronted with an absence of structure, and we can hence simply hope to find certain regularities likely to reduce "complexity" up to a certain proportion.

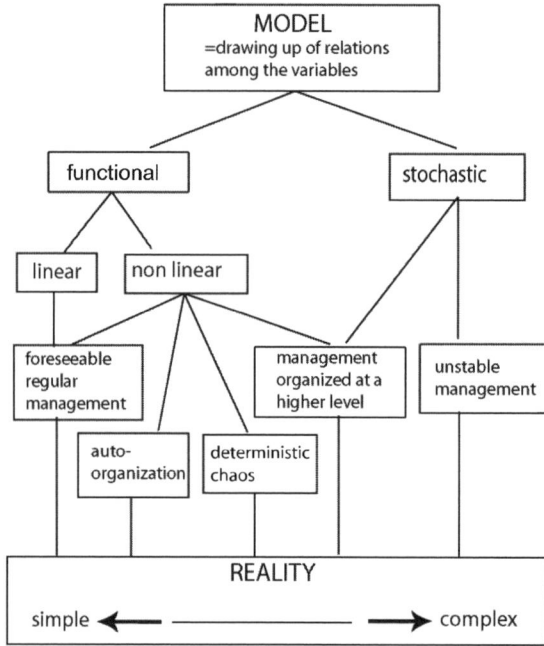

Figure 2.5. *Types of relationship in the models*

Very often, mainly in human geography, when reorganization takes place in a system, it is due to external disturbances, or to evolutions occurring on another scale, and not due to internal fluctuations. In a simulation of the spatial growth of a city during the second half of the 20th century [DUB 03] it can be seen that the major events that led to "bifurcations" were the building of residential complexes in the 1960s, followed by the creation of peripheral commercial centers in the 1970s, the "vogue" of the technological parks in the 1980s and the acknowledgement of industrial risks in the early 21st century. It must be admitted that although they could have been predicted in a sagacious intellectual reflection, their insertion in a forecasting model was problematic. It is likewise for all important "bifurcations" in human history, from the industrial revolution, introduced from the outside into the SIMPOP model [SAN 97], up to decolonization in the 1950s, and the fall of the Berlin Wall. It is evident that the exploitation of lunar resources, when it occurs, will lead to a transfer of investments into the spatial conquest (at best, or into military expenses at worst), but no model could ever ascertain which "bifurcation" will occur, since "neither the cosmic future, nor the biological future, nor the anthropo-social future can be deduced from algorithms" [MOR 82]. What we know, on the other hand, is that self-organized systems possess a behavioral adaptability, such that "the future cannot be known beforehand".

It is this adaptability which is pertinent to study, by dismissing the idea of teleological processes, according to which a system could have a project. Teleological causality is, in Kant's [KAN 65] words, a causality directed "towards ends independent of natural conditions and necessary in themselves". The gravitational attraction of a city center, for instance, or of an industrial zone, is an external finality. It lies in probabilistic dependence on a natural law. Kant thus distinguishes three kinds of finalities: that of nature, that of the organized beings, and that of man as a moral agent, which is the only one with a clearly teleological causality. The question arises as to the aptitude of the organized society having a finality corresponding to a project, independently of the material constraints. We can sense that yes, but we can also sense that it is hardly possible to apprehend this highly tenuous and elusive ability in a "geographical system", which is mainly founded on material elements. If the system integrates only material elements, it cannot scientifically isolate the role of myth in the accomplishments of societies. This is where the limitation of the positivist approach lies.

The problem arises because, when we try to introduce non-material indicators in a program, they are either subjective or uncertain, like "the sensitivity of the entrepreneurs to the advantages of location" in Allen's model, or elsewhere "the sentiment of regional identity".

The evolution may occur globally on the basis of individual deterministic hypotheses. The best known example is Schelling's model [SCH 78], where, starting from a random distribution of dissimilar populations, segregated nuclei are constituted after a certain number of iterations, on the simple basis of rules on tolerance towards social diversity, which by themselves did not imply segregation. Similarly Dominique Badariotti and Christiane Weber [BAD 02] take as decision-making units the household heads, divided into groups according to 6 social categories, 4 age classes categories and 2 statuses (proprietors/tenants). The propensity of a group to relocate (which varies from 0 to 1) depends on the degree of a group's affinity with the neighboring groups. The transition function corresponds to the initial number of households in each sector, modified by the sum of the movements between this sector and the neighborhood, for the concerned group:

$$n_G^{t+1}(s) = n_G^t(s) - \Sigma\, f_G(s,v)$$

with:

n_G: number of households of the group G;

s: concerned geographical sector;

f: flows between sector s and the neighborhood;

v: vicinity, defined as all the "nearby" sectors for which the degree of affinity of the group G is greater than that of the sector s.

These models do not however integrate retroactive phenomena. Fortunately, a lot of reasons could contribute, if not to eliminate, at least to restrain urban segregation: these could be political decisions (public construction programs), or individual reasons (family ties between diverse populations) or simply a vague "learning" phenomenon, related to the consciousness of the risks for social peace of a total apartheid. In this respect, it will always be difficult to reconcile the "scientific" vision and the "real life" subjective vision. The outcome, for every researcher, runs the risk, according to Edgar Morin [MOR 82], "of being a schizophrenic solution, that is, at two levels of thinking that never communicate. Thus the technocrat sees a society of determinisms, mechanisms, processes, but from time to time, the technocrat makes a philosophical leap, sees the society made up of fellow-citizens and subjects with their problems and needs".

To avoid facing these questions, the geographer often withdraws into the "morphological" aspect of his work, limiting himself to the study of the form, studied for itself. In the DLA (*diffusion limited aggregation*) model, where each animated particle of a Brownian movement aggregates with the mass that it encounters [DAU 03], the process may be easily oriented towards the formation of axes, or rather of nuclei. The risk is to end up in purely formal models and to take the form for the phenomenon. Speaking, for example, of a fractal city, if we try to account for a certain similarity of the urban forms at different levels (neighborhood, district, agglomeration), we are led to separate the forms from the causes. The form includes all the external features of an object, and these are not necessarily related to its nature. Let us consider the "form" of the constellations. Is it perhaps better to turn our attention directly to the causes, rather than reflect on them by using the forms that they have created as intermediaries?

2.5. Spatial agents

In order to free ourselves from a disembodied conception of the model, the idea of introducing "agents" in the modeling process naturally arose. A pioneer experiment of "multi-agent systems" was carried out by Léna Sanders, Denise Pumain *et al.* [SAN 97]. It involved reconstituting, over quite a long period, the evolution of a system of cities on a territory delimited by a grid of 236 hexagons. Each cell was characterized by a type of natural environment (plain, river, slope, etc.), it could accommodate human occupation (habitat, route, railway, etc.) and could benefit from natural resources (agricultural lands, mineral resources).

The simulation is based on an initial population pattern and natural resources. The ability of each cell to improve this initial potential depends upon the population figure, the technical level of the period, and the links with the neighboring cells. The status of each cell influences its growth potential, according to hypotheses on the relation between the levels of the functions (agricultural, commercial at level 1, 2, 3 or 4, administrative at level 5 or 6, industrial) and population growth. The state of the cells is related to qualitative variables (soil utilization) as well as to quantitative variables (population figure, proportion of traders, etc.). Information exchanges develop between spatial "agents" with a fixed position in space, contrary to what happens in non-geographical applications of multi-agent systems. The notion of "proximity" evolves during the modeling process, gradually, as the position of the cells in the urban hierarchy changes.

The term "agent" introduces some ambiguity, since it is often confused with the term "player" used when the "player strategies" are mentioned. The CNRS program "Complex Systems in Human Sciences", already cited above, thus considers "that although in physics the elements constituting the system are relatively simple and homogenous…in human and social sciences these elements are highly complex and differentiated, since they are "cognitive agents", equipped with representations, with memory capacities and intentions, able to develop individual strategies". This constitutes a simplistic conception of social sciences, perceived by physicians, unless it is the sign of a wrongly understood categorization of geography within the scientific field, related to its subject, which is firstly space and its effect on society (even if we cannot disregard the feedback of society on space).

The question does not arise in physical geography, where "agents" can be assigned a simple behavior: air density, soil permeability. In human geography, it would undoubtedly be pertinent to differentiate the geographer's work from that of the sociologist or the psychologist. The geographical modeling unit is not the individual (on which the pertinence of geographical analysis provides only partial, and sometimes even superficial results) but the portion of space that is in some way "de-anthrophized". If we consider the agent as a spatial cell, it is this cell that, having an activity (a highway, a railway station, a supermarket, a hospital), is acting simply by its spatial proximity. This action takes place independently of the "spatial strategy" (if it exists at all) of the hospital director or of the supermarket director, strategies which, regardless, do not fall within the geographer's domain. However, what the geographer can analyze is the effect of a change of strategy on space: hypermarket expansion, hospital closure, creation of a high speed railway station. Jean-Pierre Treuil *et al.* [SAN 07] opportunely distinguish "Eulerian", formulations, in which equations specify what happens in every fixed unit of space, and "Lagrangian" formulations, in which equations specify what happens in the elements. When space is perceived as a structure of relations, the researcher "thinks about a stratum of reality composed of player-type entities, and about the evolution of the perception that they have of the world, according to their own benchmarks …he tries to use the description of their behavior as the basis for explaining the phenomena described at a higher organizational level". On the contrary, when space is perceived as a reality existing in itself, we try to model the transformation of space-type entities in their forms, their properties and their relations.

We must be aware of the fact that a model can only adapt itself to an aspect of reality, and that it is impossible to take into account everything that influences the course of things, from macro-scale (geo-political evolution) to micro-scale (psychological evolution of individuals or groups), when we have to analyze, for example, at medium scale, the evolution of an urban area. Scientific research can only take on very well delimited research fields, even though public opinion drives towards an all-embracing analysis, as proven by the success of "the greenhouse effect".

2.6. Incremental modeling

The objective of modeling for a researcher is to reproduce the functioning of a geographical phenomenon in the best possible way, in order to try to use the model thereafter for testing hypotheses. The purpose is not to test "one" single model, nor to operate a system that invents itself by producing its rules in an endogenous manner. On the contrary the model should rather be considered as a research tool without normative ambition, and which the researcher constructs in a stepwise manner. Cellular automata are well suited for this "incremental modeling", which makes it possible, through successive simulations, to constitute this "virtual laboratory" referred to by Michael Batty [BAT 01] as indispensable to research in human and social sciences. In geographical modeling a cellular automaton is a structured organization of interconnected spatial cells that are organized according to a topology defining the neighborhood ties. In certain experiments carried out in the MTG laboratory, the state s of every spatial cell i at instant t+1 is a function of its state at instant t and of the state of the vicinity V at instant t:

$$s_i(t+1) = f(s_i(t), V_i(t))$$

This neighborhood V_i is defined according to the type of problem at hand. In a physical geography program on the simulation of surface runoff [DEL 04], the vicinity links are defined by the adjacency of the cells, and in a simulation of the evolution of land use in an urban area [DUB 03], these links are determined according to the distance from other cells. The basic idea is to allow a virtual experimentation, to appreciate the spatial outcome of a set of rules, and their consequences over time, in order to eventually validate the hypotheses, modify them or generalize them. In the example cited above [DUB 03], a base of rather complex "rules" has been outlined. This rule base is easily modifiable by the user, which allows him to make his rules evolve gradually with the simulations. In other cases, on a prospective basis, a single rule, such as a potential equation, is calibrated differently according to the type of land use. Thus, Françoise Dureau and Christiane Weber [SAN 07] assign a weighting of land use (for example density of dwellings) at each pixel of a satellite image, and this weighting plays the role of mass in a potential model:

$$P_i^{\alpha} = A + \Sigma_j \, M_j^{\alpha} \, f \, (d_{ij})$$

where:

P_i potential at the pixel i;

M_j^{α} weighting of land use for the pixels j;

d_{ij} distance ij, limited by a floating grid around the pixels i;

A is the value adopted for the pixel i itself, that is:

$$A = M_i^{\alpha} / (0.5 \, S_i^{0.5}/\pi)$$

S being the surface of the pixel i.

The simulation relies on the translation by rules of a certain experience. In "expert systems" reasoning is guided by general rules of logical inference ("If… then…"), and by proven events. As for the cellular automata, they contain, according to experiments by experts, rules of action, which directly influence the state transformation of the cells. Like in multi-agent systems, of which there are specific cases, the objects interact amongst themselves. Thus, three types of simulation can be summed up in the following way:

Expert systems	Multi-agent systems	Cellular automata
Rules of logical inference	Behavioral rules Reactive agents Cognitive agents (that interpret the environment)	Rules of action Reactive agents (according to the environment)
Work on an external object (an event base)	Distributed intelligence (objects that interact amongst themselves)	

Figure 2.6. *Three types of simulation*

The periodical readjustment of the modeling process is as indispensable as the rectification of the trajectory of a rocket. "Nor does any man imagine", wrote Hume [HUM 48], "that the explosion of gunpowder, or the attraction of a loadstone, could ever be discovered by arguments *a priori*". Far from just allowing the discovery of local laws, cellular automata models are a means of testing relationships between variables, that can be "corroborated" as Popper used to say [POP 59], as long as they are not "falsified" by other experiments.

In the construction of an aggregation model according to the classical laws of cluster formation, Andre Dauphine [DAU 03] is led, in order to simulate urban growth in the French Riviera, to integrate complementary geographical laws, in order to reproduce the observed evolution and the principles guiding it, in this case, the polarization towards existing cities, as well as selective attraction or repulsion processes according to relief and land use.

Jean-Louis Lemoigne [LEM 77] defined three projects for "systemography" (Figure 2.7). Conceptually, we start from the finalities, to manage an evolution of the structure in its environment, with respect to these finalities; for an analysis, we start from the functions, to interpret them with respect to some finalities, so as to infer from them the structure that can accommodate them. Finally "in the event of a simulation, we should start from a structure, obtained beforehand through conception or analysis, we should make it work and evolve in an environment, and we should compare the results of this simulated activity with conceivable finalities".

	Finalities	Environment	Structure	Functions	Evolutions
Conception	①	②	③	④	⑤
System Analysis	④	③	⑤	①	②
Simulation	⑤	④	①	②	③

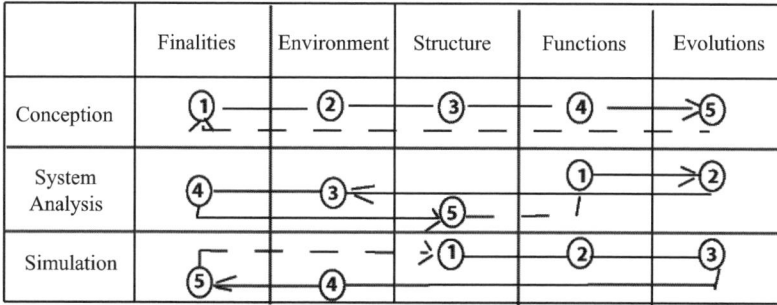

Figure 2.7. *Three working methods of systemography, according to J.L. Le Moigne [LEM 77]*

The development of a normative modeling provided an opportunity for the finalization and application of the main rules likely to govern spatial interaction. These rules are still valid, but we derive benefit from inserting them into cellular automata models, which can easily adapt themselves to the constraints of geographical space. Starting thus from the field work, in a bottom up approach, diverse simulation hypotheses may be tested by comparison with the observed evolutions, and, by restoring the behavioral complexity of the cells, we can manage to generate macroscopic spatial configurations. The consequences that may be derived from them, far from having only a local significance, are the only ones able to lead to new explanatory theories.

This reversion to an inductive approach, after a period during which deductive methods largely dominated, should not be considered as a matter of principle. The back and forth between deduction and experimentation is a part of the constant work in research and in the preparation of renewed theoretical hypotheses. The aim of continual slight reductions in the role of chance in the explanation remains the justification for research.

2.7. Bibliography

[AKI 86] AKIRI P., *Une application à Rouen du modèle de Lowry,* Cahiers Géographiques de Rouen, France, no. 25, 1986.

[ALL 81] ALLEN P., *Self Organization in Complex Systems,* Colloque Thermodynamique et Sciences de l'Homme, University of Créteil, France, 1981.

[BAD 02] BADARIOTTI D., WEBER C., "La mobilité résidentielle en ville – Modélisation par automates cellulaires et système multi-agents à Bogota", *L'Espace Géographique,* Paris, no. 2, 2002.

[BAT 01] BATTTY M., TORRENS P.M., *Modeling Complexity: the Limits to Prediction,* www.cybergeo.presse.fr

[BOU 03] BOULANGER P.L., BRECHET T., *Une analyse comparative des classes de modèles,* Institut pour un Développement Durable. Ottignies (Belgium), 2003.

[BRU 90] BRUNET R., "Le déchiffrement du Monde", *Géographie Universelle,* vol 1, Belin, Paris, 1990.

[CHA 84] CHAMUSSY H., CHARRE J., DURAND M.G., GUERIN J.P., LE BERRE M., UVIETTA P., "Une expérience de modélisation dynamique en géographie pour des interventions en aménagement du territoire: le modèle 'AMORAL', in [GUE 84].

[CLE 03] CLERO J.P., *L'Autorité du paysage,* Etudes Normandes, Rouen (France), no. 3, 2003.

[DAU 03] DAUPHINE A., *Les théories de la complexité chez les géographes,* Anthropos, Paris, 2003.

[DEL 04] DELAHAYE D., DUBOS-PAILLARD E., GUERMOND Y., LANGLOIS P., "From modeling to experiment", *GeoJournal,* 2004.

[DUB 03] DUBOS-PAILLARD E., GUERMOND Y., LANGLOIS P., "Analyse de l'évolution urbaine par automate cellulaire. Le modèle SpaCelle", *L'Espace Géographique,* Paris, no. 4. 2003.

[DUP 99] DUPUY J.P., *Aux origines des sciences cognitives,* La Découverte, Paris, 1999.

[FOR 69] FORRESTER J.W., *Urban Dynamics,* MIT Press.

[FOT 89] 1969 FOTHERINGHAM A.S., O'KELLY M.E., *Spatial Interaction Models: Formulations and Applications,* Kluwer, 1989.

[GUE 84] GUERMOND Y. (ed.), *Analyse de Système en Géographie,* Presses Universitaires de Lyon (France), 1984.

[HAG 67] HAGERSTRAND T., *Innovation Diffusion as a Spatial Process,* University of Chicago Press, 1967.

[HAG 77] HAGGETT P., CLIFF A.D., FREY A., *Locational Models,* Edward Arnold, 1977.

[HUF 63] HUFF D.L., "Defining and estimating a trading area", in Ambrose P., *Analytical Human Geography,* Longman, 1970.

[HUM 48] HUME D., *An Enquiry concerning Human Understanding* (1748; New York: Oxford University Press, 1999).

[KAN 65] KANT E., *Critique de la faculté de juger,* 1790 (Vrin, Paris, 1965).

[LAU 97] LAURENT M.A., THOMAS I., *Modèle d'interaction spatiale et agrégation des lieux: l'exemple des données criminelles,* L'Espace Géographique, Paris, no. 3, 1997.

[LEM 77] LE MOIGNE J.L., *La théorie du système général* (4th edition), PUF, Paris, 1977.

[LOW 64] LOWRY I.S., *A Model of Metropolis,* Rand Corporation, Santa Monica, 1964.

[MAX 94] MAXWELL T., COSTANZA R., "Spatial ecosystem modeling in a distributed computational environment", in VAN DEN BERGH J., VAN DER STRAATEN J., *Towards Sustainable Development,* International Society for Ecological Economics, Island Press, Washington DC, 1994.

[POP 59] POPPER K., *The Logic of Scientific Discovery*, 1959.

[PUM 89] PUMAIN D., SANDERS L., SAINT-JULIEN T., *Villes et Auto-organisation*, Economica, Paris, 1989.

[PUM 96] PUMAIN D., "Une déconnexion entre la théorie et le modèle ?", *Géographes Associés*, Paris, no. 18.

[PUM 03] PUMAIN D., "Une approche de la complexité en géographie", *Géocarrefour*, Lyon (France) no. 1, 2003.

[SAN 97] SANDERS L., PUMAIN D., MATHIAN H., GUERIN-PACE F., BURA S., "SIMPOP, a multi-agent system for the study of urbanism", *Environment and Planning B*, vol. 24, 1997.

[SAN 07] SANDERS L. (ed.), *Models in Spatial Analysis,* ISTE, 2007.

[SCH 78] SCHELLING T.C., *Micromotives and Macrobehavior,* W.W. Norton, 1978.

[TAY 75] TAYLOR P.J., "Distance decay in spatial interactions", *Catmog- Geo Abstracts* (Norwich), 1975.

[WAL 00] WALRAS L., *Eléments d'Economie Politique Pure,* Librairie Générale de Droit, Paris, 1900 (new edition 1952).

[WIL 74] WILSON A.G., *Urban & Regional Models in Geography & Planning*, John Wiley & Sons, 1974.

Chapter 3

The Formalization of Knowledge in a Reality Simplifying System

In geography, modeling is a commonly used method for capturing a part of reality and helping to understand its function. We seek to better understand the geographic pattern, by identifying the system of elements observed in a location as well as its interactions with other locations by bringing out regularities and laws and by placing particular attention on unexplained residual specificities.

The first steps of research therefore consist of formalizing the knowledge in a reality simplifying system. This simplification implies choices and imposes an abstraction. The global perspective of the research, the researcher's questions, much more than the existence of data, will influence the choices of formalization and the methods of understanding put into action. The modeling therefore poses the question of formalizing the knowledge. The last of these could take many forms, such as formalization by literary language, formalization by calculated observational criteria (raw database and creation of indicators), facilitating the transition to a mathematical equation (when we use "hard models"), or formalization in the form of sagittal drawings, chorems ("soft models"), or maps, often realized in the field or in areas unsuited to the use of an equation.

The investigation of the cultural domain of contemporary urban societies is an illuminating example in this respect. We will show the necessity of using, in order to comprehend the whole of the cultural system in place in the city, several paths of knowledge formalization. Defining the protocols of investigation in an area where

Chapter written by Françoise LUCCHINI.

the cultural measure is a question of debate proves to be delicate but nevertheless possible. We will offer perspectives regarding the progressive nature of the elaborated model, that is to say its re-formulation by successive steps of development and diverse scales of investigation. Starting with the national triangular diagnostic, "politics-equipment-practices", followed by inter-urban competition and complementarities as well as sub-communal deserts, the cultural potential of cities will therefore decline. The problem of knowing how to gauge the complexity of the cultural system in place will always be present in such research that seeks to simplify reality in order to understand it.

3.1. Formalizing a complex cultural system using a series of perspectives

Borrowing different paths of knowledge formalization is an experimental approach capable of finding new forms of understanding observed reality, comprehending it in its wholeness and above all manages its complexity. Generally, one of the first paths used is a formalization focused on written knowledge. What we call "state of the question" or "reasoned bibliography" corresponds to the initial step aiming to discern what has been developed regarding the subject in terms of information and extensive knowledge, as well as the different angles of approach dealing with the same information. The choice for the researcher is then to orient his approach according to one perspective rather than another and to organize the bulk of knowledge and information in a simplified form in direct relation with an angle of approach and a clear problematical set.

We therefore rewrite and rework this printed information in draft form, or as a simplified formalization schematic in direct relation with the problem faced. Let us highlight here that the problem set chosen to guide the research has more implications than the existence or non-existence of data.

3.1.1. An initial perspective on culture and the city: the French example

In the research relative to social and cultural activities presented here, the problems could be considered under the following terms: we wish to *verify through the presented cultural apparatus of French cities, that culture has a part in the necessary diversity of the French urban system.* The hypothesis is to state that the cultural issue for cities lies greatly in the diversity and not in the homogenity (often feared and denounced). As much as in other domains where major forms of urban change could develop with the diffusion of innovations of any type, we have widely observed a small freedom of movement for cities that conserve their relative position, in terms of dimension or of economic or social dynamics [PUM 78]. This question thereby generates conceptual choices regarding the notions of "city" and of "culture", influencing our approach and its appropriate developments.

We design our cultural mechanism in a social and territorial context. The first test pertains to the system that is in place in France, and that primarily of the urban sector, since cultural activities and equipment are services particularly represented in the city [LUC 02], where the majority of the population is located considering the general urbanization of French society. Cities are attracted to proposing a large diversity of cultural activities in the hope of educating, entertaining and satisfying the inhabitants, but also for self-promotion. In France, the city is a key geographical echelon for culture, since municipalities are the principle owners of cultural sites and principle public sponsors (more than 10% of their budget), far ahead of the Secretary of Culture and Communication [LUC 02].

In this work we share our conceptual choice to "consider the city" with urban theories that do not omit the spatial dimension. These theories also "place the urban condition as a particular way of organizing the space, to live, to use, to exploit and to control the territory, as well as the scale of the city itself and the scale of the urban networks or of the city systems" [PUM 96]. This approach to the city raises conceptions that sometimes formalize the city as an element covering a territory, sometimes as a node in a network of relations or an elementary part of a more global system. In the same way, we can borrow an idea from geographer Brian Berry, who presents the city as "a system in a system of cities." In such a way, the patterns produced by social life may be questioned, without deviating too far towards the psychic foundations of the inhabitants [SIM 03], but insisting more on the cultural foundations present in society and their implications on the global system.

Superimposed on this theoretical and conceptual choice the city is the cultural context that we will shine a light on before resorting to different paths of formalization. Culture represents one of the main issues for contemporary man. Of course, countless definitions of the word "culture" exist, but they do not manage to hide a paradox: simultaneously, there is a "unity of culture" and "diversity of cultures." "Unity" since human society cannot exist without culture, and we agree to consider that culture constitutes the heart of the human identity and that it is a complex set of knowledge, practices, rules, values, makers, as well as surveys on the presence of cultural activities, arts, myths and beliefs transmitting themselves from generation to generation, reproducing in each individual, maintaining a social complexity and at the same time evolving in time and space. We can also talk about cultural "diversity" since each culture is unique. Cultures maintain social identities with distinctive characteristics. Culture therefore precedes the individual, but equally the group of individuals. At the smallest level, it can be confused with personality, affectivity, sensibility, social adherences and therefore the deepest thoughts of human beings [CER 80]. While, on the other hand, at the level groups, culture (often called mass culture) can be imposed by national scholastic education, family setting or even proposed by the media, cultural industries and urban services. It is necessary to choose an operational concept of culture. There could be many ways to take interest in cultural expression in cities (creative forms, innovations, frequency of visits, econometrics, discourse analysis, cultural management, etc.). Cultural definitions and the relative hypotheses in the system of cities orient the tactic proposed here. It is a matter of conceiving simple and comparable information about the urban agglomerations by surveying the multiplicity of cultural forms at the heart of a society. This formalization therefore approaches culture in society and does not reduce itself to the intimate aspect of a person. We observe culture as a collective construction. This combines the analysis of the political choices of local and national decision in the cities collected from strictly artistic domains (music, dance, literature, fine arts, theatre, monumental and museographic heritage, cinema) or belonging to a register of activities close to daily culture (music cafés, contemporary music halls, radio stations, bookshops, etc.). We therefore approach the legitimist and democratizing points of view of culture at the same time (rendering the best works of humankind accessible to the greatest number of French people, borrowing from the well known formula of Malraux, and therefore valuing the "academic arts") as well as relativist democratic objectives (an advocacy for all forms of culture, choosing the "all culture" way). This attempt to measure is a test of the efficiency of the French urban network in its capacity to adopt cultural innovation (spread of equipment and services over the territory) leaning towards recognizing the cultural diversity of cities. It is also recognizing national policy in France, which is, among other things, very concerned with culture through the defense of cultural exception. This does not eclipse the vitality of local initiatives. Cities were and still are the center of an abundant cultural life which owes as much today to the search for quality of life for citizens as to the demand for an attractive image for the city.

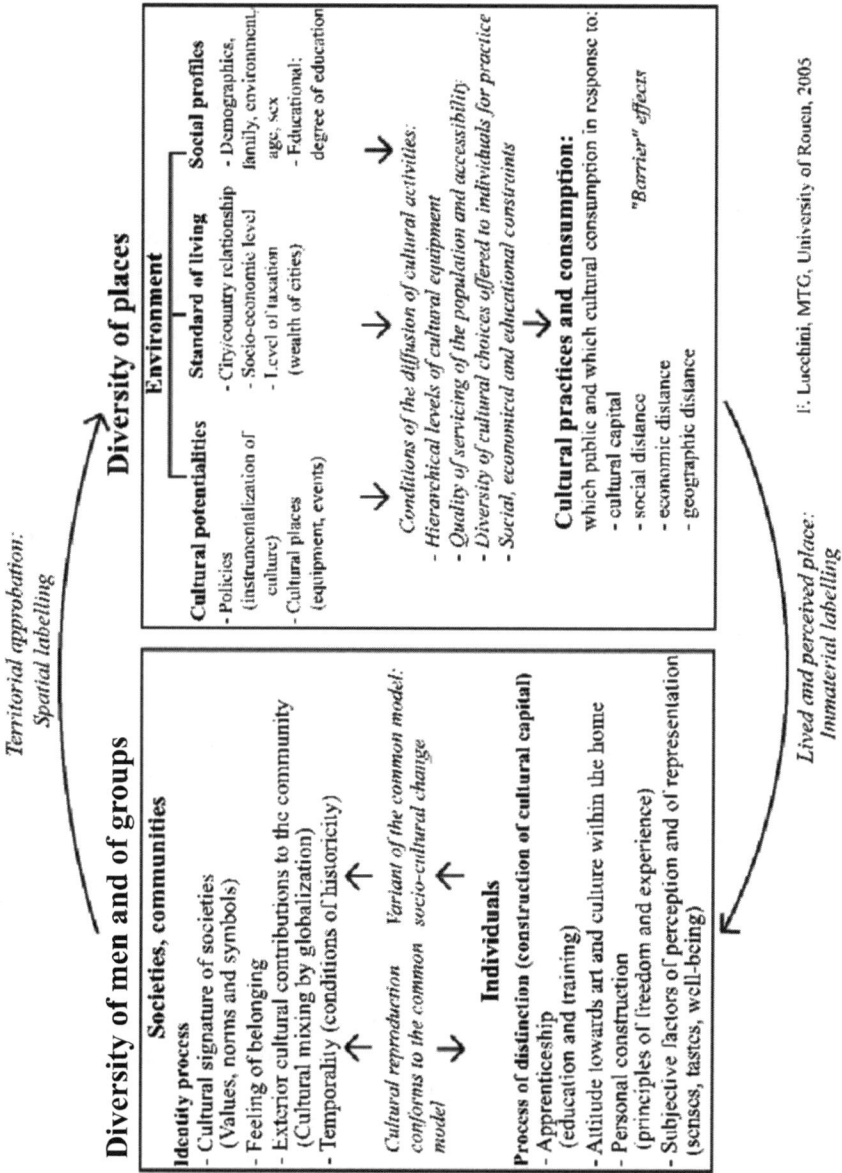

Figure 3.1. *A cultural system in society.*
Cultural consummation, social structure and
geographic space: material and immaterial highlights

3.1.2. *A simplification of the cultural system in place in France that is transposable to other countries*

Figure 3.1 proposes a reading of the cultural system in society and a simplified sagittal formalization of conjoined information and knowledge. We insist particularly on a "round-trip", that is to say a strong interaction between society and places, between people and their cities in all their diversity, with material and immaterial territorial cultural markings, and lived and perceived places of life. Three panels detach themselves quite clearly from the reading of this figure in order to understand culture in society in a more operative way (leaning towards a calculated formalization). We must be acquainted with the management of culture *at a political level* – national and local – which will condition or frame the territorial device in place. It is necessary to conduct a quantitative and qualitative investigation on the *location of cultural services and equipment* in order to evaluate the presence of cultural activities (spatial analysis). Finally, the *cultural practices* panel of France advises us on the efficiency and appropriateness of these cultural services by the population. These three panels – policies, equipment and practices – constitute a triangular approach drawn up on the French territory, but which certainly seems transposable to countries other than France, therefore opening the door to comparative international investigations.

Supplements to the book-like formalization written or outlined therefore seem necessary in light of the envisioned problematical set. This case takes the form of territory investigations, national or localized surveys, specifying certain mechanisms or a more general functioning of the cultural system. Another path of knowledge formalization is used in this case: the consideration of formalization by numbered transcription, with the constitution and organization from a database, coordinated with the elaboration of relevant indicators and metrics.

3.1.3. *Culture: possible measures*

A dilemma which the researcher must overcome arises in the cultural domain. Behind the cultural activities, the notion of cultural equipment assumes that we consider, altogether, different cultural domains. We accept the need to mix the genres, even those situated at different levels (e.g. the writings of Nietzsche and the popular dialogs of Audiard). The discussion is not centered on the divisions separating the cultural genres, but on the links that exist between the social positions and cultural choices. The cultural practices invented and diffused by groups of individuals, often enabled by the existence of urban cultural apparatus, are therefore coded by the observer and approached as a whole. The cultural phenomenon is approached in a concrete manner, attempting to explain cultural activities and practices by economic, political or social characteristics, rather than by irreducible particularities. A collection of interrelations are inferred behind the existence of a cultural system at all geographical echelons, in which individuals and political and commercial structures interact (Figure 3.1). The individuals are guided by their social position and by strategies issuing from their own cultural capital [BOU 69]. The creations and diffusions of cultural practices are perceived as terms of supply and demand. Profitable goods and services are introduced to the market by enterprises. Public authorities consider these cultural activities as favorable instruments to the self-realization of development projects and urban valorization.

The observation and consequently the collective measure of culture are situated in a direction diametrically opposed to the idea of A. Finkielkraut who proclaims that "culture senses itself, but does not measure itself" [FIN 87]. We realize here that if Finkielkraut opposed all measures of culture, it is because he defends the choice from an intimate definition of culture as a thought, reflection and experience of a person, and that he pushes the cultural legitimacy so as not to put established culture and other cultural forms on the same level, even if the two genres can be seen as existing side by side in the culture of an individual. We have decided to bypass this dilemma and to favor, instead of the intimate definition of individual culture, that of a collective construction having visible and measurable forms, even in an imperfect manner, at the heart of all society. The question of legitimacy is posed here in terms of a lack of absolute precision when measuring culture. The decision to measure must moderate itself from a critical perspective on the resource, which is but a summary or an imperfect medium, but which nonetheless allows direct qualitative advances in the understanding of the system.

3.1.4. *Culture in a centralized state: a French diagnostic turned towards the elaboration of a transposable investigation protocol*

3.1.4.1. *Cultural policies: between consensus and diversity*

The French state is very concerned with culture. As proof, since 1959 a department dedicated to cultural affairs has existed assuring a centralized and hierarchized organization, and the durability of principle objectives despite political changes. This state intervention in the cultural domain is quite old in France, although it was less explicitly sanctioned before the creation of the department (art sponsorship, royal academism, revolutionary republican principles, etc.). It constitutes one of the three models of cultural administration present in Europe [COU 00]. Recently, a decentralization of decisions has been put in place and the support of municipalities and other territorial bodies is encouraged under the slogan that "culture is everyone's concern".

The position of international defense of a "cultural exception", revered in France, is not at all foreign to this evident implication in the cultural domain (culture is not merchandise like any other; it must be protected from market law). In this context, a comparative study at the end of the 1990s of the 100 largest French cities and their political strategies shows that culture has become, for them, an issue of planning to combat the inequalities of access to culture and developing an attractive urban environment [LUC 02]. It is therefore normal that cities propose a large diversity of cultural activities in the academic and artistic universe, or in a more popularist sector. They can, through this tool, participate in the artistic and cultural training of individuals, favor social links – if necessary – in the city, value a corporate image and a local identity, and also hope for economic repercussions. Municipal cultural policies have been outlined throughout this research on a sample of about 100 cities. Faced with the diversity of situations encountered, as much the choice of the cultural domains (music, dance, theatre, cinema, circus, literature, fine arts, diverse shows, heritage, etc.) as that of the connected functions (conserve and value heritage, insure the artistic education of the citizens, enliven the neighborhood, etc.) or even for stated strategies (followed objectives, rallying of human, financial or technical resources, collaborations, listening to the people, etc.), a small number of urban cultural policies emerge from this research. Traditionally the data from the research was initially formalized as quantitative variables (frequency of choice and municipal priorities), leading to the attainment of figures simplifying the large bulk of basic information (Figure 3.2).

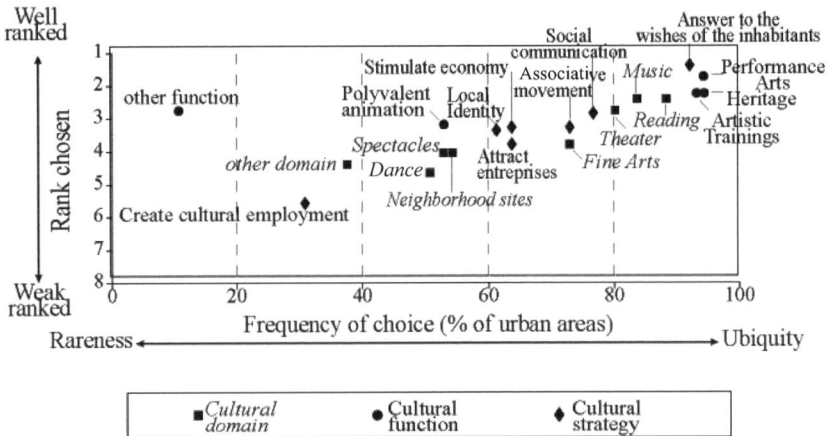

Source : F. Lucchini, 110 Surveyed municipalities in 2002, 2005.

Figure 3.2. *Cultural choices of the 110 surveyed municipalities*

Then, in such a way that we can emphasize the structures relative to cultural behaviors, the information on municipal political behaviors obtained from the research was transformed into qualitative variables (coded in complete binary) and compared with a factorial analysis of correspondences. The typology in eight categories, according to the elected choices (Figure 3.3), reveals urban behaviors opting firmly, for example, for a *cultural eclecticism* (multiplicity of cultural choices and partnerships, high cultural expenses for Amiens, Angouleme, Bourges, Chambery, Grenoble, Lyon, Tours, Toulouse, Valenciennes, etc.), contrary to other behaviors in favor of a *specialization of activities* (few choices for Elbeuf, Frejus, Le Havre, Marseille, Melun), others for a *relatively solitary behavior* in such a way as to elaborate and finance these activities (a great decision-making and financial autarky for Calais, Compiègne, Metz, Montauban, Pau, Saint-Omer, Tarbes, etc.) so that the general rule is to favor collaborations and crossed financing (city, state, region, department, DRAC (regional administration of cultural affairs), sponsorship) around cultural projects. Certain cities look more to *value a local identity* (Caen, Lorient, Meaux, Perigueux, Villefranche-sur-Saone) whereas others look adamantly for an *economic dynamism* and financial repercussions through culture (Ales, Brive-la-Gaillarde, Beziers, Douai, Dunkerque, Epinal, Sete, etc.). This non-exhaustive political painting effectively realizes two large perspectives. On one hand cities flatter social demand by trying to answer to the greatest number of cultural wishes of the inhabitants, on the other, they try to distinguish themselves by developing more original behaviors, generally more prestigious and affected by mass media, issued from a policy of cultural offering that supports the local artists' potential, and cultural professionals. Once this assessment of political strategies is drawn up, it becomes interesting to tackle the effective location of 30 or so categories of surveyed cultural equipment and services in these same cities.

3.1.4.2. *Geographical configurations of culture*

After 50 years of national political actions in favor of culture, relayed to local scales, the geographical evidence is there. There still exists, despite a territorial blanket that is becoming wider, pronounced disparities in terms of cultural services and equipment available to the population. Overall, these services and equipment respect the distribution of the French population. They therefore copy an anterior state of functional differentiation between cities: urban hierarchy. As this equipment is essentially urban, imbalances appear between rural and urban zones, but also between big and small cities, or between town centers and suburbs. The second piece of evidence to highlight is the over-equipment of the Ile-de-France region and particularly that of Paris, which is much more accentuated than a simple response to the effect of the population and to the national tradition of political centralism [MEN 94] [LUC 02].

We have not yet achieved territorial fairness in terms of access to a large selection of cultural offerings. There also exist more "cultural" territories than others, but they are not necessarily more "cultivated".

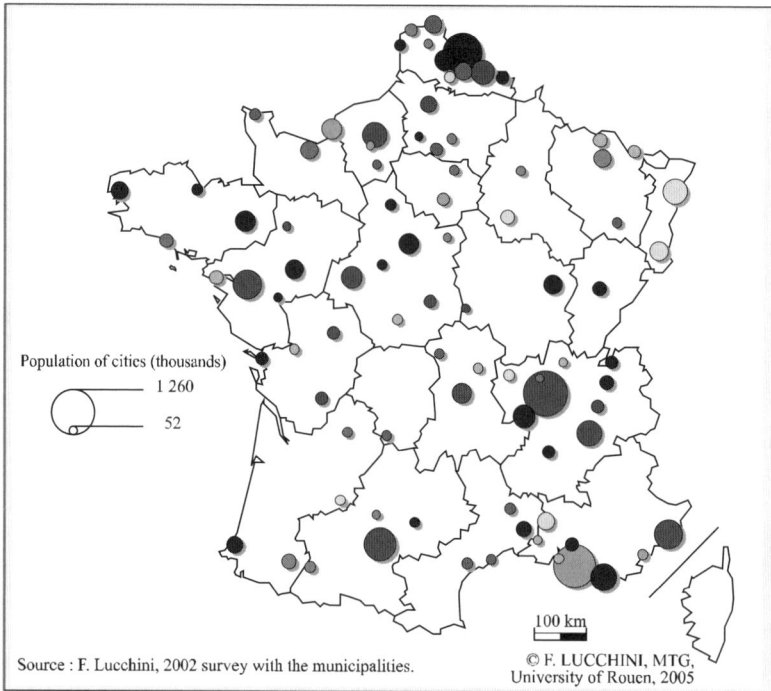

Population of cities (thousands)

1 260

52

Source : F. Lucchini, 2002 survey with the municipalities.

© F. LUCCHINI, MTG,
University of Rouen, 2005

100 km

Number of cities	Significance
16	Multiplicity of cultural choices of partnerships, and major cultural expenditures
5	Less intensive cultural profusion and choice of local identity
5	Specialization towards the internal life of the city with the help of major human resources
7	Less intensive specialization of city liveliness
9	Towards a financial and decisional autarchy, weak human and financial resources
9	Greater human resources and less autarchy
11	Small cities with fewer human resources, economic orientation and external cultural influence
25	Choice of original domains and numerous cross-subsidisation

Figure 3.3. *Differentiated urban behaviors for the 110 surveyed municipalities – results of factor analysis (see also this map in the color plate section)*

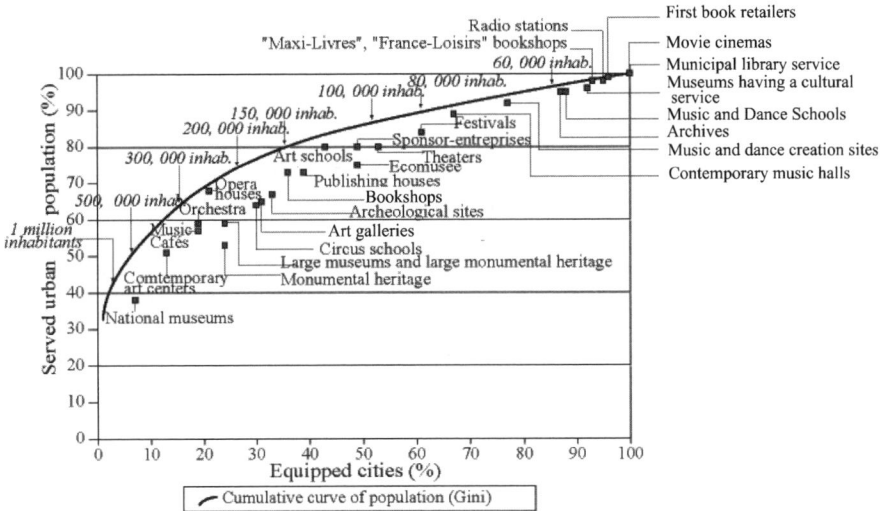

First book retailers
Movie cinemas
Municipal library service
Museums having a cultural service
Music and Dance Schools
Archives
Music and dance creation sites
Contemporary music halls

Radio stations
"Maxi-Livres", "France-Loisirs" bookshops
60, 000 inhab.
80, 000 inhab.
100, 000 inhab.
150, 000 inhab.
200, 000 inhab.
Festivals
Sponsor-entreprises
300, 000 inhab.
Art schools
Theaters
Econmusée
Opera houses
Publishing houses
500, 000 inhab.
Orchestra
Bookshops
Archeological sites
1 million inhabitants
Music Cafés
Art galleries
Circus schools
Large museums and large monumental heritage
Comtemporary art centers
Monumental heritage
National museums

Source: F. Lucchini, MTG, University of Rouen, 2005

Figure 3.4. *Service and concentration of cultural equipment of French cities*

We distinguish among the cultural equipment, those with proximities that are widely distributed over the territory (municipal libraries, cinemas, music and dance schools), those that are rarer (opera houses, archaeological sites, music-cafés, national museums), whose geographical distribution remains rather uneven (Figure 3.4). The first group tends to serve a local population, whereas the second group benefits from a larger influence than the population size interested in cultural apparatus. In addition, in terms of service-per-inhabitant and cultural choices, it is not the large cities that rank highest, but the average and small cities: Ajaccio, Arles, Bastia, Evreux and Quimper. Trends are perceivable in the cultural domain. Thus to be different, to go off the beaten path, to please a different audience, or to reclaim places differently, cities have opted over the past 30 years for festivals. The festival trend brings change and modernity, and cities therefore become independent of an urban image far too often associated with the economic environment. By way of their successively creative, educative, playful and integrative actions, festivals claim to be favorable to cultural changes. Finally, in a general manner, each French citizen finds himself no more than 30 km from a festival, which explains this success as much as the political consensus.

In addition to the problem of geographical distances, other parameters play an equal role in accessing culture: social, educational and economic selection constitute another restraint for the pastimes and cultural practices of French citizens [BOU 69]. Sociological reports established at the national level clarify this bond: the recognized cultural audiences find themselves within a population which has attained a generally high level of education, composed of executives, students and inhabitants of large cities, the most loyal of which are single and have more eclectic tastes [DON 94]. We have observed over the past 30 years different levels of cultural practices in France, from the most eclectic to the most narrow-minded, or even a complete absence of cultural practices among certain French citizens (a "non-audience"). This is why we must highlight the importance of municipalities for cultural policy, which constitutes a key level of activity, in short an alternative offer for the district's cultural network which is better adapted to the demands of the population and which opens up numerous cultural practices. At the same time, the comparison of the distribution of cultural equipment with municipal policies revealed by surveying 100 large cities [LUC 02] seems to confer only a rather moderate impact of policies on the bulk of equipment (only 20% of the equipment pattern could be accounted for by the initiated municipal policies). Nonetheless, we have every reason to believe that the impact of this policy, however small at the end of the 20th century, is continually working as a means of cultural reinforcement.

3.1.5. *The necessary re-formulation of knowledge to overcome the successive and qualitative steps of advancement*

The formalization of the inherent complexity in the cultural domain is a delicate operation that requires varied steps of advancement and the use of different paths of formalization. Revealing the cultural system in place in society as well as its different interacting elements has led us, up until now, to use the literary approach, then the formalization of soft models (sagittal schemes, typology, maps) and finally a more mathematical, quantitative and modeling method. It is appropriate to test these initial perspectives relative to the "cultural model in society" on another territorial environment and other observation scales – European, local – in order to measure the whole soldity of the elaborated cultural system.

Let us reformulate the preceding approaches as a "global cultural model". It is with the support of a large panorama of cultural services or activities offered to the population in Europe that we can define the cultural component of the European territories, as *a full extended day-to-day service that participates in local development and contributes in reality to territorial competitions*. The cultural model reformulated as such (a system susceptible to further evolutions) will be compared with other territories and other geographical scales of investigation.

3.2. Differentiation of the system of cities by culture: contribution of the spatial analysis for testing the "global cultural model"

One aspect of territorial competition in Europe asserts itself through European diversity in the domain of *the cultural offering of the European territories* and the paradoxes that it reveals. It is fair to speak of *cultural diversity* at first since the situations appear multiple: national debates on the specificity of culture (what definition to choose), generated national and local cultural policies (from the defense of a "cultural exception" to a simple financial participation), the commercial domain of culture, the circulation of cultural goods and services, without forgetting the followers (and their levels of practice) generate a complex cultural system. It is equally right to talk about *paradoxes* insofar as we notice, despite diverse national cultural situations, indeed the opposite in political terms, a *convergence* in the territorial device of cultural offerings proposed to the population.

Since these cultural services and activities are essentially located in town, they are hereafter studied on the inter-urban echelon (comparability of cultural offerings between large French cities and those of the UK) in the same way as at the intra-urban echelon (comparability of cultural offerings inside two test cities with divergent political contexts: Rouen and Brighton). A change of observation location (another national context), then a change of scale of observation (passing from the inter-urban level to an intra-communal level) are therefore determined to test the efficiency of the model.

3.2.1. *A methodological investigation to define the cultural potential of British and French cities and their competitive capacity*

The question is to know if the two opposing contexts, in terms of national cultural policies, being about a spatial organization, based on the different cultural offering, on the territories in terms of potential. These cultural services and activities could also follow a spatial organization similar to that of Europe, despite an acquired specificity in the cultural sector. We could equally question the perspectives over the long-term, that is to find out if we are heading towards *an emerging cultural differential* or on the contrary a homogenization which is often praised in the context of globalization.

Answering these questions requires the use of localized calculated measures, of the cultural capacities of French and British cities, in terms of equipment and services offered to the population. The question of the calculated measure will not be discussed here (see above, on the choice of rendering collective practices and not personal practices of culture). It is more the operational utility of measuring culture and particularly "the current cultural offering" which will be at the heart of this study. The difficulties encountered for the collection of data were numerous. The criteria compatibility needed for this study requires a general determination of the nature of cultural equipment, in French and British cities, belonging to the academic and artistic field but also to a more popular field. The existence of sparse data, or sometimes a complete lack thereof, leads us to create our own database from land surveys and paper-based information (Figures 3.5 and 3.6). The great complexity of the British territorial division has exacerbated the difficulties encountered in terms of realizing spatial and above all cartographic information.

- *Municipal public libraries*, Direction du Livre et de la Lecture, Ministère de la Culture et de la Communication, Annuaire 1994
- *Cinemas*, Centre National de la Cinématographie, 1990, 1998, 2000
- *Dance groups*, Performing Arts Yearbook for Europe, 2000
- *Summer festivals*, Association Dclic, Ministère de la Culture et de la Communication, Guide des 10 000 manifestations Festivals et Expositions 1994.
- *Publishing houses*, Livre-Hebdo, guide 1994.
- *Museums having a cultural service*, Direction des Musées de France, Ministère de la Culture et de la Communication, 1993
- *Operas*, Performing Arts Yearbook for Europe, 2000
- *Large orchestras*, Performing Arts Yearbook for Europe, 2000
- *Associative or commercial radio stations*, Département d'Etudes et de Prospective, Ministère de la Culture et de la Communication, 1995

Figure 3.5. *Retained cultural equipment for French cities*

It is in the monumental domain of historical heritage, particularly for the British side of the study, where the difficulties of the lack of local censuses were most notable; that is why the constitution of the localizable data is still underway for the current database.

3.2.1.1. *Cultural service in the city*

The first observed convergence between French and British cities is their cultural service and their level of concentration in terms of offered equipment. We expect that the equipment of large cities will be, at the same time, varied and numerous to meet the urban demand.

- *Public libraries* in the UK and the Republic of Ireland, Libraries Association, yearbook 2003.
- *Cinemas*, Dodonna Research, Cinemagoing 10th, March 2002
- *Dance groups*, Performing Arts Yearbook for Europe, 2000
- *Summer festivals*, Performing Arts Yearbook for Europe, 2000
- *Publishing houses* in the UK and the Republic of Ireland, http://www.lights.com/publisher/db/country-United-Kingdom.html, 2003.
- *Museums* in the UK and the Republic of Ireland, Museums Association, yearbook 2002.
- *Museums* in the UK surveyed as *Designated Collection* (collections ranked as remarkable), Museums Association, yearbook 2002.
- *Art Galleries* in the UK and the Republic of Ireland, Museums Association, yearbook 2002.
- *Operas*, Performing Arts Yearbook for Europe, 2000
- *Large orchestras*, Performing Arts Yearbook for Europe, 2000
- *Independent radio stations*, RAJAR/ Ipsos-RSL, Media Pocket Book 2001.

Figure 3.6. *Retained cultural equipment for British cities*

In studying the large French and British urban areas exceeding 100,000 inhabitants (57 in France; 69 in the UK and Ireland), we notice the existence of urban levels of cultural service, organized in a hierarchal manner. The geographical configuration of cultural equipment in French and British cities follows a distribution conforming to a location of *ordinary services*. Some cultural equipment can be qualified as frequent, since they are present in all the cities studied. Others are rarer and appear less frequently, especially in the bottom of urban hierarchy. We distinguish several steps between the service level of rare equipment, uniquely present in some cities (generally large in size and more diverse), and a larger accessibility for daily equipment, ubiquitous with the cities. We can also identify levels of cultural services, from the most daily to the rarest. In the French and British cases, *libraries*, *cinemas*, and *museums* are very common equipment (often ubiquitous), with 9 out of 10 cities equipped. *Operas*, *orchestras* and *dance groups* can be regarded as rare (present in a maximum of 4 in 10 cities). Between the two we observe some other types of equipment quite common in cities, such as *radio stations*, *publishing houses* or *festivals* (Figures 3.7 and 3.8). Some gaps are noticeable between the service offered by the British and French cities: more cities offer festival type activities, radio stations, dance and orchestras in France, whereas more cities have publishing houses and museums in the UK and Ireland.

Nevertheless, cultural equipment as a whole takes part in urban centric thinking and their localization conforms to the theory of central locations: cities differentiate themselves in cultural matter in contrasted ways, contingent on the quantity and variety of the cultural equipment they accommodate. We discover economic phenomena of agglomerations by the distribution of cultural services. Also, the French and British cities, which are central places in a system of interdependent cities, differentiate themselves by culture, evenly conserving their relative positions at the national level.

Nature of the cultural equipment	% of equipped cities[1]	% of urban population served[2]
Municipal public library	100	100
Cinema	100	100
Radio station	96	99
Museum	95	97
Festival	70	89
Publishing house	56	81
Dance group	42	73
Orchestra	39	70
Opera houses	35	73

Figure 3.7. *Frequency of urban cultural equipment in France*

3.2.1.2. *Cultural primacy*

There is a phenomenon of over-concentration of culture in the main city, which is perceivable in France as well as in the UK. This phenomenon can not be explained with such a simple answer as a higher population in capital cities. We see here a specific outcome of cultural activity, which manifests itself spatially in an analogous manner (a geographic concentration pushed to the extreme in the capital city), no matter what system of cultural administration chosen at the national level (Britain's liberal system compared with France's prioritized and centralized system) (Figures 3.9 and 3.10). The indications of primacy, however, are less pronounced for the UK (from 9 to 1), revealing a less intense geographic concentration of cultural equipment in the French case (from 44 to 2.5).

---

1 "Equipped cities": % of urban units possessing at least one cultural structure over the total number of urban units.
2 "Served population": % of urban population served among the total of the population of the 57 French agglomerations and the 69 British agglomerations.

Nature of the cultural equipment	% of equipped cities[3]	% of urban population served
Municipal public library	100	100
Museum	99	99
Cinema	99	99
Gallery	77	91
Publishing house	72	89
Radio station	62	80
Festival	43	75
Ranked museum – designated collection	36	63
Orchestra	26	64
Dance group	19	52
Opera	12	53

Figure 3.8. *Frequency of urban cultural equipment in the UK and Ireland*

3.2.1.3. *A differentiated reading of the global levels of equipment*

Despite the urban weight which it seems necessary to underline in order to explain the distribution of cultural equipment, we can explore the phenomenon further by investigating the services available per inhabitant and the notion of cultural efficiency. Let us take the British case as example. Figure 3.11 presents the level of cultural equipment observed for 10,000 British citizens, and therefore realizes the relative good position of small sized cities for offering a cultural experience to their inhabitants. Oxford and Cambridge, followed by the cities of York, Dundee, Norwich, Belfast and Brighton obtain the best level of service-per-inhabitant, ahead of London and other major cities.

---

3 See notes on previous page.

Nature of the cultural equipment	Portion of the 1st city in the total number of cities (%)	Portion of the 2nd city in the total number of cities (%)	Name of 1st city	Name of 2nd city	Primacy indicator
Publishing houses	83.8	2.0	Paris	Lyon	44.4
Municipal libraries	38.3	5.5	Paris	Lyon	7.0
Opera houses	29.2	4.2	Paris	All other cities	7.0
Cinemas	30.4	4.5	Paris	Lyon	6.8
Orchestras	40.6	6.3	Paris	Lyon, Marseille, Toulouse	6.5
Museums – cultural services	30.8	4.9	Paris	Marseille	6.3
Festivals	26.6	5.8	Paris	Grasse-Cannes-Antibes	4.6
Dance Groups	34.2	13.2	Paris	Marseille	2.6
Radio stations	11.7	4.8	Paris	Lyon	2.5

Figure 3.9. *Primacy index[4] of French cities*

We can also observe the quality of service in terms of variety: the diversity offered, the *palette* of activities and cultural services available? This is no longer concerned with information linked to the quantity of equipment present (gross quantity available to the population), but rather a glimpse at the possible choices opening a widened or more eclectic cultural practice. Figure 3.12 therefore proposes a reading in terms of cultural diversity of British and French cities (Figure 3.12).

If the large British cities offer a good level of diversity in cultural activities, we can particularly note a sustained position in the pattern of average cities, which seems to accord, as with their French counterparts, to a particular focus on the diversification of activities above the quantity of services offered (for example, Avignon on the French side is classed at the same level of cultural variety as Paris; the same, on the British side, as Cardiff and Brighton).

---

4 "Primacy index": ratio between the percentage of the 1st city and that of the 2nd city.

Nature of the cultural equipment	Portion of the 1st city in the total number of cities (%)	Portion of the 2nd city in the total number of cities (%)	Name of 1st city	Name of 2nd city	Primacy indicator
Publishing houses	55.1	6.0	London	Oxford	9.0
Orchestras	56.7	6.7	London	Glasgow, Cambridge	8.5
Dance groups	66.7	8.0	London	Dublin	8.3
Opera houses	60.9	8.7	London	Dublin, Cardiff	7.0
Museums	22.1	4.9	London	Leeds, Manchester	4.5
Festivals	29.8	7.1	London	Edinburgh	4.2
Cinemas	21.9	7.7	London	Birmingham	2.9
Galleries	20.5	7.5	London	Manchester	2.8
Public libraries	24.7	9.6	London	Birmingham	2.6
Radio stations	15.1	6.5	London	Birmingham	2.3
Ranked museums	12.5	11.3	London	Manchester	1.1

Figure 3.10. *Primacy index of British cities*

Beyond the cartography of this choice of cultural services and activities, the study of French cities allowed us to understand that the cultural variety of a city is better explained by the educational level of the inhabitants than by the social or economic composition of the city (coefficient of correlation greater than +0.7). Cities having the most higher-level graduates therefore offer a larger selection of cultural equipment, whereas those where the level of education is lower have a much more limited cultural choice.

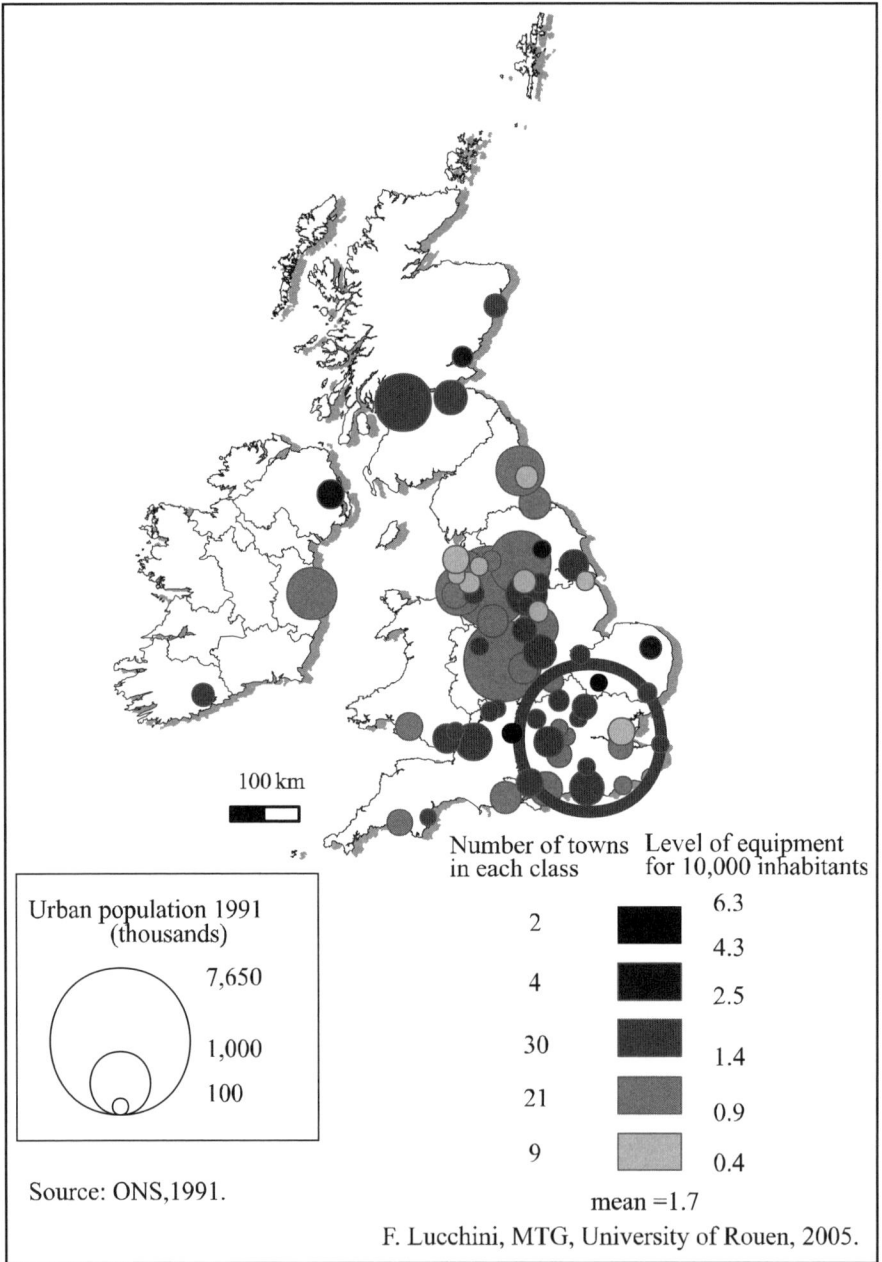

Figure 3.11. *Cultural service per inhabitant in the UK and Ireland, 2002*
(see also this map in the color plate section)

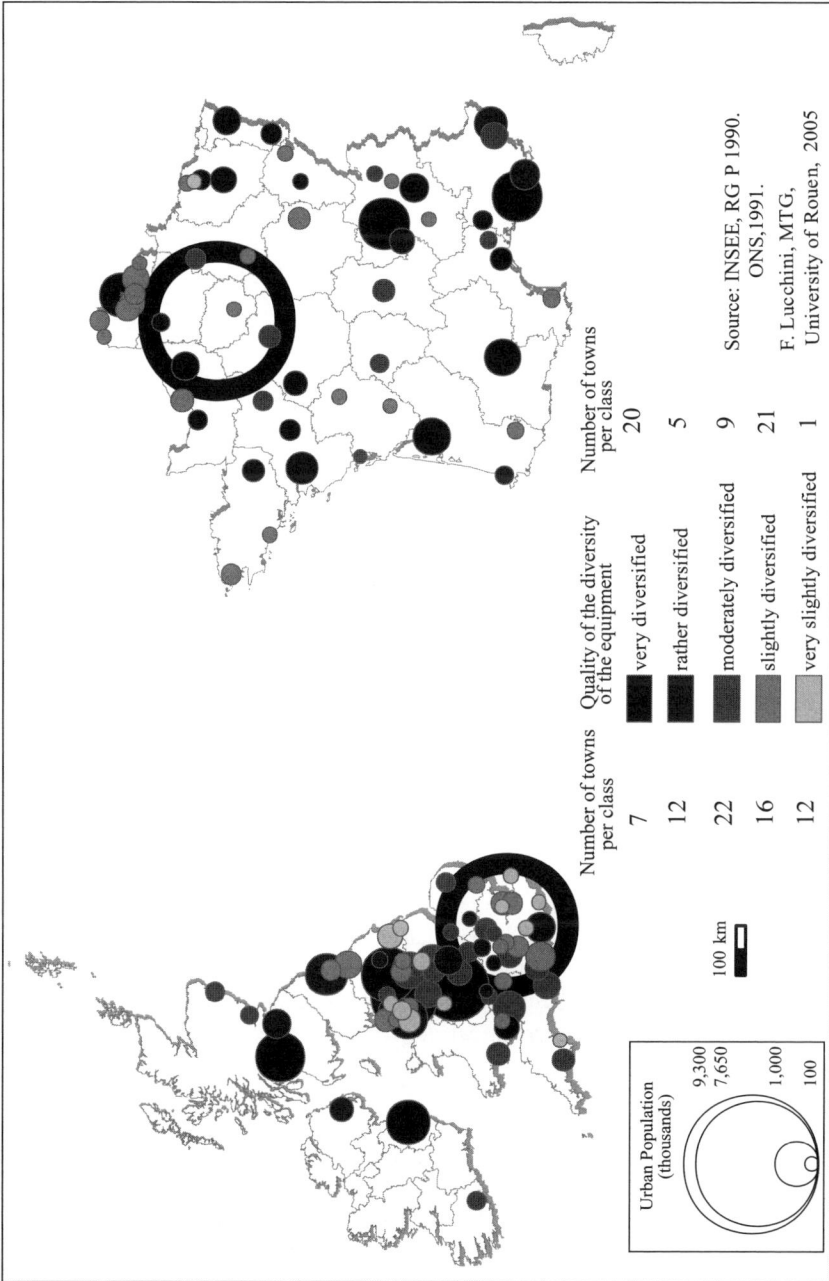

Figure 3.12. *Cultural diversity of British and French cities, 2005*
(see also this map in the color plate section)

After having constituted the French and British databases, the initial urban analysis and subsequent courses of investigation introduce surprisingly similar situations in very different political contexts. Through tests at the national scale, we can observe the resistance of the global cultural model, that is to say the pertinence to comprehend cultural activities as innovative services that were diffused and are still diffused in cities in France, the UK and Ireland. The presence of cultural services and equipment, in quantity and variety, is an advantage developed by cities. The investigation, however, certainly remains open for an in-depth comparison between French and British cities, for the above presented global cultural indicators, and for others to add to this echelon of investigation. If a convergence of national situations is observed in the cultural sector, it does not signify a cultural homogenization for all. In reality the solidity of the system in place, hardly perturbed by national cultural politics and by the influences of globalization, simply has the capacity to integrate a freedom that is noticed in the existence of a varied cultural selection that one can look at on a state or international level.

3.2.2. *A comparative intra-urban study of two cities: similar disparities at the heart of the urban areas of Rouen and Brighton*

With an understanding at the national and inter-urban scale, we then use fine geographical localization to test the global cultural model defined above. This consists of first comprehending the cultural equipment at the heart of a local dimension and valuing the spatial functioning and dynamic, before having the power to launch analyses on individual cultural practices. Two cities were chosen, each situated in a national context of different cultural supervision. The political and cultural supervision for Brighton is liberal, with little participation in cultural life from the state – essentially financial and regulatory – whereas the supervision is centralized and hierarchical for Rouen, with a state that plans, orients and finances cultural life. Local powers must, in both cases, apply a cultural policy that takes into consideration the dimension of the agglomeration or the basin. What is it about the spatial device of the cultural equipment in both contexts? Are we also witnessing at the intra-urban a certain geographical convergence already observed at the inter-urban level?

To answer these questions, the cultural services and equipment were inventoried in an almost exhaustive manner in these two agglomerations both of around 300,000 inhabitants, situated a hundred kilometers from their national capital and enjoying a notable cultural image. A strong differential in terms of the quantity of cultural structures (Figures 3.5 and 3.6) favors Rouen in comparison to Brighton. We count 223 places devoted to culture and leisure (equipment, services, art schools) in Brighton compared to 629 in Rouen. These places can, among other things, bring together several cultural activities, for example, a music and dance school, a library and a cinema in the same complex. Once this assessment of the strength of infrastructure is put forward, we can notice a common dynamic of distribution for the cultural sector in the two cities.

3.2.2.1. *First convergence: a tendency for concentration benefiting the town center*

There exists a tendency for the concentration of cultural equipment that favors the town center at the expense of more peripheral areas in both agglomerations (remoteness in relation to the town center, see Figures 3.13 and 3.14). This concentrated layout of cultural services follows, in a surprising way, the law of urban density enacted by Clark in 1941 for the distribution of population: very concentrated in the center and less and less dense the further we go from the heart of the city. This law has since been used to explain the distribution of urban activities and services that conform to it (decreasing density of these activities from the center according to an exponentially negative curve).

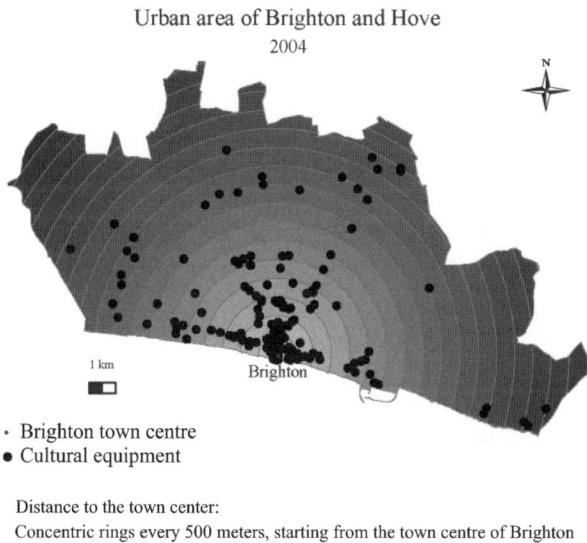

Urban area of Brighton and Hove
2004

· Brighton town centre
● Cultural equipment

Distance to the town center:
Concentric rings every 500 meters, starting from the town centre of Brighton

F. Lucchini, MTG, University of Rouen, 2005

Figure 3.13. *Cultural equipment and distance to the town center of Brighton and Hove, 2004 (see also this map in the color plate section)*

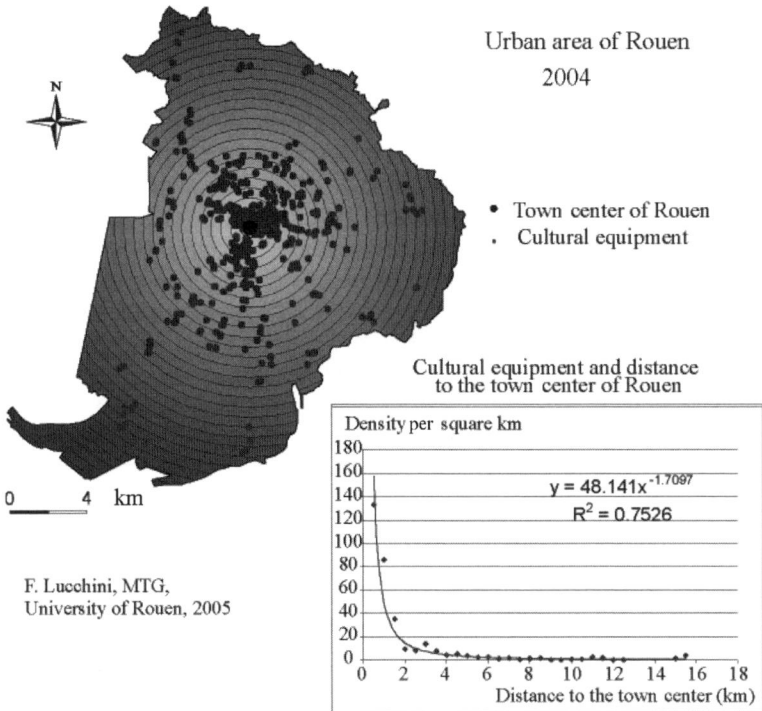

Urban area of Rouen
2004

• Town center of Rouen
. Cultural equipment

Cultural equipment and distance
to the town center of Rouen

0   4   km

F. Lucchini, MTG,
University of Rouen, 2005

Identification of the ring	Surface of the ring km²	Distance to town center km	Number of services	Density per square km
1	0.78	0.5	104	133.09
2	2.34	1	202	86.17
3	3.91	1.5	137	35.06
4	5.47	2	52	9.51
5	7.03	2.5	58	8.25
6	8.59	3	119	13.84
7	10.16	3.5	78	7.68
8	11.72	4	50	4.27
9	13.28	4.5	65	4.89
10	14.85	5	58	3.91
11	16.18	5.5	39	2.41
12	16.97	6	41	2.42
13	17.85	6.5	22	1.23
14	19.11	7	30	1.57
15	19.69	7.5	14	0.71
16	18.61	8	19	1.02
17	17.72	8.5	33	1.86
18	16.72	9	3	0.18
19	14.59	9.5	4	0.27
20	13.44	10	9	0.67
21	11.92	10.5	7	0.59
22	8.52	11	20	2.35
23	6.49	11.5	14	2.16
24	4.51	12	1	0.22
25	3.55	12.5	1	0.28
30	1.97	15	3	1.52
31	1.54	15.5	6	3.89

Figure 3.14. *Cultural equipment and distance to the town center of Rouen, 2004
(see also this map in the color plate section)*

This means that cultural sites have a tendency to arrange themselves like any other service to the population, close to the consumer, like a commonplace service. This inter-urban assessment brings together here some already formulated conclusions with a scale of inter-urban investigation in France and notably in the UK.

3.2.2.2. *Second convergence: a distribution by determined character*

The method of "quadrat counting" [UNW 81] makes it possible to demonstrate that this distribution of cultural activities in Rouen is *not at all the result of a random distribution*. In fact, there is a significant gap between the distribution of cultural activities observed in reality and the theoretical distribution that the cultural activities should follow if they were located at random (Poisson distribution). This implies that the geography of these activities calls on dynamics other than that of chance. It does not conform either to a regular distribution, like that of public schools in the city that are more evenly arranged over the urban territory. In reality, it is instead sports infrastructures that have a more even distribution in the city, because practicing sports is an educational obligation in France. This brings to light the fact that, for many among them, sports complexes are found near schools and evenly distributed in the city, following the example of schools (the study of distance matrixes for cultural/school equipment and sports/school equipment confirms these distribution assessments).

By resorting to a spatial analysis these advances make it possible to account for the efficiency of the formalization of the cultural system, as much for its comparison with other political contexts as for other scales of investigation (inter-urban, intra-urban). The same methodology applied to Brighton accounts for intra-urban conclusions similar to those drawn up for Rouen.

Of course, other spatial analysis tools can be used, particularly when we are confronted with a near lack of information, and therefore with a major difficulty of knowledge formalization.

3.3. Alternative formalizations

To further these observations regarding distribution of cultural services, we ask ourselves how we could carry out investigations when we find ourselves faced with a near absolute lack of information on the subject. This is particularly the case for the sector of individual cultural practices, studied almost uniquely on the national scale. We have practically no finely localized (for each separate cultural site) statistical information relative to practices at our disposal, and furthermore, what results from the action of associations is not always known or measurable. What is more, taking this thought beyond observations revealing paradoxical situations found at the heart of the global cultural model requires that we take a step back, so as to better understand the global functioning of the system at the same time as managing its complexity. Two alternative formalizations are hereafter contemplated; the first is relative to the expression of an urban cultural potential to better define the link between cultural offering and individual practices; the second is oriented towards a reading of the global functioning of the system and its complexity.

3.3.1. *Measuring urban cultural potential*

For the two cities observed above, the knowledge of the cultural public contained in these structures would be an aid to better understanding the operation and dynamic of distribution of cultural activities, and of their real impact at the levels of the inhabitants' practices (areas of influence for cultural structures and equipment). Now, we have already highlighted the thinness, or even absence of localized and comparable knowledge on the subject. As well, beginning with the example of Rouen, investigations were elaborated to understand the public and geographic influence of different types of cultural equipment better (Rouen's Opera, municipal library of a peripheral district).

Let us take the example of the Rouen Opera and its frequency of use (Figure 3.15). Our interpretation of this artistic field situates itself below the communal geographic echelon, so as to make out, by an empirical test, the capacity of influence possessed by cultural structure that we can qualify as "rare", as opposed to a public municipal library for example. For the 2003 season, the Rouen Opera had 2,710 season-ticket holders in an area of less than 5 km around the center of Rouen (also corresponding to the location of the Opera house in the city – 1,792 holders belong to a single district of Rouen), 934 holders reside between 5 and 10 km from the center, 27 between 10 and 15 km and still a few holders distanced by nearly 200 km. We are certainly in the presence of an area of cultural influence that surpasses the local dimension, or even the regional. On the other hand, with a cultural equipment of great locality, the empirical test conducted on the members, in 2002-2003, of the Elsa Triolet library in a suburb of Rouen, situated on the left bank of the Seine, shows a relatively concentric area of influence compared to that of the Opera house, which covers the local area without exerting an influence on the whole agglomeration of Rouen.

Stepping back from these empirical tests and taking the perspective of urban cultural management that causes modifications to the situations of unequal distribution in the city, we can underline the evident action of three simple principles concerning the distribution of equipment in order to better administer a cultural potential on the whole of the urban agglomeration. In the first place, there is the principle of *oneness*: there exists but one cultural structure of this nature in all the agglomeration. We can also find ourselves faced with a principle of *solidarity* when the structures are networked at the heart of the agglomeration to allow the inhabitants of a non-equipped district to benefit from a service.

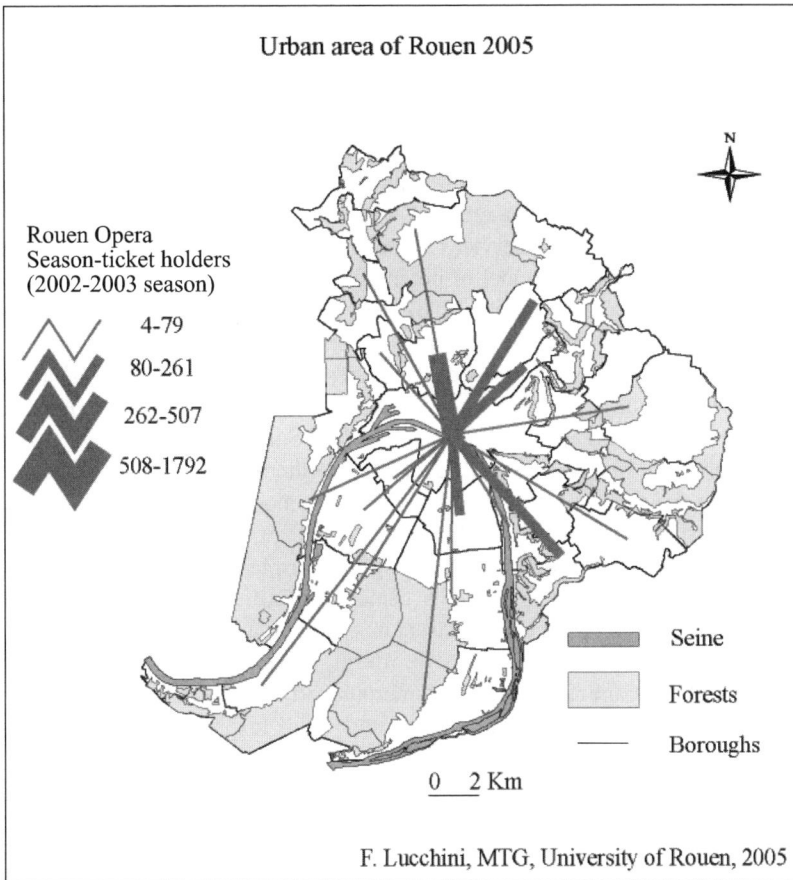

Figure 3.15. *Area of influence of the Rouen Opera, France, 2005*
(see also this map in the color plate section)

Finally, a principle of *geographic influence* of cultural structures allows us to discern the more or less extensive areas of influence for the equipment (according to the location of members, supporters and clients). These principles mix together and are at the heart of research into community interests for political decision makers at the level of the entire urban area concerned with internal rebalancing. These well-timed investigations lead us to consider alternative measures of the intra-urban cultural potential (since the information on the number of visits to structures is very difficult to obtain, when it exists at all) in order to paint a more polished cultural picture in terms of the quality of services offered: where are the most culturally well-equipped areas – many present services – which are less well-equipped, which neighborhoods have a wide variety of cultural activities – quality of service throughout the proposed diversity – and which offer only a small range of different activities?

Figure 3.16 shows two measures of cultural potential in the city, the one formulating the quantitative offer of services, the other the diversity of this offer. The two maps reveal the existence of "cultural nodes" in the city and more or less enclosed zones. A regular grid pattern of Rouen's urban space made it possible for us to tally the level of cultural services for each square of 500 m^2 in terms of volume and variety. Then using the model of potential established by Stewart [WAR 58], a linear interpolation gave us the image of these levels of cultural services, high (in red) and low (in blue). In both cartographic approaches (quantitative offering and diversified offering of cultural services) we generally observed the same locations of elevated levels of cultural nodes: the central part of the agglomeration, particularly the district of Rouen, the university center of Mont-Saint-Aignan, the area closest to the center on the left bank of the Seine, the northern area of the agglomeration (Houppeville) and the south-western area of the city along the left bank of the Seine. For the weakest cultural levels, we observed that the largely urban east was endowed with a particularly low cultural service along the periphery, and likewise a belt that surrounds some poor areas in Rouen's center (the cultural potential of the town center would push the presence of structures further away). It must however be noted that for the diversity of the cultural offering (Figure 3.16a), the cultural nodes greater than the average appear more numerous and are distributed in a more even manner for the whole of the urban area. Cities seem to play the cultural diversity card when competition in terms of the volume of offered services heats up. Let us underline in the meantime that compared to the number of inhabitants per block, we are surprised to find a cultural service lower than the average in the most populated urban neighborhoods, there being the eastern and northern parts of the city where it is not rare to observe peaks of more than 5,000 inhabitants within a 0.25 km^2 radius. A positive aspect however is that we found a more numerous and diversified cultural offering on the left bank of the Seine, where the population is more working-class and less wealthy.

Urban area of Rouen

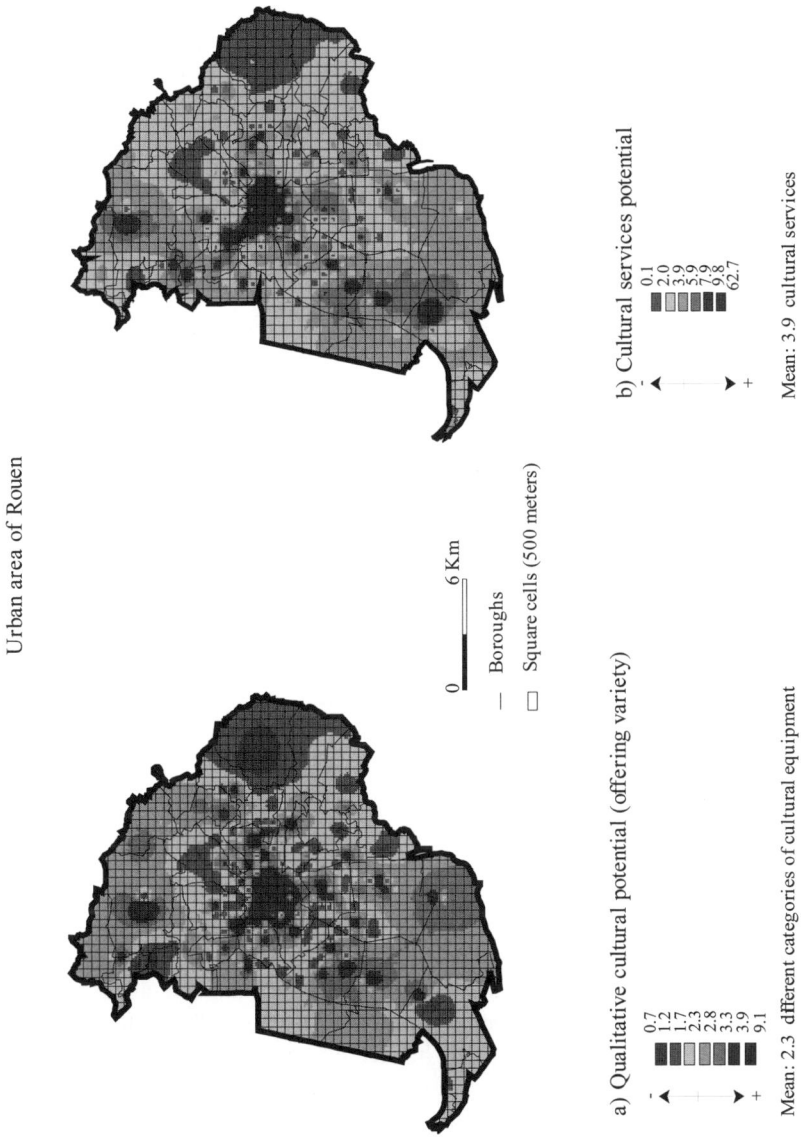

0   6 Km

— Boroughs

☐ Square cells (500 meters)

a) Qualitative cultural potential (offering variety)

0.7
1.2
1.7
2.3
2.8
3.3
3.9
9.1

Mean: 2.3 different categories of cultural equipment

b) Cultural services potential

0.1
2.0
3.9
5.9
7.9
9.8
62.7

Mean: 3.9 cultural services

F. Lucchini, MTG, University of Rouen, 2005

Figure 3.16. *Two measures of cultural potential in Rouen*
(see also this map in the color plate section)

These investigations shine a light on certain mechanisms involved in the spatial distribution dynamic of cultural services and activities at the heart of the city. This knowledge is the first step to understanding the cultural practices of the inhabitants, as far as this is allowed by the potentials for equipment and present cultural services and their variety. Yet, the dynamic cannot be totally controlled, which is due to the confidentiality of demographic information at the finest intra-communal level, or so long as our understanding of the equipment's actual use by the public remains so weak.

3.3.2. A way to better define the global operation of the cultural system

Despite cultural diversity which cannot be denied throughout Europe, we still observed a convergence of situations of cultural offering over space at the same time on part of the inter-urban comparison, and on the intra-urban spatiality of these activities. We are certainly in the presence of a global system, in a state of equilibrium, and it seems clear now, to borrow an expression from Durkheim, that the "whole is more than the sum of the parts". This systemic vision of culture in society becomes attached to identifying the *elementary operators* as well as the organization of these *operators* (their relations), which constitute a new entity. This organization possesses qualities that its elements do not at an interior level. We can say that the system is itself constituted of sub-systems when we identify qualities which are not present at a lower level. What is more, the *interaction* between the different elements makes the simple relations more complex with the loops of positive or negative retroactions that can correct variations of the global system and finally maintain an equilibrium, or, from a more pessimistic point of view, lead to the destruction of the system (emergence of areas of poverty, deterioration of certain urban zones, etc.).

Resorting to a more "systemic" formalization also makes it possible for us to ask questions about the elements composing the cultural system, its functioning, and about the levels of complexity in place. Figure 3.1, formalizing the cultural system in society, reminds us of the implication of different agents in the cultural societal environment: the basic individual (with his/her educational, familial, social and professional components), social groups (with congruous or modified reproductive mechanisms), national and local political authorities and their rules, cultural professionals and amateur associations, cities and their capacity for development, cultural equipment and services with their levels of practice and cultural consummation (protected or not by the law of supply and demand). Observation of the French and English-speaking cultural systems has brought to light the fact that a system composed of different elements can lead to a geographic situation relatively equivalent to the structures of cultural offering in the two countries. The differential is well situated in the composition of the elements, while it is their interaction more than in their qualities of global organization that stand out. It is on the different levels of complexity (elements, systems, sub-systems) that interesting paths of investigation are opened up, and a decisive step will consist of leading to a clear definition of the territorial or functional agents composing the system. The viewpoint is deliberately simplistic, thereby improving our perception of the myriad components in an obscure reality.

3.4. Conclusion

Following the painter Paul Klee, our research efforts adhere completely to the famous formula, "art does not reproduce the visible, it renders visible". Simplifying the geography of cultural activities for better understanding does not furnish an identical reproduction of reality, but rather a vector of comprehension for the complex system that is in place, modeling it by different formalizations that all converge to a more informed reading of the forms of culture in society. It is clear that we cannot be content with just one formalization of knowledge and it is necessary to incessantly sweep the domain of investigation in such a way as to produce a major qualitative change in the knowledge. The methodology employed is interrogated and renewed constantly, which is the engine that drives advances in knowledge. This seems to be an approach particularly suited to a contemporary reading of the world, which itself could be understood as a complex system of interrelations.

3.5. Bibliography

[BOU 69] BOURDIEU P., DARBEL A., 1969, *L'amour de l'Art, les musées d'art européens et leur public*, Edition de Minuit.

[COU 00] Council of Europe, 2000, *Cultural Policies in Europe, a Compendium of Basic Facts and Trends*, Bonn, ed. Du Conseil de l'Europe.

[CER 80] DE CERTEAU M., 1980, *La culture au pluriel*, Christian Bourgois Editeur.

[DON 94] DONNAT O., 1994, *Les Français face à la culture. De l'exclusion à l'éclectisme*, Collection Textes à l'appui-Sociologie, La Découverte.

[FIN 87] FINKIELKRAUT A., 1987, *La défaite de la pensée*, essai, Paris, Gallimard.

[LUC 02] LUCCHINI F., 2002, *La culture au service des villes*, Collection Villes, Anthropos, Economica.

[MEN 94] MENGER P.-M., 1994, "L'offre culturelle française: une concentration dictée par le marché de l'emploi", Center de sociologie des arts, EHESS-CNRS, in *Problèmes Économiques*, 2.381.

[PUM 96] PUMAIN D., ROBIC M.-C., 1996, "Théoriser la ville", in Derycke P-H, Huriot J-M, Pumain D, 1996, *Penser la ville, Théories et modèles*, Collection Villes, Anthropos, Economica.

[PUM 78] PUMAIN D., SAINT-JULIEN TH., 1978, Les dimensions du changement urbain, CNRS, Paris.

[SIM 03] SIMMEL GEORG, 1903, "Métropoles et mentalité", in RONCAYOLO M., PAQUOT TH., 1992, *Villes et civilisation urbaine XVIIIe-XXe siècle*, Larousse.

[UNW 81] UNWIN D., 1981, *Introductory Spatial Analysis*, Methuen & Co Ldt, London and New York.

[WAR 58] WARNTZ W., 1958, "Macrogeography and Social Science", *Geographical Review*, no. 48.

Chapter 4

Modeling and Territorial Forecasting: Issues at Stake in the Modeling of Réunion's Spatial System

4.1. Introduction

While Eratosthenes and others before him have reflected at length on questions pertaining to the complexity of geographic space and its evolution, the ambition to establish geographic models of such evolutions has strangely taken a long time to see the light of day.

Over the last few decades, theoretical contributions to understanding this complexity, which is both spatial and temporal at the same time, have experienced a manifold increase in the field of geographic science, and with them the appropriate computer tools. These breakthroughs made it possible to render the hypotheses on the working principles of the spatial systems explicit, hypotheses that have remained implicit in the geographic reasoning for too long a time. They have perhaps not been sufficiently exported outside the domain of the discipline.

In addition, after recalling a few major theoretical breakthroughs and the expectations they gave rise to in geography and beyond, we present a few territorial forecasting processes for which a socio-spatial system of modeling becomes primarily a scientific guarantee masking a discourse that is more ideological than scientific. We will then propose two modelings of Réunion's socio-spatial system to

---

Chapter written by Gilles LAJOIE.

emphasize the essential contributions made by the territory modeling process when the intention is only to strive towards a certain "spatial understanding".

4.2. A few major theoretical breakthroughs for modeling spatial complexity

Since the advent of new geography, the major preoccupation of an increasing number of geographers is to arrive at a theoretical construction that would explain what is henceforth called a "spatial system". Based on the observation that geographic spaces are shaped by spatial systems, much effort has been expended in recent years in an attempt to understand the structure of these systems at all levels, from the global system to urban agglomerations, with the concept of a region coming between the two. By doing so, we have redefined the main purpose of geography as the act of throwing light on principles of organization that generate spatial complexity, with the theoretical reflection henceforth at the heart of the geographic process.

Ambitions for such a project were evidently considerable and its consequences quickly overshot the single field of geographic science notably by affecting the subject of geography at the secondary education level. As is recalled in this respect by Jean-Pierre Renard [REN 00], "the model of nomothetic geography, even if it is not the only one, is adopted by authors of several manuals, erasing or considerably reducing the gulf that once used to separate research from geography in school. (…) The development in school geography as in geographic science, forces us to question the geographic facts and places, domains of knowledge; to formulate working and interpretative hypotheses of space in order to be able to go from a particular spatial analysis, an oft-criticized descriptive tool, but one which is often interesting and necessary, to a formalization and a generalization of knowledge as well as interpretation of the world. The issue is the movement between the particular and the general, the specific and the common, the detail and the law, etc. It is clearly this constant back and forth movement that makes it possible not only to construct geographic science and increase our knowledge of the complexity of spaces, but also to give the keys for understanding the world to the students, which is the primary purpose of geography education."

There were evidently several paths taken to provide keys for understanding the world while at the same time constructing a new geographic science, with that followed by Roger Brunet [BRU 92] in the 1980s becoming widely known and enjoying particular success. Defined by its author as the science or art of interpretation and subsequently of composition of the elementary structures of geographic space, chorematics made it possible "to resolve in geography the fundamental contradiction between the general and the particular, the law and the individual, the nomothetic and the idiographic". To remove this contradiction, the author proceeded to a theoretical construction of what a spatial system is with the central concepts of complexity, retroactions and systems as a backdrop, while at the same time looking into questions on the societal nature of this system.

According to Brunet's [BRU 92] analysis, "if there is a system, there is energy: what can the energy of this system be? If there is energy and a system, there are stakeholders. If there are stakeholders, there are strategies, interests, representations that can go to the extent of being myths. This system is open, at least to nature, which is in itself an open system: exchanges take place with the outside. This system does not fall from the sky, it is the work of man, hence it incorporates all sorts of legacies, denying or modifying them. This system is not alone, it has neighbors, an environment in the broadest sense of the word, it is in interaction with other places and systems." This analysis superbly highlights the complexity of spatial systems and at the same time the theoretical difficulty in imagining the interactions within these systems for establishing a model.

During the 1980s, this ambition for modeling spatial systems and more specifically their interactions also led to the development of modeling via systemic analysis. As emphasized by J.L. Le Moigne [MOI 78] in an important book, this theoretical approach is based on the structuralist paradigm wherein the object is considered in the totality of its functioning structure and its evolution, that is, in the diachronic and synchronic dimensions, as a system in constant transformation. The theoretical innovation was thus to imagine the spatial system at the preferred level of its interactions and in its scale interlocking, since, as the author also stated, "a system is not irreducible: due to its construction, it is always made up of systems that are themselves made up of systems".

This innovative theoretical approach to geography gave rise to the highest of expectations, evinced by this almost conclusive statement from Henry Chamussy [CHA 86]: "in any case, I do not see a problem before a geographer, in any domain of this science, that cannot be studied by using the concepts and the methods of the theory of the system: from the problems that the most physical of geographers is confronted with (erosion, type of climate, hydrology, etc.) to those that the most specialized of geographers in social domains studies (urban ghettos, location of the services, port traffic, etc.)".

The conception of a model for making the dynamics of a system explicit would then consist of first defining the frontier between the system and its environment, and then specifying the relationships between the essential variables that determine the state of the system at an instant. The modeler subsequently plunged right into the complexity of the system by introducing new intermediary variables before mathematically formulating all the relationships. One of the major breakthroughs of this type of modeling was the obligation for the modeler to render explicit, for each relationship between variables, a certain number of hypotheses on the working principle of the concerned spatial system by way of modeling, hypotheses that have remained implicit for a long time in geographic reasoning. A number of geographers thus saw an essential instrument of geographic reflection in systemic modeling. According to M. Le Berre [BER 86], in the mid-1980s, this new approach "forces one to consider the space differently and with this act alone it questions the discipline and evokes an important step forward in theoretical reflection. This advantage is valuable, whatever the conception that one might have of the systemic, whatever the degree of formalization used in the representation of spatial complexity and whatever the manner in which geography is considered."

Allegiance to these new conceptual tools by a considerable number of geographers in the 1980s can also be explained by the possibility that computer tools gave the researchers the possibility to finally go into the experimental phase of the research. As has been highlighted in this respect by M. Le Berre, "the simulation models make possible in a new way, a new form of experimentation in geography. They help in the retrospective comprehension of the studied phenomena and in the theoretical exploration of the possible futures when used discerningly, and if they are constructed solidly, some of them can provide valuable assistance in decision-making in the matter of planning".

We can affirm that while the siren song of systematics was largely heard and appreciated by the community of modeling geographers, notably at the time of presentation of the first results of the AMORAL model [CHA 83], the breakthroughs as well as the limits of these procedures were highlighted and assumed from the premise of the authors' own reflections. While prerequisite mathematic formalization makes it possible to model a large number of interactions between several variables in the form of equations, the questions on separation, hierarchy of the variables and relationships clearly constitute difficulties. In fact, before transcription in the form of equations, the construction of a graph depicting the interactions between variables, then the weighing of the relationships as per their functional implications in a model, constitute stages that are always delicate depending somewhat on the accuracy of the modeler's choices. Even more obviously, the difficulties related to the integration of space and time in these models were also discussed and assumed by the authors of these modelings. For example, in the case of modeling the space of the southern Prealps (AMORAL model), the division into "countries" constituting the elementary spatial unities of the model was supposed to be in line with the emergence of a spatialized collective conscience corresponding at the same time to the elementary units of territorial planning. Thus, the complexity of the functioning of a territory evoked in Figure 4.1 and modeled mathematically by 150 differential equations was considered at a unique scale and as was then highlighted by J-P. Guérin [GUE 84], if by chance, "the evolution leads to the production of a new space, then it is necessary to start all over again…".

Furthermore, as highlighted in Figure 4.1, this type of modeling enabled the acknowledgment of the spatial constraints only by integrating them in the form of variables. The direct influence of geographic space on the functioning of the model was not raised, with the space moreover supposed to be homogenous at the level of elementary spatial units of the model. Finally and above all, the different variables pertaining to the populations (social groups, forms of habitat, type of employment, etc.) were within these supposedly homogenous elementary spatial units, with the model reflecting a level of overall analysis.

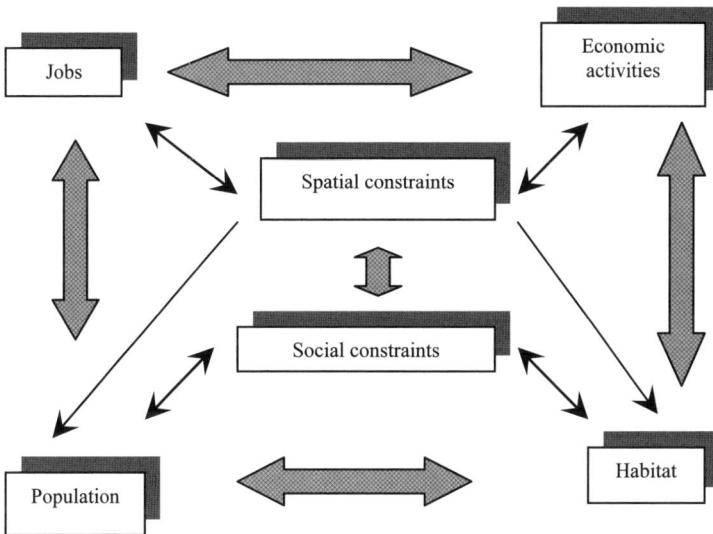

Figure 4.1. *The spatial complexity at the rich times of systemic modeling: the thematic sub-sets of the AMORAL model taken from [GUE 84]*

The integration of time in this systemic modeling posed a certain number of problems recognized by the authors. Other than the act of considering a discreet and non-continuous time in the model, which immediately leads us to choose the pace of time, the mathematic formalization in differential equations informed the modeler as to the state of the model at the instant t, then at the instant t+1, which led *ipso facto* to make the hypothesis that a phenomenon likely to affect the model did not unfold at a slower pace of time.

These so-called continuous models relying on a system of differential equations first had the advantage of being simple thanks to the use of pure mathematical formalism, with their computerized version hardly presenting any particular problems. Translating cause and effect relationships between input variables and output variables, their success also depended on their robustness. Inherent to the tradition of modeling of the dynamic systems initiated by Forrester [FOR 79], the simulations, which this type of model led to, remained perfectly mechanistic, with the state of the system at T+1 already contained in its state at the instant T. There is perhaps one essential difference with the new families of modeling which give pride of place to the concept of emergence, including the soon-to-be discussed multi-agent modeling.

Other more general difficulties can be mentioned against continuous models, difficulties that are demonstrated by Cal[unfilled]roni [CAL 02] thus: "the major criticism that can be made with regard to current mathematical models pertains to the difficulty, even the impossibility, of taking into account the actions of the individuals and thus the effective modifications of the environment resulting from their behavior. The majority of collective phenomena are the result of a set of individual decisions that take into account behaviors of the other stakeholders of the system. By considering actions only by their measurable consequences at the global level, it proves to be difficult to explain the phenomena emerging from the interaction of these individual behaviors... The continuous approach is, by nature, all-embracing; in fact, it supposes that inside a system, the properties are homogenous and isotropic. Insofar as the individual and no longer the population is considered as the modeling unit, the individuals are in interaction in space and time, and the effect of the distribution of the individuals and distances at which they interact, becomes of primary importance."

These criticisms led a certain number of modelers to despair of equations being able to describe the relationships between too many variables and led them to explore discrete models. Among the discrete simulation techniques, cellular automations remain some of the most used. Recognized in its own right as a new family of models after the first few successes of the famous "game of life" in the 1980s [BAU 99], they were exploited primarily in physics to explore chaotic processes in fluid dynamics. Very soon, the advantages of this type of model were recognized for dealing with the difficult question of emergent collective phenomena in domains that are as varied as population dynamics in biology, simulations of the forest fires in ecology (see Figure 4.2) or the study of the development of the urban systems in geography [LAN 97].

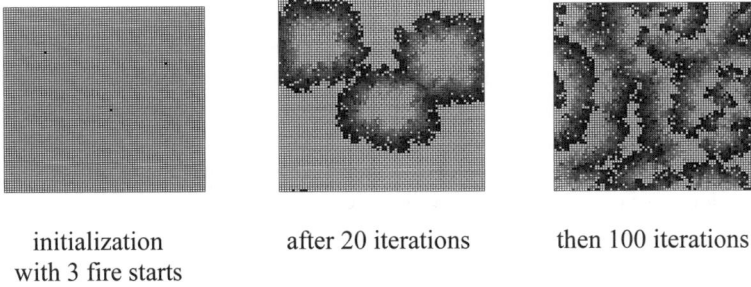

initialization   after 20 iterations   then 100 iterations
with 3 fire starts

Figure 4.2. *Simulation of a forest fire, taken from [FAT 01]*

As is maintained by N. Fatès [FAT 01] following an in-depth analysis of cellular automations, "it seems as though a new scientific paradigm has been developed, its main characteristics being to deal with problems as per a parallel ascending approach, from the simple to the complex, and by determining the behavior of the elementary entities locally".

This change of approach can also be roughly found in the multi-agent modeling, which allows direct representation of the individuals, whether they are social or spatial, through their behaviors and interactions. From this perspective, a phenomenon can be modeled as resulting from a set of the interactions between a certain number of autonomous individuals. These so-called "individual-centered" models make it possible to integrate not only quantitative parameters (the stock variables of the old models) but also qualitative parameters (differentiated behavior of the agents).

In addition, the "multi-agent" simulation comes in at all the system levels, from the single agent to the set of agents that constitute the society or the environment of the model. According to Ferber [FER 95] an agent can be defined as a special object that perceives the environment in which it evolves and where it acts in accordance with a certain number of objectives, called trends (a function of satisfaction, reproduction, survival, etc.). This freedom of action is also translated into the power to communicate with other agents, all of them endowed with specific trends, their own resources and competencies that trace behaviors aiming for satisfaction of specific objectives. This research can lead its "active" agents to modify their environment, that is, a space that is generally equipped with a metric, by exploiting "passive" objects, from their creation to their destruction. Relationships thus unite agents and objects that inhabit the environment in which operations take place (perception, creation, transformation, destruction, etc.).

Thus defined, this virtual world is what we call a multi-agent system (MAS). This type of modeling was largely developed around 10 years ago in the domain of life sciences and social sciences. Geography is concerned very directly since at the heart of the MAS, the question of the interactional system connecting the agent to his environment reappears through a set of perceptions followed by actions. This clearly explains the large volume of research on the environment proposing multi-agent modeling of ecosystems or agrarian systems.

4.3. Modeling and territorial forecasting of the socio-spatial system of Réunion

4.3.1. *Spatial complexity and social urgency in Réunion or future deviations*

In the scientific field, these successive breakthroughs in realizing the spatial complexity have led to continual lively debates, the essential element being effectively to remain within the frontiers of science. As was recalled by H. Chamussy in the case of the AMORAL model that has already been discussed, "the complexity of the space in dynamics is necessarily simplified; we know it and the consequences of this oversimplification are assumed; but we can think that the essence of the operation has been understood and could be called the spatial understanding".

This research into spatial intelligence evidently motivates all geographers and, well beyond this microcosm, all those who "have anything to do with space" including all the local stakeholders who are found notably in other reflection groups carrying out territorial forecasting. In Réunion as elsewhere, the socio-spatial system has thus been the subject of a certain number of studies whose declared objective was to attain this famous spatial intelligence. This objective has been clearly presented as a necessary preliminary to the exercises of territorial forecasting that have increased dramatically in a few years, with prospective reflection finding perhaps a more favorable terrain for its growth with the elaboration of the Schéma d'Aménagement Régional (regional planning scheme) by the Conseil Regional, adopted by the Conseil d'Etat in 1995, which fixed the broad lines of planning for the decade.

Thus, since 1994, "songes sur La Réunion (dreams for Réunion)" have been published in the local press and present three scenarios that are quite traditional: European integration, the Indian island chain and sustainable development of the Creole-speaking areas [PRE 94]. From 1998 onwards, the association ODR (Observatoire du Développement de la Réunion – Réunion development observatory) has assembled a group of experts to debate on "l'avenir et les enjeux stratégiques de La Réunion à 2030 (Réunion's future and the strategic issues in 2030)". Based on a more constructed methodology, the assembly has led to the drafting of a dozen scenarios that are often iconoclastic [ODR 02]. More recently, "the Cahiers réunionnais du développement" published a special issue on "La Réunion à l'heure des choix" drawing up three scenarios for 2020: the economic and social *status quo*, social and rural reform, economic and urban reform [AKO 02]. Finally, in September 2002, the Conseil Économique et Social (economic and social council) published its reflections on the "issues and challenges for Réunion by 2020."

This list of the main exercises in territorial forecasting work carried out recently on Réunion sends back several messages and primarily, that of abundance born from the urgency of the responses needed for planning issues of the territory "falling under the public domain". In a few years, about 20 scenarios have been imagined, with their construction acting as some sort of first response given urgently to a territorial crisis in everyday life that was evident to all the inhabitants of Réunion (unemployment and social instability, lack of space for agricultural activities and for the habitat, difficult traffic, etc.).

These urgencies truly constitute the driver for local reflection with respect to the future of the territories. Figure 4.3 taken from the "La Réunion à l'heure des choix" [AKO 02] perfectly illustrates this situation because the three scenarios with 2020 as the objective are judged by the yardstick of results obtained in three domains: unemployment, useful agricultural area, i.e. for sugarcane cultivation, and the housing density per hectare.

Given that the intention is obviously not to comment on the figures for 2020 ,we simply underline that scenario 1, which extends the curves, represents only the current anguish of exploding unemployment, diminishing areas of sugarcane cultivation – coming down to the ominous threshold of 30,000 ha (profitability threshold) – and an urban pressure that is increasing.

	UNEMPLOYED	Unemployed *in % of employed population*	Agricultural area *including sugarcane*	Housing density per ha (urban areas 1999)
2000	130,400	36.5	43,500 ha *incl. 25,500 for sugarcane*	6.6
2020 Scenario 1 *Status quo*	192,000	45	41,000 ha *incl. 22,000 for sugarcane*	8.6
2020 Scenario 2 *Social and rural reform*	200,000	50	28,000 ha *incl. 15,000 for sugarcane*	8
2020 Scenario 3 *Economic and urban reform*	50,000	12	42,000 ha *incl. 25,000 for sugarcane*	10.6

Figure 4.3. *The scenarios in figures taken from [AKO 02]*

Another image reflected by certain forecasting studies is that of science applied to the territory. In this respect, the forecasting process undertaken by the Futurs Réunion group that set out to understand the interactions between different spheres (territorial, societal, economic, etc.) has heuristic objectives so ambitious that they deserve attention.

Based on "a structural analysis of the factors that determine Réunion's evolution, the Réunion system is divided into a matrix with eight sub-systems governed by a combination of key variables: population, employment and occupation, production system, infrastructure, environment, ways of life, politico-institutional system, international context... The identification of the variables that have a role to play in Réunion 's evolution by 2030 and the present and potential reports referring to one another is a vital step in the forecasting process. The tool retained was the cross-impact matrix for highlighting the motivity and dependence relationships between the variables". In short, the Réunion system is thus modeled with 39 variables that constitute the eight sub-systems presented in Figure 4.4 [ODR 99]. Upon perusal of such a table, the reader cannot avoid being impressed, all the more so because the Futurs Réunion group makes it clear to us that its forecasting process is based on a principle of the "Delphi" method, which indicates that "experts are better placed than other citizens for explaining the logic of social phenomena and movements". "Major personalities in the current functioning of the island", whose caliber as experts and competency areas are not specified, were thus invited to estimate the assets, constraints and margin for maneuver for the variables grouped together as sub-systems. However, the methodological effort produced imagining Réunion's possible futures does call for some comments before discussing the scenarios.

In the first place, it is not futile to recall that the empirical and inductive process that relies on the identification of trends was largely used in the sociology of recent times for identifying the variable factors of social change [DIR 90]. It implies that we have to be in a position to conclude whether a link exists or not between all the trends taken two by two, with the cause and effect graph processed using a network analysis algorithm. Without entering into the details of the criticisms formulated against this method and relayed by the very significant Revue Française de Sociologie [RFS 97], let us first highlight the fact that recourse to the experts is sometimes satisfying only for organizers of the reflection on the forecasting.

For the average reader, the thought process of the *expert* who recognized a cause and effect relationship between two variables remains irreducibly obscure. The reader has in front of him a *"cross-impact matrix"* which to him looks like an immense black box composed of all the unexplained cause and effect relationships. Thus a number of questions remain in abeyance.

How many of the 1,521 possible cause and effect relationships (39 x 39 variables) were explored given that many are not applicable? In addition, does the matrix not cross completely heterogenous objects, some corresponding to real empirical observations (demographic indicators), others lending themselves less easily to measurement and instead constituting theoretical objects, which are in fact widely discussed (cultural identity; value and behavior, etc.)? In short, which relationships are considered as motivating or dependent?

DEMOGRAPHY	1	Natural population growth
	2	Net immigration
	3	Structure per age
	4	Geographic distribution
EMPLOYMENT AND OCCUPATION	5	Rate of activity
	6	Professional qualifications
	7	Labor costs of the factors
	8	Subsidized employments
	9	Unemployment
	10	Social dialog
PRODUCTION SYSTEM	11	Economic growth per sector
	12	Emerging sectors
	13	Productivity of the factors
	14	Know-how and poles of excellence
	15	Mastery of the new technologies and norms
	16	Research and development
INFRASTRUCTURE	17	Land occupation
	18	Housing
	19	Public facilities and services
	20	Internal transport
	21	External transport
	22	Telecommunication system
ENVIRONMENT	23	Water distribution and treatment
	24	Resource management and conservation
	25	Treatment of waste
	26	Major natural hazards
WAYS OF LIFE	27	Available income
	28	Structures and ways of consumption
	29	Level of education
	30	Cultural identity
	31	Values and behaviors
	32	Citizenship and civil society
POLITICO-INSTITUTIONAL SYSTEM	33	Distribution of the competencies
	34	Distribution of the public resources
	35	Stature of the public sector
INTERNATIONAL CONTEXT	36	Image and attractiveness
	37	Dynamics of European integration
	38	Inter-regional relationships
	39	Growth of the world economy

Figure 4.4. *"The Réunion system and its eight sub-systems" taken from [ODR 99]*

Finally we must say something concerning the search for the potential rupture points of certain key variables presented in Figure 4.5? The experts have ruled that a rate of activity divided by 2 would represent a rupture, just as the external transport costs divided by 2 (?) or multiplied by 3 or even the price of water multiplied by 20. Obviously, we can speculate as to whether these figures are significant or not.

Again, we must determine whether any sense can be accorded to these figures other than the fact that they exist for a short while during the debate of the experts. As for the societal variables that elude all quantifications, their rupture points sometimes seem to have been attained a long time ago (values and citizenship: commonplace delinquency). The typology of the scenarios which follows must provide, if not responses to the questions, at least some instructive light on the finalities of the forecasting processes.

4.3.2. *The trend scenarios or the probable future*

In the forecasting exercise proposed in "La Réunion à l'heure des choix", the status quo or real time scenario obviously remains an unacceptable scenario. It depicts with precision an apocalypse which weds planning problems (explosion of the individual habitat, precarious constructions without permit, land speculation, sacrificed sugarcane cultivation area, etc.), social problems (constantly growing unemployment, deepening social chasms, explosion of criminality and delinquency, etc.) and political problems (institutional status quo, absence of national project for overseas).

Rate of activity	Divided by 2
Land occupation	Land speculation Commoditization
External transport	Price divided by 2 or multiplied by 3
Water distribution	Price multiplied by 20
Cultural identity Values and citizenship	Community isolation, monoculture, globalization Commonplace delinquency

Figure 4.5. *Examples of the potential ruptures for a few key variables taken from [ODR 99]*

However, apart from the problems mentioned, it is their sequencing that deserves attention. While the relationship with land is clearly addressed and a part of the text is titled "planning system: a territory in precarious and fluctuating equilibrium", the angle of attack, in particular, seems narrow and simplistic. In fact, the only land planning tools that see the light of day on the period are the Parc National des Hauts, along with "coastline protection measures" that are not elaborated. These tools, which are supposed to make it possible to control spatial organization prove to be inefficient due to the aggravation of unemployment which leads *ipso facto* to degradation of living conditions in the urban zone. The imagined result is the progressive loss of control in the territory, which leads to "a return to the land" with the development of precarious habitats on agricultural or natural spaces occupied without permits.

The movement between social and spatial, between the territory and its social, economic and political dimensions is broadly simplified to the extreme. It is particularly distorted when passing through what could be called "an ideological prism". The cause and effect chain and the succession of subsequent events can be expressed in the following manner: in the absence of a decisive aid from the State ("a national political project for the overseas areas"), a social problem (unemployment) takes on a spatial dimension (return to the land), as if this recourse were always possible in Réunion despite national regulations (Parc National des Hauts). The scenario goes to the extent of predicting "disrespect for the law that becomes a cultural norm", with the re-appropriation of land thus taking place without ownership deeds.

This condemnation without appeal deserves our attention. In fact, the current situation can already be deemed difficult with respect to the above-mentioned indicators (unemployment, precariousness, etc.) and can partly explain the sense of urgency. For all that, it is doubtful whether an observer will see the premises of a major crisis affecting all aspects of social life (delinquency, return to the land and reign of lawlessness) and forever modifying the living environment of the inhabitants of Réunion (deteriorated environment, pollution, full-scale blockages in the transport networks, etc.).

In other words, we propose a hypothesis that, while taking a simple look without any nuances at the present and the future that awaits us "if we do not do anything" is also a part of the exercise. The objective analysis of the strengths and the weaknesses of the territory quickly becomes transformed into a sort of catharsis that highlights in particular, the stigma of the predicted crisis, which leads to an imbalance between the "reasons for believing" in the coming catastrophe and the "reasons for not believing in it" that are logically kept secret.

Since the exercise must start with the drafting of a consensual mobilization project based on the idea of reform (positive or voluntarist scenarios), the current trends table will always resemble the work of Jérôme Bosch...

4.3.3. *Catastrophic scenarios/unacceptable futures*

As per our hypothesis, the hidden objective of the forecasting exercise is to impose the reform. We will thus not be surprised with the tenor of the unacceptable scenarios, whose headings speak for themselves (inter-ethnic war, independence, world company, Singapore on Réunion, etc.). Contrary to trend scenarios that cannot be expected to separate the wheat from the chaff in the current trends, these unacceptable scenarios are within their logic when they predict parlous futures. In fact, the general plot traces the disappearance of current trends, or at least, their brutal bifurcation leading to ultimate chaos. The message is clear: reform or chaos, and through creative efforts by the experts, it takes on a more or less political and often cultural tone, with "economic horror" obvious as the permanent feature.

Thus, in the scenario of independence described by the "Futurs Réunion" group, the "Répiblik Reyonez" is proclaimed in 2007 and causes an unprecedented emigration (150,000 departures/year, largely civil servants from the mainland). As a result, unemployment recedes thanks to the jobs released during the honeymoon period, followed quickly by a serious crisis due to the end of French and European financial aid, exit from the euro zone and the return of the CFA Frank, finally creating an unprecedented monetary instability. With the decrease in household purchasing power, consumption crumbles and the local businesses are severely affected. As a result, "in 2025, the situation has become equivalent to that Madagascar experienced in 1999, malaria is present once again, infant mortality has increased tenfold, lack of education has given way to illiteracy".

At the heart of the crisis, the people of Réunion accept by referendum the Indian proposal of becoming in 2031 an "overseas island district administered by a governor taking his orders from Delhi". However, "Indian assistance" is only an illusion. In the chapter on deep-rooted disruptions, it is noted that the land is redistributed to the peasants by the Indian National Office while at the social level, the caste system appears, English becomes the official language and Hindi and Urdu are made obligatory. Having become, against its wishes, a back-up military base for India in the Indian Ocean, Réunion sees its youth sacrificed in the battalions of *tirailleurs mascarins* (mascarin infantry) sent to the front in an Indo-Pakistan war.

Among the catastrophic scenarios, the independence scenario is by far the most developed and most detailed as regards its cause and effect chain, perhaps commensurate with the anguish that this future with its conclusive rupture from the French mainland generates. It is also archetypical of unacceptable futures in more ways than one. In fact, the political dimension is omnipresent from the beginning of the scenario (creation of a National Réunion Congress, adoption of the constitution of the "Répiblik Reyonez", etc.), but the general disturbance is economic in nature, its consequences on the desintegration of the social body being endless (explosion of delinquency, psychological withdrawal, development of sects, etc.). While, in short, the international geopolitical context decides the fate of an island that has become a pawn on the geostrategic arena for India, the initial dysfunction is clearly economic in nature.

The importance of the economics in the emergence of "catastrophic scenarios" is sometimes denoted in the title itself (World Company, Singapore on Réunion). In "World Company", the global movement for concentration of enterprises has a violent impact on Réunion and a single company is then in a situation of perfect monopoly. The society is then recomposed into three distinct groups: unemployed, employees of the "World Company" and government servants. In a context with vivid social tensions and confronted with the unemployed who do not have access to the global model of mass communication, the government servants and the rare private sector employees take refuge in "small colonies" protected by urban militias.

Once again, economics and its associated events win over the political and the social, which suggests the necessity of the reforms proposed in voluntarist scenarios tracing out desirable futures.

The last characteristic of these catastrophic scenarios is that they are purely theoretical forecasting exercises, without precise connections to the actual situation of the region. While the methodological tool always decomposes Réunion's system into sub-systems with a few variables that are truly spatial (geographic distribution of the demographic variables, land occupation, etc.), the precise references to the territory are absent.

It thus follows in the inter-ethnic war scenario, wherein Réunion resembles the Beirut of the 1980s with its armed militia, its heightened inter-ethnic tensions against the backdrop of economic crisis, albeit one from where territorial conflict is absent. This resurfaces however in the scenario of the Natural Reserve wherein Réunion becomes one of the 15 sensitive zones classified as "heritage of humanity" and turned into a sanctuary by an all-powerful European Union, which has become owner of the land. In this unacceptable future, Réunion's population is concentrated in three cities situated far from the coastal areas and connected by a common transport network.

Unfortunately, when the territory is re-injected into forecasting reflection for this scenario, it is done with great liberty and forgetting that the territory has particularly pregnant memories that explain its inertia in the long term. This is clearly recognized by the authors of this scenario, who make the understatement that "*the financial and social costs of such a situation are hardly realistic*".

4.3.4. *Reformist scenarios/desirable futures*

After having explored trend scenarios leading to crisis situations that prefigure the catastrophic scenarios, the forecasting process naturally opens out into the drafting of a voluntarist or reforming scenario likely to become a genuine territorial project. This prerequisite reform sometimes appears in the heading as well as in the scenarios imagined in "La Réunion à l'heure des choix" [AKO 02] or in its two rural and urban variants.

Thus in the first scenario, "social and rural reform" is imposed on all as "degradation of the social situation and increases in criminality led to the economic and institutional status quo being questioned". In this desirable future, help comes in fact from the cultural resources of Réunion's Creole-speaking society. For example, individual initiative is encouraged to manage the land resources and the more traditional resources of national unity in an optimal manner. Thus inhabitants can construct their own house, using aid and building in controlled spaces and they can live off the fruits of the formal and informal economy, with the food resources produced in the family gardens constituting an additional income for the households.

As for planning programs in the territory, concentration and densification of the habitat and the work areas are given up in order to prevent social tensions and environmental damage. With urban sprawl increasing, the agricultural area recedes but we finally know how to reserve the best lands for agricultural production and farming. The scenario however predicts a few difficulties in containing the dispersion of the habitat and the costs of this dispersion in terms of equipment at the end of the period (2020).

In its urban variant, the voluntarist scenario described in "La Réunion à l'heure des choix" also relies on the idea of "economic and urban reform", but everything opposes it to "the social and rural reform", with the term to term opposition evidently not being fortuitous (urban *versus* rural; economic *versus* social). While the preceding reform started from the base (individual initiative encouraged), this one comes from the top, with the European Union finally authorizing massive aid programs for production in the territories suffering from heavy structural handicaps (i.e. application of the waiver clauses of Article 299-2 of the Amsterdam Treaty). Thus, thanks to a reduction in costs (work, production, transport, etc.) associated with these tax incentives, economic development takes place on time and unemployment regresses strongly while illegal work and subsidization of income gradually disappears.

As for planning schemes in the territory, this desirable future provides for genuine development strategies championing economic activities with strong potential (tourism, agriculture). The quality urban areas become more densely populated without deterioration of the social climate due to a social elevator that functions once again, with the range of accommodation getting diversified. New cities appear mid-slope on lands with limited agricultural potential and without environmental interest. In short, this voluntarist densification is a success and human pressure is controlled better.

The top-down idea of reform opening out to a desirable future for Réunion is also at the heart of the scenario entitled "Common law" and imagined by the Futurs Réunion Group. In fact, after "the inhabitants of Réunion achieve a real and full integration of their island with the national whole, without any particularism", the French State, aided by Europe, launches a vast "Réunion Plan" that will establish vital "structural and temporary recovery measures and subsequently support measures for local development". Paradoxically, this voluntarist scenario which traces a future, both is desirable and believable at the same time, does not have any precise reference to the land despite evoking local development. However, in the majority of desirable futures imagined by the Futurs Réunion group ("the high tech paradise, the alchemy of the success, Creole romances") help comes from science and technology. Within a few decades and thanks to particularly ambitious training plans, the island is equipped with specialized human resources not only in the NTIC, but also in "biotechnologies, neuronal systems, meteorology, geothermal science, artificial intelligence, human redress, modeling, after-sale technological service, fundamental research, etc". Finally, it should be noted that these different desirable futures often go through the prerequisite reforms. Whether it is a "Réunion Plan" or a European recovery plan, whether it is economic and urban or social and rural, basic reform seems to be the fundamental element of all desirable futures as an essential bifurcation in the pursuit of current trends.

At the end of this typology of the scenarios, a few remarks on the form and substance can be formulated. As for the form, geographers can only rejoice that a public debate has started on the future of the territory, with the forecasting reflection henceforth "decentralized". Réunion has not resisted the surge of territorial forecasting, thus following a general trend. Starting with the premise that it is *good to think of the future of one's land*, especially in a situation of crisis, everyone tries his hand at methods with more or less theoretical baggage, and creativity. Yet, while "understanding of the territory" always seems to the declared objective, the means to attain it do not systematically come within the field of science. In reality, in crisis situations, only the activity of reflection on the future of the territory sends back a positive image and the forecast can locally become a therapy. However, a few comments must be made on the substance of the forecasting process.

A first paradox with respect to the great diversity of the productions can be highlighted. In fact, to open out the reflection to a greater number of people and launch the public debate on the future of the territory, the detailed presentation of the heavy methodologies used in forecasting is often avoided. At the same time, we do not wish "to leave science" and its positive aura entirely, and this is why a few principles and scientific methods used in the forecasts (structural analysis, motivating, dependent variables, etc.) are often recalled at the beginning.

In the end, the attentive reader who is expecting an objective and scientific analysis of the functioning of the territory remains unsatisfied. He finds himself confronted instead with a set of black boxes (relationships between variables), or even questions without response (spatial inscriptions of the variables). In the same way, the curious reader will not hesitate to wonder about the competency areas of the experts who are supposed to understand the functioning of a territory better than the others. From this methodological and conceptual uncertainty, what follows is a surprising diversity of productions that could, in short, be detrimental to the forecasting reflection of the whole. The reader discovers not only traditional forecasting exercises that provide hypotheses and propose a clear path but also purely literary exercises that are closer to the short story than to the territorial forecasting scenario. Some will probably argue that the creativity of the experts in forecasts should not be curbed and they are not wrong, but in that case recourse to scientific vocabulary is no longer necessary. A science-fiction novel cannot be called a scientific book even if its contents are drawn from it. They are intellectual productions of a different kind.

A final comment can be made with regard to the problematic movement between the social and the spatial, between the forecast and the territory. Here as elsewhere, these measures are presented as reflections on the territory and its possible futures, while the scenarios most often ignore the geographic space, its levels, its constraints and its memories.

The trends spotted in the present and whose curves are traced over 20 to 30 years are generally aspatial and when their geographic dimension is mentioned, it is with a level of such high generalization that it is no longer pertinent (rural, urban, coastline area, useful agricultural area, etc.). Upon perusal of the scenarios, it is observed that regionalization of the trends is not attempted, as if the territory was a single homogenous whole. This movement between the social and the spatial is not a recent development and remains a major theoretical difficulty. For all that, territorial forecasting does not ignore it because, in Durkheim's words, social facts generate physical substrates that constitute the territory. As is highlighted by the sociologist Grafmeyer [GRA 94], "between the material devices, structural facts, institutions, free currents of the social life, there are no differences of nature, but only unequal degrees of crystallization of this social life".

4.4. Modeling of Réunion's socio-spatial system

4.4.1. *Graphic modeling of Réunion's complexity*

Faced with the socio-spatial complexity, the siren song of modeling blended with the approach of territorial forecasting can lead to certain problems as we have just seen for Réunion's system. While it is true that we can think that territorial forecasting has dissimilar rules to spatial modeling, the objective for both is the same: to arrive at spatial understanding. In fact, how can we write the possible futures of a given space without stating beforehand the assumption that we have apprehended most of its the main working rules in the present instant?

In the case of Réunion's socio-spatial system, a general feeling of urgency can perhaps explain such deviations: urgency in the matters of housing, means of transport, employment, reduction of social inequalities, etc. This sentiment, shared by the politicians and the various social players who are involved in the planning of territory is in fact that there is an urgent need for understanding the present and its main trends in order to have better control over an uncertain future as much from the social point of view as from the spatial point of view. Faced with this dual observation of complexity and urgency, the contributions made by modeling deserves to be highlighted, given that they make it possible to strive towards a real intelligence of the space.

An important contribution is that of chorematics. As has been perfectly summarized by Grataloup [GRA 93], "at the base of the chorematic project, there is a desire to deconstruct spatial complexity". Expressed differently by the initiator of this scientific project [BRU 90], "chorematics helps in researching; imagining; understanding; followed by presenting and making others understand". Construction of the graphic model is thus not the result of the research, but the research itself.

For Réunion's socio-spatial system, this deconstruction of spatial complexity can start with a comparison. As proposed in Figure 4.6, the simultaneous vision of the chorotype of the tropical island, a great favorite of the genre [GOD 97] and the chorotype of Réunion island makes it possible to organize the reflection. A great number of elements for structuring Réunion's space can obviously be found in the traditional model.

CLASSIC MODEL OF "SUGAR ISLAND"

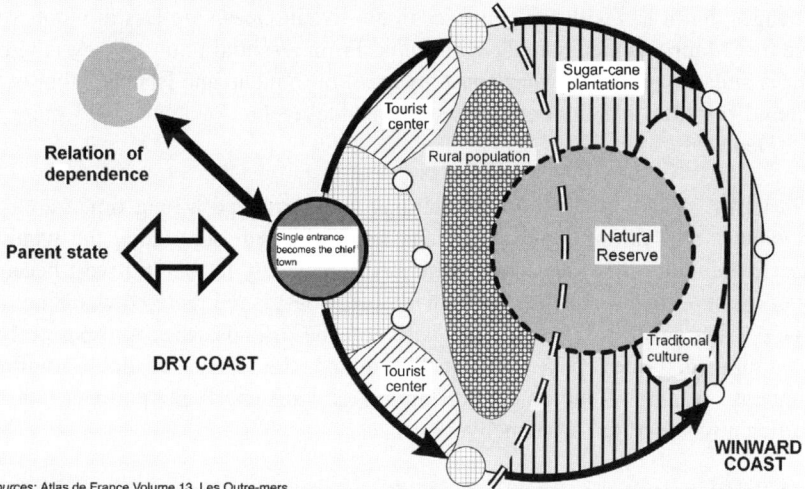

Tourist center

Sugar-cane plantations

Relation of dependence

Rural population

Parent state

Single entrance becomes the chief town

Natural Reserve

DRY COAST

Tourist center

Traditonal culture

WINWARD COAST

Sources: Atlas de France, Volume 13, Les Outre-mers

LOCAL ADAPTATION AND TERRITORIAL DISPARITIES

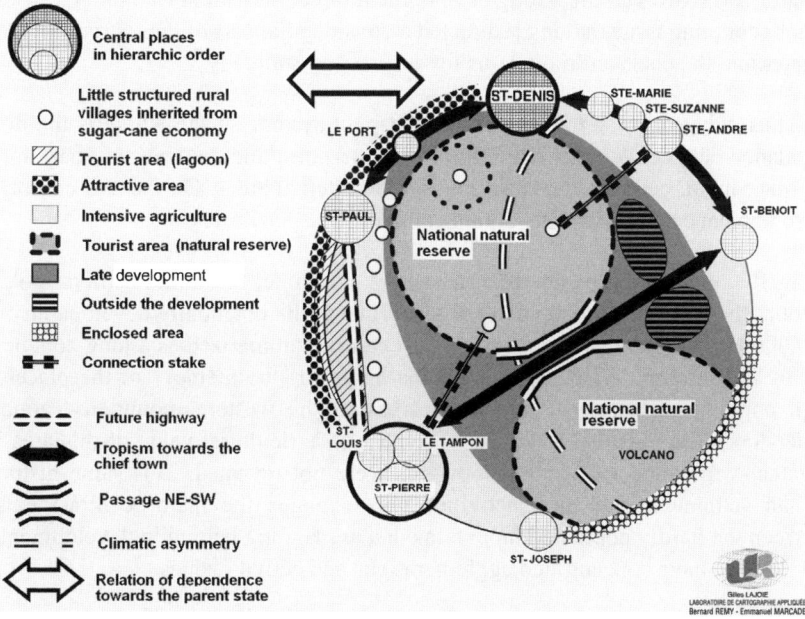

Central places in hierarchic order

Little structured rural villages inherited from sugar-cane economy

Tourist area (lagoon)

Attractive area

Intensive agriculture

Tourist area (natural reserve)

Late development

Outside the development

Enclosed area

Connection stake

Future highway

Tropism towards the chief town

Passage NE-SW

Climatic asymmetry

Relation of dependence towards the parent state

ST-DENIS

STE-MARIE
STE-SUZANNE
STE-ANDRE

LE PORT

ST-PAUL

National natural reserve

ST-BENOIT

National natural reserve

VOLCANO

ST-LOUIS

LE TAMPON

ST-PIERRE

ST-JOSEPH

Gilles LAJOIE
LABORATOIRE DE CARTOGRAPHIE APPLIQUÉE
Bernard REMY - Emmanuel MARCADE

Figure 4.6. *The chorotype of "the sugarcane island" and its local adaptation*

It is the environment of the system that needs to be highlighted first. The island is dependent on the mainland and its exchange flows are deeply imbalanced. This dependency, born from history, is perhaps more sensitive in Réunion and the exchanges that it has with the countries in the Indian Ocean are less in number than those that Mauritius has, notably with India. From a spatial point of view, the entry point is situated on the leeward coast, as in all the "sugarcane islands", that is, in a sheltered position that the ancient historic capital (Saint-Paul) affirmed more than Saint-Denis.

In terms of land occupation, plantations were arranged a long time ago as per pluviometry and altitude, with sugarcane fields situated more along the windward coast. In Réunion, it is henceforth cultivated intensively in the arid west because a part of the superficial overflow of the windward coast is redirected towards deficient areas of the west. In the interior parts of the islands, the agro-pastoral and silvicultural activities have historically shared the space available, with the mountains of the Réunion appearing as an oft-enclaved periphery to the insufficiently structured urban network.

In Réunion as in the general model, the touristic spaces are concentrated along the sheltered coast and at a reasonable distance from the main town of the department. However, the strong development of ecotourism in Réunion represents a major economic issue, having redirected a considerable part of the tourist flow to the cirques and the enclaved mountains for several years already.

Thus, what characterizes this insular capital system on the whole is the double imbalance between, on one hand, the windward coast and the leeward coast and, on the other hand, between the coastline and the internal areas. On the two chorotypes, these imbalances underpin the organization of these territories.

In Réunion, while occupying the narrow coastline, the initial settlers took into account the natural constraints of the site (100,000 ha potentially developable out of 250,000 ha) and the history of settlements can be summarized as a long conquest of slopes and cirques. What remains of this history is the memory of the places; the high population density at mid-slope areas of the western mountains cannot be understood if the history of the enhancement of agricultural value of this area with sugarcane from the 19th century onwards were not revealed. This same history of human settlement and their activities thus explains the major contrast existing between the hardly populated mountains that are lagging behind in development and the Bas that have concentrated agglomerations and activity zones.

The affirmation of a polynuclear urban pole in the south constitutes an important difference from the general model, with this economic counterweight having in fact taken on an almost institutional dimension with the bi-departmentalization project proposed a little rashly by the Jospin government (the project of two departments, one in the north and one in the south, was quickly buried). Between the two economic poles of the north and the south, a zone of very high human density was thus formed and causes crucial traveling problems on the island, all the more because the coastal tourism along the lagoon is at the heart of this narrow coastline area.

This spatial concentration of human beings and flows leads locally to a situation of "circulatory coma" since 75% of travel on the island has only six districts as a destination. Despite the various problems that it brings about (increase in land prices, travel difficulties, etc.), this dynamic is continued and nourished partly by external factors: 60% of the new arrivals at Réunion settle down in one of the four big districts of the island: Saint-Denis, Saint-Paul, Saint-Pierre and Le Tampon.

To deal with this concentration of men and activities and to resolve the problem of travel, a mid-slope road is under construction in the western mountains (the tamarind road) that risks reinforcing the tropism of the big north-western quarter of the island while opening up a part of the mountains at the same time. This reinforcement of the pull exercised by the main town of the department will perhaps be further accentuated when one of the big projects of the Regional Council – a common transport on its own site (TCSP) – will be functional from Saint-Benoît to Saint-Paul in around 2012.

Finally, the coast protected from the trade winds appears destined for exposure to "economic flows" for a long time yet while a dead angle is being formed in the south-east, along a coast exposed to the trade winds but protected from "economic flows".

The chorotype of the Réunion's system finally highlights the fact that help may come from the east, the zone between Sainte-Marie and Saint-Benoît with several advantages including the space available in the mountains and especially the proximity and rapid accessibility with regard to the main town of the department.

4.5. Towards a modeling of the dynamics of Réunion's system

Since 1990, Réunion's population has increased at an average of 1.8% per year whereas the population of mainland France progressed at a rate of 0.4% per year. This natural difference alone explains the total growth rate of nearly 86%. At this rate, the population of the island, which was 740,000 in 2002 will have gone beyond 832,000 in 2010, with the million mark possibly attained around 2030.

Such demographic projections immediately take us back to the spatial problematic in a micro-insular context where space has become rare for all the socio-economic players. From the spatial point of view, this growth in the demand and the pressure that it exerts on land leads from this present moment onwards to one of the most troubling problems of the past few years: anarchic development of cities and urban sprawl. Despite the priorities announced in a Schéma d'Aménagement Régional (regional development scheme) in 1995, urban densification has not happened as planned.

"Constructing a city on a city" remains a desire shared by everyone, but at the local level the sum total of individual decisions leads to urban sprawl. In 1995, the SAR advocated a minimum density of 30 housings per hectare for the urban extension zones. At the level of rather big neighborhoods, this number is attained only in the two agglomerations of Saint-Denis and Port while the percentage of housings in collective buildings is also the highest there.

In 1999, there were 45 agglomerations of at least 2,000 inhabitants that occupied 31,000 hectares, that is, one-third of the developable 100,000 hectares of the island (with the remaining 150,000 hectares including protected natural spaces, steep slopes, volcano, etc.). Their surface area has increased by half since 1990. Today close to 86% of Réunion's inhabitants live in the urban agglomerations with low population density.

Far removed from the model of a European city, Réunion's city is a "garden-city" [LAJ 01] and its average density is 6.6 houses per hectare. Finally, like the agglomerations, the burgs (200 to 2,000 inhabitants) have also spread considerably in 10 years while their average density is half that of the agglomerations (3.4 houses per hectare). These burgs constitute one-third of the island's rural population of nearly 100,000 inhabitants. Half of this rural population lives in the southern micro-region while the rural population is only 4% of the northern micro-region population. With most of Bas having been urbanized, the rural people mainly occupy the mountains, but the structuring of the burgs, also advocated in the SAR, still faces some difficulties. Given these observations, the location of urban growth in the future is evidently a sensitive topic for Réunion and the modeling of this growth can be a tool that could provide useful information for reflection. It is from this perspective that we propose as a conclusion to this reflection, a modeling of the urban growth based on a cellular automation developed at the University of Rouen by Patrice Langlois [LAN 01].

Constructed with a regular kilometric grid as the base, this automation defines the states of the 2,500 cells it is composed of and indicates the lifespan as well as the rules for transition of the states of the cells. With the objective being to model the development of urban sprawl, we have chosen to define the states of the cells as population density levels and not as types of land occupation as is generally done. In fact, at this kilometric scale, the rules for transition between two types of land occupation seems to us as being difficult to "model", given the number of parameters that come into play. Generally speaking, the question of the level of analysis also comes up.

Modelling of the evolution of the density of population in La Réunion

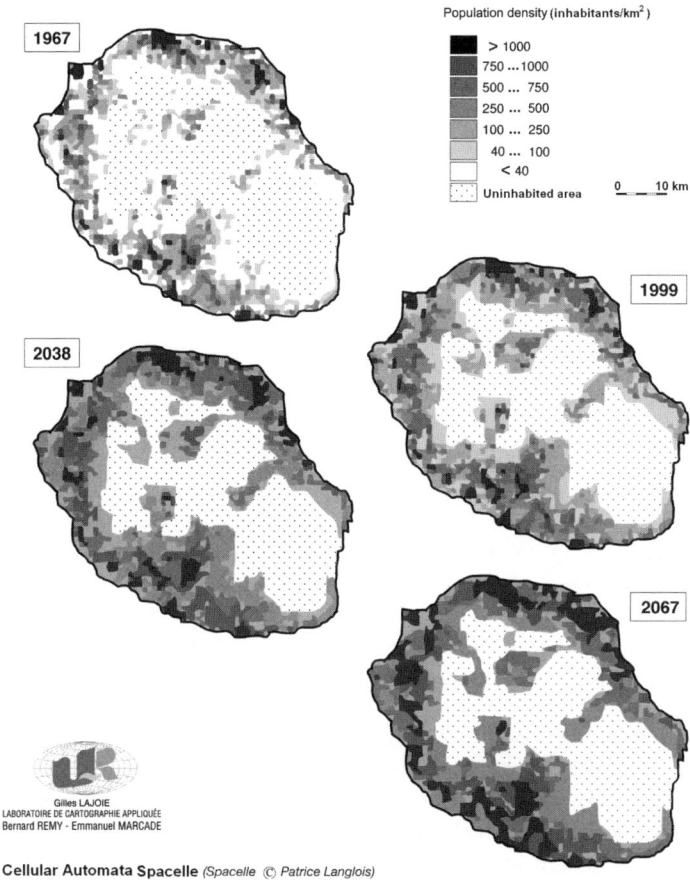

Figure 4.7. *The dynamics of urban sprawl in Réunion as per a simulation of a cellular automation [LAN 01]*

On the scale of a kilometric grid and on an island of around 1,000 km², the aggregation of different modes of land occupation leading to a single state seems hardly pertinent, particularly in the eyes of the local players.

For these reasons, we opted to classify states into seven population density levels, with one particular state reserved for the cells that cannot be developed (naturally protected areas, volcano, steep slopes, etc.). In fact, a density level refers us to a powerful reality, which can be directly interpreted by the largest number of people. While it is true that this particular level aggregates very diverse types of land occupation that are not mentioned in the model, the question of the transition from one state to another is greatly simplified and it is clearly this question which is at the basis of the functioning of a cellular automation.

Once this principle is admitted, the change in the state of a cell simply corresponds to a variation of the density of human beings and activities at a given point of time.

In the proposed model, the time zero (T0) of the simulation corresponds to the kilometric grid of population densities in 1967 taken from the Atlas of the overseas departments published by the Centre d'Études de Géographie Tropicale (CNRS) in 1975 [CEG 75]. The model calibration relies in fact on a second kilometric grid established this time on the 9,700 sectors of the INSEE at the time of the 1999 census. By starting with the 1967 grid which aggregates 416,000 inhabitants, the objective of this calibration is only to distribute the 706,000 inhabitants of the 1999 census within the kilometric grid by applying relatively simple rules for the transition in state. These neighborhood rules aim to reproduce a phenomenon of spatial diffusion of the densities which we could "seriously" monitor from 1967 to 1999.

Several simulations were launched by integrating positive (coastline, employment zone, etc.) and negative (volcano, steep slopes, disconnection of urbanization from the Regional Development Scheme, etc.) proximities and Figure 4.7 presents the most basic simulation (density diffusion as per a neighborhood principle).

In this modeling, by continuing with the phenomenon of urban sprawl, the mark of one million inhabitants is attained in 2038 while the demographic projections by INSEE foresee this symbolic event occurring around 2030. Going still further with these trends, at the end of 100 iterations, in 2067, the model projects an alarming spatial distribution image of 1.2 million inhabitants in Réunion, albeit one which does not integrate the outflow of the demographic transition that should stabilize Réunion's population to around a million inhabitants if, of course, the immigration remains at a reasonable level.

The innovation provided by this modeling is evidently the fact that it results in a cartography that has the merit of sustaining the debate and providing material for the forecasting reflection on quite a sound modeling base for which the hypotheses are formulated, just like the stages of reasoning that have led to the model, its results and its limits.

We will have understood that the interest of such an exercise is evidently not to predict in detail the distribution of Réunion's population but to sustain the debate on urban sprawl and the necessity of making Réunion's cities more dense. In fact, the attachment to the Creole case and the Kour (the Creole garden) is still well entrenched in the local culture and 80% of Réunion's population lives today in individual houses. As a result, building plot prices soared in the last decade (85% increase on an average with local hotspots 200% as per the property observatory of Réunion's Urbanism Agency). This modeling, which is mathematical and graphic at the same time, thus has the advantage of impressing people by producing the image of a general urban sprawl on the scale of the island if the densification effort is not more active in the next few years.

In such a context, this type of spatial modeling can constitute an extra tool for participating in the civic debate on the evolution of Réunion's cities, which are not only spreading and becoming denser but this in very uneven proportions.

4.6. Conclusion

Whether it is chorematics or modeling of urban diffusion based on a cellular automation, the objective of using models is still to strive towards a certain spatial understanding. While this is fortunately not entirely limited to modeling, a few elements refer to it in the scientific field where hypotheses on the working principle of the spatial system are always explicit, with the limitations of this type of intellectual exercise having to be recognized and assumed.

In such a case, the geographer is in a position to contribute to the large civic debate on the future of the territories. While the multiple advantages of the chorematics are no longer to be demonstrated by deconstructing spatial complexity and then simply realizing it, the contributions of modelings that go from the simple to the complex (cellular automations, multi-agent systems, etc.) perhaps deserve to be better known and diffused among the numerous stakeholders of the territory.

Constituting a tool among other tools for exploring the spatial complexity and its development in the long term, modeling can shed light on the reflection on territorial forecasting, which we hope will emerge stronger from it.

4.7. Bibliography

[AKO 02], "La Réunion à l'heure des choix: trois scénarios pour 2020", *Les cahiers réunionnais du développement no.14, Revue Akoz - espace public no.14*, 2002

[BER 86] LE BERRE M. *et al*, Cheminements systémiques: du modèle AMORAL à une réflexion théorique en Géographie, L.A.M.A., U.S.T.M.G., 1986.

[BRU 90] BRUNET R., *Les modèles graphiques engographie*, Anthropos, 1990.

[BRU 92] BRUNET R., *Les mots de la géographie: dictionnaire critique*, Collection Dynamique du territoire, Reclus La Documentation Française (First edition), 1992.

[CAL 02] CALDERONI S., Éthologie artificielle et contrôle auto-adaptatif dans les systèmes d'agents réactifs: de la modélisation à la simulation, PhD Thesis, Réunion University, IREMIA, Ecole doctorale de mathématique et informatique de Marseille, 2002.

[CEG 75] CEGET – Centre d'Études de Géographie Tropicale du CNRS, Atlas des départements d'outre-mers.

[CHA 83] CHAMUSSY H., CHARRE J., DURAND M.-G., GUERIN J.-P., Le BERRE M., UVIETTA P., "Le modèle AMORAL; analyse systémique et modélisation régionale des Préalpes du Sud", DATAR-GREP-USMG Report, Grenoble, 1983.

[CHA 86] CHAMUSSY H. et al., Cheminements systémiques: du modèle AMORAL à une réflexion théorique en Géographie, L.A.M.A., U.S.T.M.G., Grenoble, June 1986.

[DIR 90] DIRN L., *La société française en tendances*, Paris, PUF, 1990.

[FAT 01] FATÈS N., Les automates cellulaires: vers une nouvelle épistémologie?, MPhil thesis in History and Philosophy of Science, Paris 1 Sorbonne, 2001.

[FER 95] FERBER J., *Les Systèmes Multi-Agents, vers une intelligence collective.* Informatique intelligence artificielle. Inter Éditions, Paris, 1995.

[FOR 95] FORSE M., LANGLOIS S., *Tendances comparées des sociétés post-industrielles*, Paris, PUF, 1995.

[FOR 79] J.W FORRESTER, "Dynamique urbaine", *Économica, no.6*, Paris, 1979.

[GOD 97] GODARD H. (ed.), *Atlas de France: les Outre-mers*, volume 13, Reclus, 1997.

[GRA 94] GRAFMEYER Y., "Sociologie urbaine", *Sociologie 128*, Nathan Université, 1994.

[GRA 93] GRATALOUP C., "Le même et l'autre: le renouvellement de la chorématique". *Espace Temps,* no.51-52, Les apories du territoires, 1993.

[GUE 84] GUERIN J.-P., "Statut de l'espace et modélisation régionale en analyse de système", *Colloque européen d'économie régionale*, Poitiers, December 1984.

[LAJ 01] LAJOIE G., ACTIF N., "Des villes-jardins qui s'étalent depuis dix ans", *Économie de La Réunion*, no. 108, Saint-Denis, INSEE, 2001.

[LAN 97] LANGLOIS A., PHIPPS M., *Automates Cellulaires – Application à la simulation urbaine*, Paris: Hermès, 1997.

[LAN 01] LANGLOIS P., Automate cellulaire SpaCelle (Système de Production d'Automate CELLulaire Environnemental), CNRS UMR 6063 – IDEES, University of Rouen – Laboratoire MTG, 2001.

[MOI 78] Le MOIGNE J.-L., *La théorie du système général*, P.U.F., Paris, 1978.

[ODR 99] "La Réunion à l'horizon 2030", ODR – Observatoire du développement de La Réunion, 1999.

[ODR 02] Scenarios can be consulted at the website address www.ODR.NET, documents can be downloaded in PDF format.

[PRE 94] PRELL B., G. BLANC G., GRANVAUX J.-L., *Songes sur La Réunion,* AGORAH, 1994.

[REN 00] RENARD J.-P., "Description et modélisation en géographie", *Hommes et Terres du Nord*, 2000.

[RFS 97] *Revue Française de Sociologie*, XXXVIII, p. 651-56, 1997.

Chapter 5

One Model May Conceal Another: Models of Health Geographies

Health geographers are sometimes criticized for their descriptive approach due to the lack of a general theory providing the key to interpretation of the observed phenomena or reference models. Atlases and maps are tools for disseminating geographical information, but evidently, they are not sufficient to understand the complex processes constituting the spatiality and the territoriality of health-related events.

The *Health and Place* review received credit for radically mooting theoretical questions. Are spatial disparities such as health inequalities that show up on maps conditioned by the places themselves as a contextual consequence, or by social and spatial groups as a consequence of composition? Would the same people have the same experience as regards their health irrespective of their living environments [CUR 98]?

The answers to these questions cannot be devoid of interest. In fact, they make it possible to evaluate whether health-related actions and policies that take space and places can indeed contribute to reducing health inequalities. The various health models that we have chosen to explore here, revolving around epistemological questions, constitute only a part of the possible schools of thought on this vast issue.

We could thus consider, for example, the epidemiological transition model and the debates that set its partisans against those who would defend the new concept of

---

Chapter written by Alain VAGUET.

emerging diseases [VAG 00]. Likewise, faced with the development of Europe and its slow progress from the health point of view, the national models continue to prevail and oppose each other more on ideological grounds than on material modalities [VAG 02]. What will the European model of the future be: curative, preventive, Bismarckian, Beveridgian?

If a geographer were to dispute a well-established epidemiological model, it could very well revive the controversy regarding its validity. Thus, the model of the valleys depopulated due to onchocerciasis as per the distance gradient from the river and hence from the carrier mosquitoes was contended by J.P. Hervouet [HER 90]. Not only are all riverbanks of West Africa not devoid of human beings, but several of the scarcely populated territories are actually a result of pre-colonial wars.

Finally, certain works could very well become models themselves, due to the fact that comparisons can be made on the basis of the reference that they constitute. The proximity or the deviance with respect to this reference system already constitutes an appreciable measure. In this way the number of potential models required to ensure comprehension can be extended.

5.1. Modeling in order to surpass descriptions?

In order to approach the complexity of spatial organization, recourse to modeling makes it possible to simplify matters, to differentiate between the superfluous and the essential while at the same time retaining the general pattern that covers a large number of cases observed [DUR 01]. Indeed, information loss must then be accepted, but this intellectually stimulating methodology attempts to circumvent the descriptive block that health geography is often criticized for and at the same time makes it possible to obtain a better understanding. Just as grammar rules, however fraught with exceptions, help us to learn languages, the construction of models in geography helps us to codify the world as it appears to us [HAG 2002]. Even if this cognitive mode of intellectual construction is made credible, we still have to question, as the epistemologists invite us to, as to the intentions of the modeler and to resolve the essential question: who decides the reference models?

In the foreword of a book dedicated to models, L. Sanders mentions "modeling families" and the plural nature of the notion. In 1962 Ackoff proposed a typology with three categories, depending on the degree of abstraction: iconic, analogical and symbolic. This in turn made it possible in 1964 for Chorley to construct a model of models with the stages of idealization, theorization and simplification that made it possible to transcend the real world in order to move towards a more "elegant" mathematical interpretation. In his book *Explanation in Geography* [HAR 69],

Harvey recounts his pleasure in delving into this form of scientific geography that enables him to work for nothing less than the joy of humanity.

This new so-called quantitative geography comprises a considerable number of models that could be of use to health geographers, but it would evidently be impossible to enumerate them here. Cliff, Hagett and Gould have thus intensively explored the trends of epidemics in space, via diffusion and interaction models.

By adapting Moseley's [BRA 73] concentration graph, Picheral [PIC 00] has, for instance, proposed a U-shaped curve that enables us to go beyond simply describing the evolution of the location of healthcare centers. It clearly shows how health facilities, after an initial historical phase of concentration in the big cities, spread out in space, only to ultimately resume their tendency to remain grouped together in large-scale structures. The advantage of this kind of "model" lies in the fact that, in principle, it remains relevant, irrespective of time and place.

Even these examples of so-called descriptive modeling, that simplify in order to be accountable, are more rigorous than a simple statement of facts. Understandably, conceptual modeling, seeking more profound causalities, remains rarer, but is often more durable. Christaller's hypothesis of central places further stimulates the theses put forward by health geographers.

For all that, we are aware that scientific geography, with its penchant for models, does not find full favor with postmodern geography. Human beings cannot be very easily reduced to a purely economic behavior, implicitly put forth, for example, when the principle of least effort is evoked with a view to optimizing healthcare locations. These limitations experienced clearly by the partisans of quantitative analysis, such as Gatrell and Contandriopoulos, push them to surpass the results provided by geographical information systems, by looking for new developments in the methods of complexity. Indeed, numerous research studies have brilliantly demonstrated that the sum of small individual preferences could have significant effects on societies, such as residential segregation. In the matter of health geography, these powerful demonstrations are yet to be established.

In the meantime, partisans of postmedical health geography (see definition below) consider spatial modeling of services as a simple description (who goes where and when?). In other words, with quantitative geography, we would have become caught in the rut of description. Several authors are in favor of combining new and old concepts so as to form an "epistemological pluralism" [CUR 98], where medical and postmedical geography models complement rather than contradict each other.

5.2. Mode of the models and models in vogue

Progressive broadening of the definition of health has led geographers to formulate new research objectives. Initially, geography dealing with the spatiality of diseases and healthcare provision was introduced as an evolution of medical geography, exclusively revolving around pathologies. According to Curtis, this period made it possible to determine the foundations of the discipline and to establish the first spatial models. Subsequently, the preoccupations that are essentially naturalistic – diseases are natural and only health professionals can treat them – gave way to humanist perspectives, that is, the social and political construction of health issues. The postmedical paradigm results from a vacillation between core questions, which paves the way for new models. Indeed, if modeling involves accepting generalization as a structured representation of the real, why not consider, as Gatrell does, all generalizing abstractions as models?

He counts up to five models of health geographies. As for the origin of the different approaches, he refers to Giddens and Bourdieu for the "structurationists", to Foucault and Beck for the "post-structuralists", and to the "Lumières" theorists for the positivists. All these positions do not preclude each other. They split the research objectives differently so as to finally produce complementary models with different fundamental concepts. This acknowledgement of the plurality led Gatrell to title his book *Geographies of Health*.

"Family" of models	Methods	Tools	Spatial objects	Concepts
Humanist Social geography Constructivism	Qualitative Explicative Inductive	Samples Interviews Experience	Places Countries	Identity, Culture Actors, Social, Landscape
Structuralist Marxist	Critical Qualitative	Macro level Dialectic	Territory	Policies
Structurationism (Giddens)	Qualitative	Reflexivity	Spatial Processes	Time geography Local structures Social constraints
Post-structuralism Postmodern Post-colonial	Distancing Critical	Deconstructi on	World	Modernity Spatial check Genre
Neo-positivist	Quantitative Informatics	Law, order Mathematics	Space	Distance Interaction Discontinuity
Naturalist	Quantitative	Statistics	Zones	Epidemiological landscape

Figure 5.1. *Main models of health geographies*

This kind of list, comprising different types of geographies, is sometimes criticized for giving the regrettable impression of a methodological supermarket that we can indiscriminately pick from. In reality, each of the families relies on philosophical choices, which in turn condition the quasi-hardliner positions.

The table, summarized and voluntarily simplified (Figure 5.1), inspired by the books by Gatrell [GAT 02] and Curtis [CUR 96], does not claim in any way to be exhaustive and even less an orderly qualitative or period-wise compartmentalization of these families. Strictly speaking, it simply helps us to better understand that health geography is in no way merely a specialized branch of geography, but that it encompasses geography in its entirety. In the process, the antinomic terms of past debates on the methods that are never entirely closed recur: nomothetic or idiographic, inductive or hypothetical-deductive, qualitative or quantitative methods. If we were to observe the evolutions in the anglophone geography, it can be seen that through sedimentation, all these intellectual propositions coexist but that the most recent notions sometimes come back to simple techniques, such as interviews, or explorations of specimen cases – with or without geographical information systems – that at other moments would have been referred to as monographs.

As for research objectives, Curtis proposed in 1996 a typology comprising various perspectives based on the opposition between traditional and contemporary health geographies; during the first stage, it included spatial modeling of mortalities, morbidities, healthcare service provision and health services, and during the second stage, humanist, materialist, critical, structuralist and lastly cultural movements. In doing so, medical geography gave way to postmedical geography.

In 2004, the same author proposed another formulation of these diverse families, with the help of the notion of conceptual landscape. The relevance of this term lies in its capacity to evoke a dynamic system, in which the various components are closely interrelated and have to be understood as a whole.

Theoretical models	Landscapes	Objectives
Sense of place/identity	Therapeutic	Places and their symbolic value
Social/political check	Power or resistance	Domination/subordination
Production/structuration	Poverty/prosperity	Social inequalities of health
Consumption/lifestyle	Consumption	Natural settings and social situations
Ecological/epidemiological	Ecological	Environment

This 2004 panorama of the main theoretical models that can be used for research in the field of health once again demonstrates the parallels between successive new geographies and the evolution of diverse human sciences. Anglophone health geography, in a remarkable epistemological endeavor, finds great inspiration from philosophers and sociologists (who are often French).

This new formulation has integrated the linguistic approach and employs the term landscape metaphorically. This makes it possible to fully reinstate the idea of a system of factors active in particular environments in order to explain geographical variations. These landscapes can be grouped into five different "ways of looking" or models. Such a point of view contributes to affirming the role of space and places in the production of inequalities and confirms the complementary nature of quantitative methods (when an objective benchmark seems possible) and qualitative otherwise.

5.2.1. *Modeling of healthcare provision*

In order to illustrate these multiple types of modeling without a wearisome list, let us take examples from health care provision, an aspect that is particularly well studied.

Health planning relies implicitly on a well-known "positivist" organizational model, that of distance friction. In order to try to provide services equally to all the citizens and at the same time avoid wastage of budgetary resources, the partitioning into health sectors is based on distances and population volumes.

In the background, the concept of centrality intervenes actively. In fact, the polarization of populations observed in urban areas should help in facilitating the provision of services. Through a ladder economy, it should be possible to reduce the number of professionals required or to redeploy them in isolated rural zones.

We know that the situation appears to be completely different. When the disposition of the inhabitants becomes scattered, when the populations show a strong indication of dispersion, its equipment poses a problem and paves the way for famous debates on land planning. Maybe in France, it is rather about a simple mutation in a liberal professional universe?

If we were to look at it more closely, several logics come into play. In the majority of cases, the gravity model constitutes a good general descriptor: the bigger and more concentrated the population, the greater number of health diagrams and maps proposed by the services. However, the exceptions also constitute fully valid lines of interpretation. These can be presented as residues of modeling (an off-putting term, it is true) in addition to the main tendency, making it possible to perceive other logics not often mentioned but which must be made explicit.

It is known, for example, that huge swathes of population in the metropolitan peripheries do not receive the services that they have a right to, often because of composition and social distance. In the same way, some cities have CHU at their disposal, while others, bigger but more working-class, are still waiting for them. On the other hand, some insular spaces have, for strategic and contextual reasons a relative plethora of equipment. Likewise, when a bridge is constructed on the estuary of the Seine, experts plan to organize a synergy between the institutions situated on both sides. For all that, several years later, the doublet effect persists. Finally, beyond the barrier of the river, the otherwise more formidable administrative discontinuity cracks down.

The division into these compositional and contextual effects does not always prove very simple and after a decade of debates in the anglophone world, Curtis suggests we may transcend these notions [CUR 04]. For her, as for many authors, physical distance is not the dominant factor among the determinants of accessibility. In any case, only the "positivist" model has difficulty accounting for other realities that the other models make it possible to process. If we subscribe to a modeling process, it is pertinent to model all azimuths and not hesitate to acknowledge the diversity in the types of distances to be taken into account.

For their part, the explanations of the "structuralist" model boil down to a comprehensive interpretation of inequalities in location: domination, class conflict. As per this model, medical science favors the curative dimension, which brings in revenue, but thus reproduces social inequalities. Indeed, observing socio-spatial configurations and distribution of social groups in their spaces, throws a lot of light on the disposition of health professionals. The plethora of specialists in the fine neighborhoods of Paris and their quasi-absence in the industrial-port cities are well-known.

As for the "structurationist" models, they enable us to understand that health-related decisions are taken under several constraints and that the practices of the people involved in turn recreate new structures.

The big upheavals in domicile trends such as metropolization and urban expansion have not led to a progressive adaptation of health service provision, which still depends greatly on locations inherited from the past. The walls of the hospital cost less than its personnel, but weightiness renders structures perennial in places that are often ill-suited. As proof of the above, we can take the big difference between setting-up big commercial stores that optimize their location in order to tap consumers, and setting-up healthcare establishments that try to cater to the demand of the same people, only when they are ill. Megastores and clinics are established in the periphery of the grouped city, with the hospitals that were created sometimes several centuries ago being rather closer to the city centers.

Partisans of the neo-positivist models, given the explosion of individual mobility, will tend to be stunned at the gap between the big commercial stores and hospitals. Partisans of humanist, structuralist and structurationist models will try to say in what way other macro logics are at play. We would be tempted to put forward the idea that the various explanations re-establish by themselves, the depth and the complexity of the objective. However, the objective itself changes depending on the writers and their favorite explicative models. The methodology of the atomic researches propounded by those who give precedence to experience lived by individuals, revolves around very small qualitative specimens that aim to explain processes rather than describing them. This seems, of course, intolerable to those who use the ecological perspective, the long series of statistics to detect an order, a spatial analysis model.

The methods of these various researchers contrast sharply with one another, in concert with their values and beliefs. A neo-positivist has faith in modernity and the achievements of science. He will develop research destined to improve locations, hospitals, ambulances, service accessibility, for example, on the basis of optimal location-allocation models [PET 01]. Rushton [RUS 91], who has intensively explored these algorithms in the poor countries from the 1970s onwards, calculated that in an Indian district, the accessibility of various services varied considerably. While the services had to move towards the people, their "spontaneous" locations were confused with the optimum locations, defined through calculations. On the contrary, services seemed to remain insensitive to the question of accessibility when the users had to ensure the mobility. He concluded flexibly by emphasizing that the accessibility to the services had to be adapted in accordance with the services planned and also in accordance with the sites, opening the door to other interpretive models integrating the effects of domination [FIS 77]. More than 30 years after these calculations had been put at the disposal of planners, evaluation remains very fragile. This is due to the fact that primary health care centers, however well located they might be, remain strongly discredited, on one hand due to a lack of medical personnel, but also because of competition from an emerging private sector. This unheralded spatial change has managed to disturb a spatial ideology that is henceforth obsolete.

Structuralists believe in science as well, but looking instead for the causes of pathologies in political systems, they will look at colonialism, liberalism, imperialism, etc. with a critical eye. They will bring to the fore obstacles to equality in the matter of social accessibility and denounce reductions in public investments.

The hospital-centered model based on the affirmation of the prowess of bio-medicine, will be denigrated by the post-modernists. They will denounce a desire for social regulation through discourses that medicalize societies. They will show that the success of the bio-medical model causes a "chronicization" of the pathologies, which the professionals who created it cannot control [BEC 01].

Can these debates cater to anyone other than the theoreticians? Evidently, yet political discourses and actions endorse a few research models and disregard certain others. In what follows, we will take a few examples to show that various models do not play an equal part, which is sometimes to the detriment of the citizen.

5.2.2. *The models put to comprehension and action testing*

Comprehension precedes action. An ecological model, which is too determinist and too descriptive, runs the risk of intervening clumsily in complex realities. Models aimed at risk comprehension propose totalitarian or even sectarian explanations. If we were to observe the evolution of locations of private health care institutions in France, their tendency for regrouping is evident everywhere. We must then learn the lessons the various models teach us.

Clinics that are more committed to research of critical dimensions in terms of the number of liberal practitioners [CLA 03], are at present relinquishing their original locations in bourgeois neighborhoods and are opting to relocate. Henceforth, they give precedence to rapid service distribution, possibilities of parking, that is, the spaces that are found easily in neighborhoods under urban renewal near industrial wastelands (Figure 5.2).

A geographer with Hotelling's spatial economic model in mind will notice that in the diagram, the location strategies led the firms competing for the healthcare market to come closer without attempting to allocate the space fairly among them. A person attracted by game theory will conclude that the individualistic behavior of these players does not help the community. We can see here an example of the "free rider" problem who will act only according to his interests, which legitimizes the regulations of a State, which is the guarantor of public assets and which here seems incapable of taking steps on locations.

The analysis of the final locations (T2) can be interpreted in several ways, either as a spatial competition event, purely organized in terms of distance (Hotelling model); or as the outcome of tensions between various forms of dominations. We have on our side opted for the latter proposition by demonstrating that the spatiality of this collective service was construed from several logics.

T1

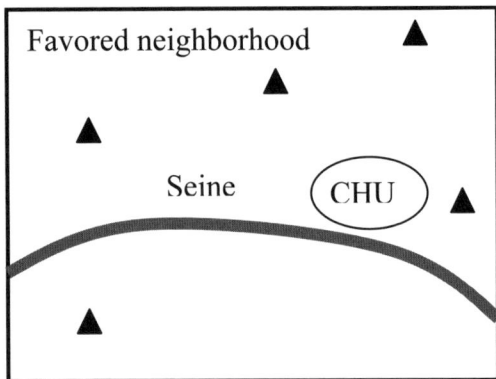

T2

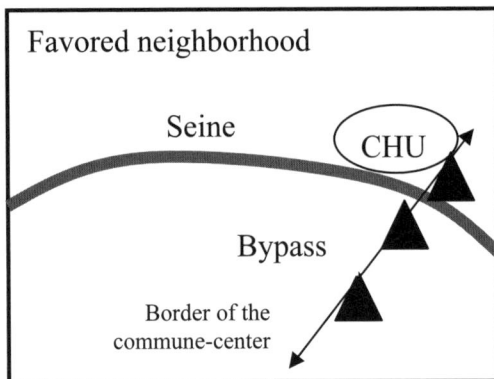

Figure 5.2. *Diagram on the recompositions of private healthcare services (City of Rouen)*

Understandably, it firstly serves the interests of professionals, who are very restricted by regulations attempting to ensure quality and security to the service users. However, the implantation site depends rather on the municipal authorities who wish to retain professional taxes on their territory. Moreover, they can also use the construction of a new health care institution as a stimulus for urban renewal [VAG 01]; another way of serving the community, since an architect from London (Lubetkin) was able to assert that clinics located in sensitive neighborhoods constituted "megaphones for health" [CUR 04], spreading their message in their vicinity. All models are welcome as a means of giving depth to a complex subject.

The second example chosen will bring the heuristic power of the new concept of a therapeutic landscape to the fore and, if need be, transforms itself into a tool in the struggle for democracy.

At present, the UK is experiencing an unprecedented wave of hospital construction. The originality lies in the mode of private funding and in the publicized ambition to produce "holistic" hospital environments, as per the words used by the Prince of Wales himself. The experts of the National Health Service (NHS) have evaluated these new buildings positively. A team of postmedical healthcare geographers have, on their part, propounded a critical analysis of the evaluation criteria and produced an alternative table, thus making it possible to contribute to the debate [GES 04].

According to Gesler, the "experts" implicitly promote criteria originating from the bio-medical model bringing to the fore the environment type, which interests the professional but disregards the social and symbolic environment that the patients/clients are especially preoccupied with. According to him, bringing the therapeutic landscape concept to the fore consists of not merely being content with measures that can be carried out from a distance or custom-made, but consists of generating qualitative work through detailed discussions with the people concerned including the public user. We can recognize here the distinction between the modernist hospital model of the early 20th century, which corresponded to the principles of medical science and hygiene alone, and the postmodern institutions that also aim to become sites for living, or even for recreation.

5.3. Conclusion

We have shown that a rhetoric which implicitly over-determines the health policies through space can be observed at the same time as an incapability to act verily on the profound causalities producing persistent differentiations.

In order to execute this project, it would be more pertinent to take into account all the families of models for health geographies. We have thus opted for a presentation that is not as descriptive of the main spatial analysis models; after all, all of these have a bearing on the health-related field but with macro-models that enable formalization of the acquired knowledge. Henceforth, the geography models can be improved in line with other researches in social sciences.

5.4. Bibliography

[BEC 01] BECK U., *La société du risque*, Aubier, Paris 2001.

[BRA 77] BRADFORD M.G. and KENT W.A., *Human Geography, Theories and their Applications*, Oxford University Press, 1977.

[CLA 03] CLAVERANNE J.P. *et al.*: "Les restructurations des cliniques privées", in *Revue Française des Affaires Sociales*, Paris, no. 3, 2003.

[CUR 98] CURTIS S. and TACKETT A., *Health and Societies*, Arnold, 1998.

[CUR 04] CURTIS S., *Health and Inequality*, Sage, 2004.

[DUR 07] DURAND-DASTÈS F., "Modeling concepts in spatial analysis", in *Models in Spatial Analysis*, SANDERS L. (ed.), ISTE Ltd., 2007.

[FIS 77] FISCHER H.B. and RUSHTON G., 1977. "Integrated rural planning in India: experiences of the pilot research Project in growth centers 1969-1974", in *Man, Culture, and Settlements* (eds. R.C. EIDT, K.N. SINGH and R.P.B. SINGH), Kalyani Publishers, 1977.

[GAT 02] GATRELL A., *Geographies of Health*, Blackwell, 2002.

[GES 04] G. WIL, M. BELL, S. CURTIS, P. HUBBARD and S. FRANCIS. "Therapy by design: evaluating the UK hospital building program", *Health and Place*, vol. 10, 2004.

[HAG 02] HAGETT P., "Locational analysis in human geography", in DEAR M.J. and FLUSTY S., *The Spaces of Postmodernity*, Blackwell, 2002.

[HAR 69] HARVEY D., *Explanation in Human Geography*, St. Martin Press, 1969.

[HER 90] HERVOUET J.P., "Le mythe des vallées dépeuplées par l'onchocercose", *Cahiers GEOS*, Paris, 1990.

[JOH 00] JOHNSTON R.J., DEREK G., PRATT G., WATTS M., *The Dictionary of Human Geography*, Blackwell, 2000.

[PEE 07] PEETERS D., THOMAS I., "Location of public services: from theory to application", in *Models in Spatial Analysis*, SANDERS L. (ed.), ISTE Ltd., 2007.

[PIC 01] PICHERAL H., *Dictionnaire raisonné de géographie de la santé*, University of Montpellier III, 2001.

[RUS 91] RUSHTON G., "Use of location-allocation models for improving the geographical accessibility of rural services in developing countries", in AKHTAR R. (ed.), *Health Care Patterns and Planning in Developing Countries*, Greenwood, 1991.

[VAG 00] VAGUET A., "Maladies émergentes et reviviscentes", *Espace, Populations, Sociétés*, Lille (France), no. 2, 2000.

[VAG 01] VAGUET A., "Du bon usage de l'analyse spatiale et de l'évaluation territoriale", *Natures, Sciences, Sociétés*, Elsevier, no. 4, 2001.

[VAG 02] VAGUET A., "Les systèmes de santé en territoire, matérialité des ideologies", *Géopoint*, Avignon (France), 2002.

Chapter 6

Operational Models in HMO

Since 2000, medical care resources have become increasingly specialized, leaving no place for small-sized hospitals. People feel the need to have access to immediate and good quality support. One of the primary aims of the National Health Insurance System (NHIS) is to constantly evaluate the quality of healthcare services, using epidemiological surveys for specific geographical areas. Considering a variety of realistic scales, some medical services may feel somewhat threatened. Moreover, in certain cases, although a minimum level of quality is maintained in accordance with the accepted criteria, small hospitals tend to limit the development of new types of therapy. In this situation, the government is in a very tricky position: either the service supplied will completely disappear due to a low level of activity or it will overspend its budget to maintain the minimum expected standard.

By determining how patients are geographically distributed, as well as evaluating their individual healthcare requirements, authorities may be able to provide an adequate healthcare structure: i.e. determining which medical services (specific specialties, type of equipment, the number of medical practitioners, beds, out-patient consultations, etc.), and to what degree, meet the requirements of the urban population. This study attempts to supply guidelines for a better balanced metropolitan area. This could be achieved not only by closing specific services, reducing the number of employees, but also by helping health professionals to set up their practice in deserted and underprivileged areas.

Other solutions include concentrating or developing partnerships between structures (e.g. cancer centers), thereby creating links between professionals,

Chapter written by Jean-François MARY and Jean-Manuel TOUSSAINT.

structures or hospitals thanks to newly emergent healthcare networks. Thus the hospital network will develop toward an effective medical care network through a co-ordination of scarce resources (co-ordination of the emergency medical service centers by on duty physicians).

In principle, as Peters and Thomas [PEE 01] both observed, "all location models reveal the fundamental dilemma existing between financial cost and accessibility". In other words, they suggest we avoid any overstretching of resources. They promote, instead, a better balanced approach which would take into account the financial requirements while remaining ready to provide patients with high quality support. The following examples concern the use of standard models of gravity centers and Thiessen's polygons, to which, thanks to geographic data systems, we can add the calculation of channels optimizing accessibility conditions.

6.1. Buffer and barycenter to determine the location of cardiac defibrillation

This following is an illustration of using a buffer and barycenter as regards to the implantation of cardiac pacemaker centers. The notion of a barycenter was first introduced by Archimedes (287-212 BC). This following equation shows a basic relation between points projected in a system and their masses. The gravity center, which is called "G" point, is the center of the system; an average point weighted according to mass or some other attribute.

$$X_G = \frac{\sum_{2\text{ to }n}^{i} \mu_i(X_i - X_1)}{\sum_{1\text{ to }n}^{i} \mu_i}$$

- If we have points
  $M_1(X_1,Y_1)$, $M_2(X_2,Y_2)$, ..., $M_i(X_i,Y_i)$, ..., $M_n(X_n,Y_n)$,

- With their masses
  μ_1, μ_2, ..., μ_i, ..., μ_n,

$$Y_G = \frac{\sum_{2\text{ to }n}^{i} \mu_i(Y_i - Y_1)}{\sum_{1\text{ to }n}^{i} \mu_i}$$

- where $G(X_G,Y_G)$ is the barycenter of the system.

The gravity center of a system of particles is defined as the average of their positions weighted by their masses. The gravity center is closer to the larger object.

If the density of an object is uniform, then its gravity center is the same as the centroid of its shape. A structured program in Avenue® Language was developed (S. Freiré-Diaz, J.M. Toussaint, Rouen University), based on these equations, to run on ArcView® software calculating the barycenter for a system of points or for a specific surface.

This barycenter could be used for geomarketing applications: for decision makers, the barycenter defines new locations without personal interest and promoting general interest. A case study is cardiac defibrillation centers. In France this therapy is underdeveloped and presents an incomplete spatial distribution. There still exist wide deserted areas without any center. However, the cardiac defibrillation makes it possible to keep alive 50,000 people per year (source Brady-Tachy) by protecting them from cardiovascular disease. Defibrillation is a process in which an electronic device gives an electric shock to the heart. This helps to re-establish normal contraction rhythms in a heart with dangerous arrhythmia or in cardiac arrest. The implants of automatic internal cardiac defibrillators constitute a prevention for a population at risk.

However, in France, National Health Authorities did not promote this therapy till recently. Some patients must travel more than a hundred kilometers to be examined. In Italy, twice as many implants use an automatic internal cardiac defibrillator (ICD), in Germany four times as many and nine times as many in USA.

In geographic information systems (GIS), the buffers compared to the cardiac defibrillation centers show the theoretical attraction area. The regions, which do not belong to these buffers, are considered as deserted areas. These areas become target areas, where a cardiac defibrillation center could be established to improve the health facilities for the local population.

All the centers (Q=66) were geo-referenced and GIS-integrated by their coordinates (XY). Four buffers were generated with a radius respectively 80, 90, 100 and 110 kilometers. With the ArcView® GIS software, this analysis runs easily. The underlying buffers can be aggregated.

The boroughs where the centroid is included in a buffer zone must be examined in a second step. Cities of up to 10,000 inhabitants are added in the selection because they are likely to host a cardiac defibrillation center. The gravity center program (Avenue® script) uses the population data as mass parameters for the cities. This GIS process has found more than ten isolated areas, particularly near the national border and Corsica, where the distances are extensive.

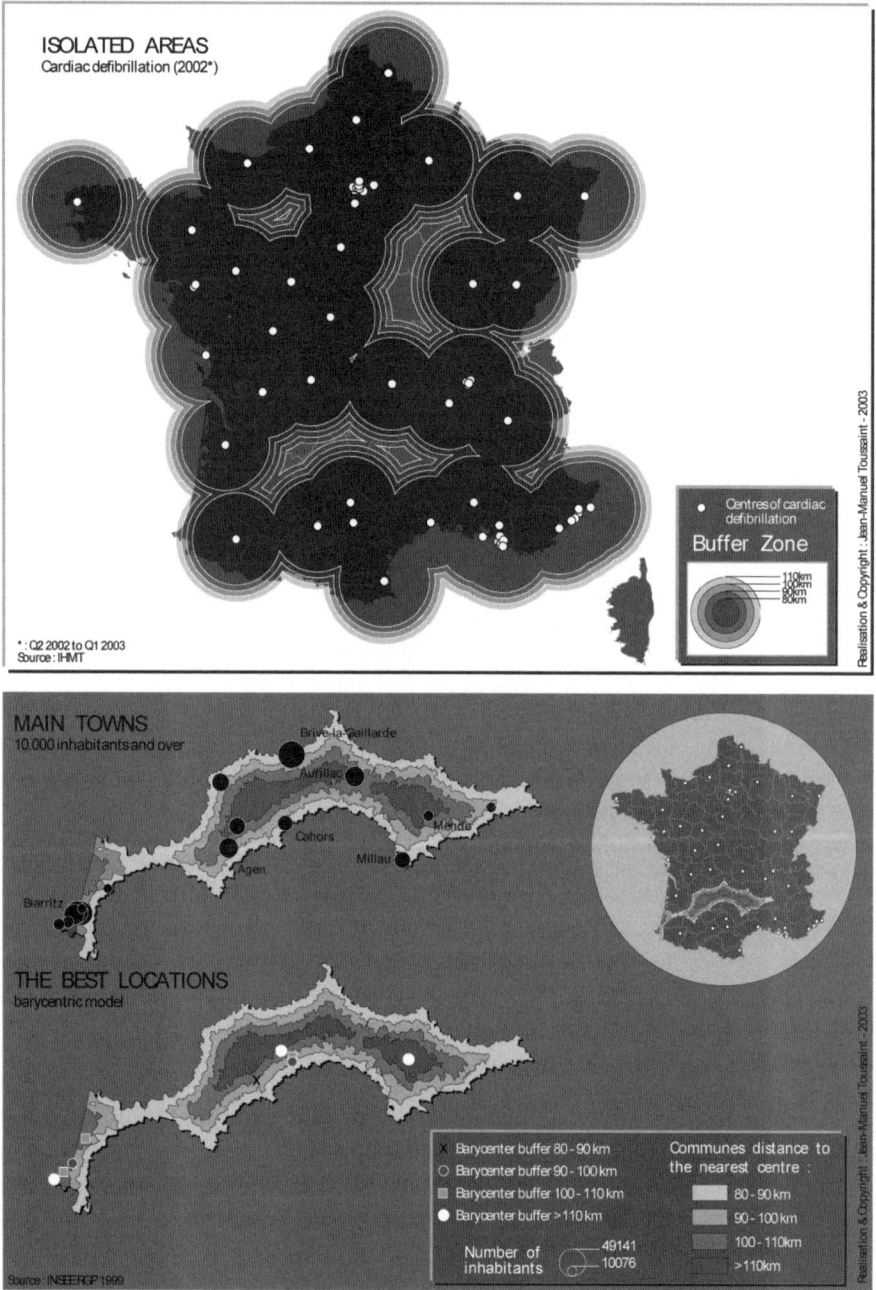

Figure 6.1. *Buffer to determine deserted sectors and barycenter to create the facility*

Stretching across two large isolated regions, from Biarritz to Mende the "south-west zone" is larger than 500 km. The gravity center (see the cross on the map) is particularly near the centroid, between Agen and Cahors. One gravity center may be not efficient enough for this zone. This region could be divided into two sectors with their own gravity center. In the middle, Cahors could be chosen to cover Brive, Agen and Aurillac (zone 4 is far from the other centers). For Biarritz, the demographic size of the city wins against its location at the border. Mende is the last gravity center, better situated than Millau.

In this approach, the use of a barycentric model and the GIS tools offer a complete support for objective measurements and territorial information about new services.

6.2. Thiessen's accessibility formula

Emergency medical assistance (EMA) is based on the same medical regulation as the emergency medical service (EMS). This regulates and offers a complete and immediate assessment, and provides the adequate and proportionate back-up according to the severity of the emergency situation occurring outside the hospital. If required, it can also allow admissions to an adequate private or NHS hospital which is ready to accommodate the patient [MAR 01]. It supplies medical advice and may decide to send out specific medical equipment such as intensive care mobile units (ICMU), on duty doctor or services providing essential medical care (emergency health transport, firemen). In certain cases, the emergency medical service may simply advise the patient to go to the nearest hospital emergency service ("H" on map 2). Since the beginning of the 21st century, the paramedic support system has evolved and the emergency medical admissions can also advise the patient, in specific areas, to consult, without appointment, at a GP (general practitioner) center where a doctor is on duty in the late evening (see grey triangle in Figure 6.2). This means usually work between 8.00 pm and midnight as well as during non-working hours.

No appointment is needed to consult the general practitioner on duty and within a specific structure called a "medical center" (*domus medica*) or in medical emergency center (with written agreement signed by the referent or hospital manager). These new types of structures have not yet found their specific territories and their fixed locations.

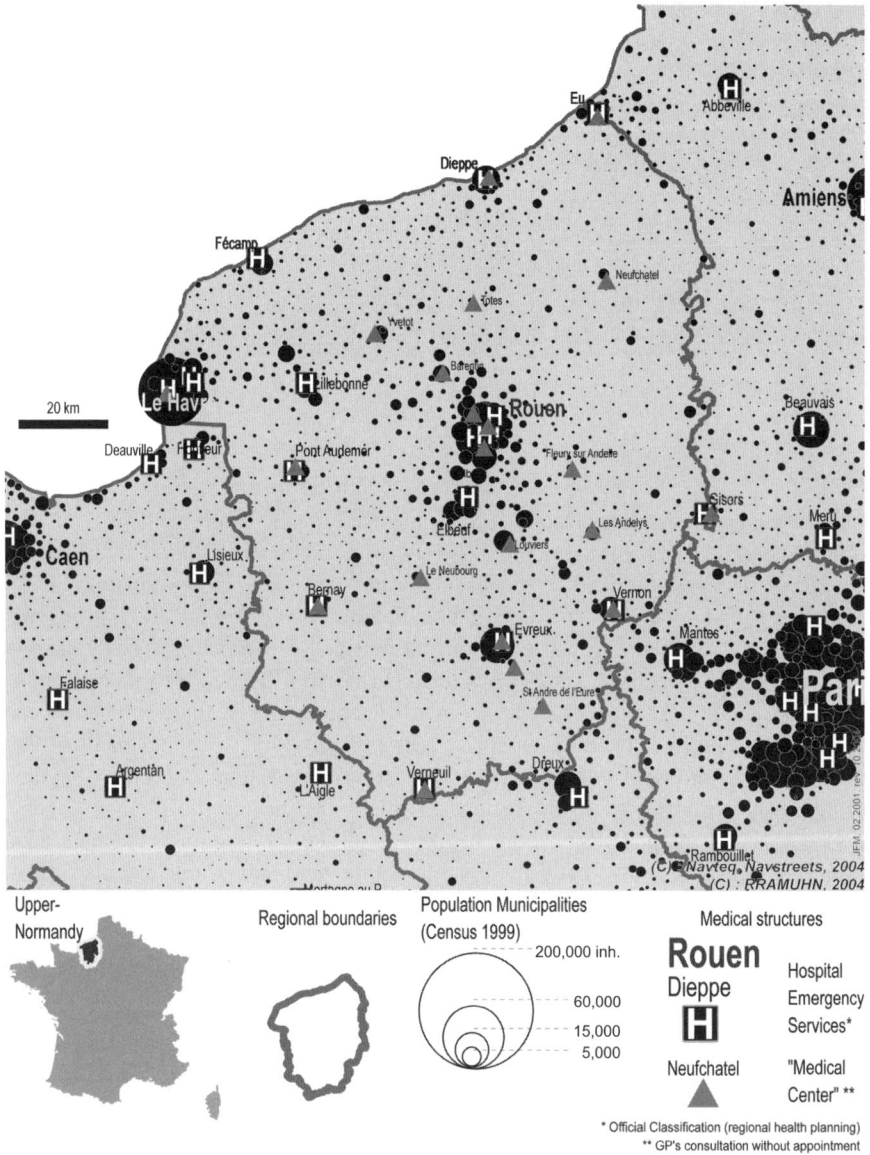

Figure 6.2. *Spatial distribution: population,
hospital emergency services and "medical centers"*

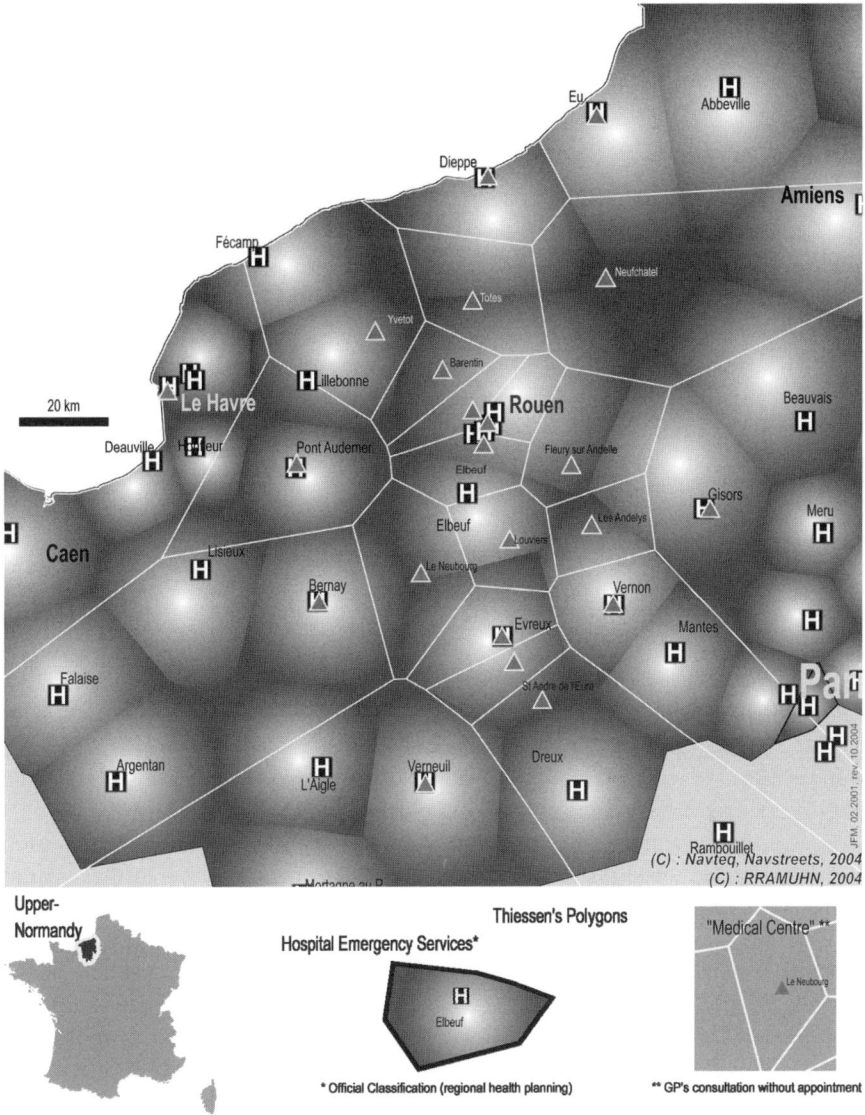

Figure 6.3. *Thiessen's polygons for two spatial distributions:
hospital emergency services and medical centers*

Moreover, people, particularly in rural areas, are not accustomed to this type of facility. In the best cases, the referent medical practitioners have agreed to aggregate the former on duty sectors. The first generation of areas may be defined by the

Thiessen's polygon method, based on more impartial and geometrical factors, which provide a more objective basis for complementary analysis.

This method is based on the use of ArcView™ geographic information system (GIS) software, upgraded by Thiessen's plug-in (Thiessen.avx file available on-line: http://support.esrifrance.fr). From the center layer, Thiessen's GIS function allows for a precise definition of Thiessen's polygons (regions). The edges of the polygons are medians of the segments linking the centers at the same level. These polygons take into account the centers and their neighbors.

The patient can be accommodated by two co-existing networks: they can either go to the hospital for medical care or to the general practitioner's surgery. Over the past few years, GPs have also undertaken home visits in cases of emergency. However, it has been observed that GPs who accept home visits for emergencies are becoming less and less numerous. This is the reason why these complementary networks must be regulated by the emergency medical service (EMS).

Even if Thiessen's polygons, which are based on the distribution of emergency medical services (see black boundaries on the map), are quite well known (e.g. undeniable polygons for the largest suburbs which are the cities of Rouen and Le Havre, see Figure 6.3), Thiessen's polygons based on the location of on-duty GPs (see white boundaries on the map) lead to a more surprising result. It was found that the area can either be portioned (around Rouen and even Evreux, Figure 6.3) or be vaguely limited (such as around Yvetot). The superposition of the two previous boundaries has led to different conclusions. Some of them cut suburbs in two parts (like Fecamp or Lillebonne, Figure 6.3), where no "medical center" is available, whereas emergency medical hospital structures exist. Consequently, patients go to hospital to be provided with all kinds of treatment out of hours. On the contrary "medical centers" are sometimes situated close together, showing a lack of coordination. Thus, between outlying towns, Tôtes, Barentin and Yvetot, three "medical centers" are located less than 8 miles (12.5 km) apart. This is obviously a waste of valuable medical resources. Even if the situation seems unrealistic, it does exist in practice. Independent GPs are organized based on their own personal criteria, sometimes without taking into account the patients' needs in terms of National Health Services (NHS). It is important at this stage to carry out an accurate analysis of the situation which has evolved over the years. In order to do so, Thiessen's polygons represent a new concept which makes it possible to measure the number of people provided with emergency medical assistance (topological analysis). It also attempts to improve knowledge of health care public centers and to obtain more relevant figures in terms of accessibility, the latter originating from the analysis of traffic networks and spatial inequalities.

6.3. Accessibility: the direct added-value of the GIS

If time is measurable in terms of space (transport), space can also be measured by time. The space-time or distance-time concepts may become the more significant analysis approach. The mathematical data are then integrated into a geographical system, closer to the actual perception. Instead of measuring metrical data, a new approach could be considered in terms of time, costs and moreover outside interactions which interfere with the space-time criteria.

Accessibility is measured in terms of time and costs. It can even be seen as the global result of distance on costs. Therefore this factor should be taken into account when a new emergency medical center is about to be located or for evaluating the performance of an existing structure. Moreover, this type of emergencies facility cannot be established without a previous analysis of the network (roads, accommodation, bus, rail, and aerial networks), the unit of measurement (time, cost, etc.), and the size and nature of the geographic area (if it belongs to an existing network or scattered reference points).

Although the concept seems easy to explain, these ideas are not easy to put into practice without a step by step structured approach.

As a practical example, consider that since 2001 the regional medical geographic information system (GIS of the emergency regional medical network in Upper-Normandy) has carried out a detailed regional survey in order to create a representative graph of the road network and of the individual boroughs which are closely interconnected. The geographical areas concerned are: (1) the Upper Normandy Region; (2) the Le Havre Estuary (see Figure 6.4). This last area is illustrated in great detail, including the municipal subdivisions and National Economic and Statistical Study Institute (INSEE) units (city blocks). The purpose of the annual referential survey is to expand its data analysis to other professionals. These figures are used by the Ministry of Transport and supply professionals with statistics with immediate and constantly updated information. Among the different sources of information, Navteq® (Navigation technologies SAS) has also been used.

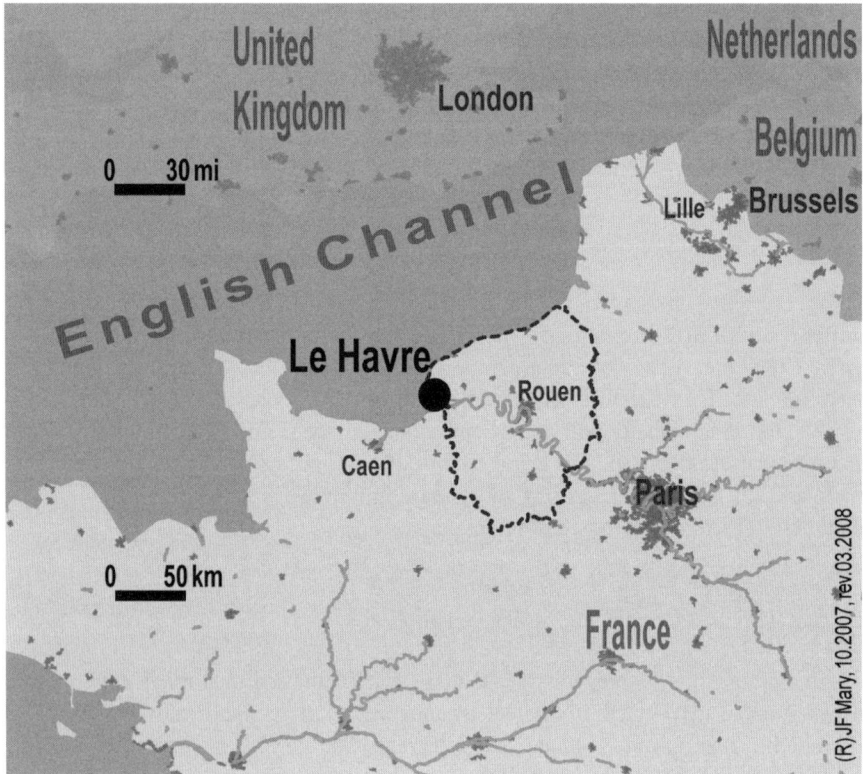

Figure 6.4. *Localization of "the Upper-Normandy" and "Le Havre Estuary"*

Emergency medical assistance is above all the first practice to be measured in the health organization, because each inhabitant must be situated at less than 30 minutes from an intensive care mobile unit (ICMU). Upper Normandy's emergency medical assistance regional network with a GIS system checked this time limit for the region. This approach could change the regional health planning. Moreover, territories where emergent medical services can operate must be defined and referred to as the basic areas in terms of public health. This consists of defining which easily accessible emergency medical service each borough depends on, without taking into account administrative criteria. The method is based upon a geographical data system built up to be able to calculate a regional accessibly matrix devoted to emergency and intensive care mobile units. This regional matrix is theoretically defined according to the road network. The area concerned is Upper Normandy and a surrounding buffer area of 30 miles (50 km) all around reaching large neighboring cities (Caen, Amiens and Paris). Nearly 3,800 municipalities are represented by their geometric centers (which must be connected to the road network and belonging to the settlement area). The established intensive care mobile units settled are represented by their pinpoint location, related to the hospitals' locations. The speed limits in more than 250,000 edges have been recorded (road network). They are based on navigation attributes (type of road, characteristics or morphology) and on the emergency medical experience on the ground. The speeds are thus modulated according to the priority level given to the vehicle used. Consequently, it is necessary to think in terms of health deprived areas to determine the emergency medical mobile unit covering each municipality.

Moreover, space-time knowledge (level of road accessibility) makes it possible to define the optimal emergency medical service location. This must be the emergency medical service of reference. Currently it should be underlined that in 2004 no national information system existed to attribute a referent emergency mobile unit to each individual municipality. The emergency administrative boundaries still remain limited to regional daily usage.

6.4. A regional database of road accessibility devoted to emergency

The emergency medical service area can be defined as "the reference area for vital emergencies". Its boundaries must not be determined by administrative criteria but on the contrary by its own specific spatial characteristics. The intensive care medical mobile unit (ICMU) must necessarily be located in the most convenient hospital for accommodating unexpected patients. The time requested to intervene remains the first criterion. The time intervention means how long it takes between the call and the moment the medical unit reaches the locality. The following map shows the spatial distribution of population with legal boundaries of emergency medical mobile units (if known) dividing the regional area in 12 health emergency districts.

The logic of the scale could be compared with the "half an hour employment areas". The poles of the emergency areas are ranged according to the number of medical teams (practitioners, nurses, ambulance drivers) which is normally proportional to the number of inhabitants to be provided with cares (one team per 100,000 habitants, corresponding to around 1,000 interventions per year).

The GIS has gathered, from the regional space-time data base, more than 130,000 flows defined geographically as segments determined by two points, and semantically as couples (municipal reference point, emergency mobile unit center). The nearest emergency medical service is thus determined for each municipality, taking into account the average time to reach the location.

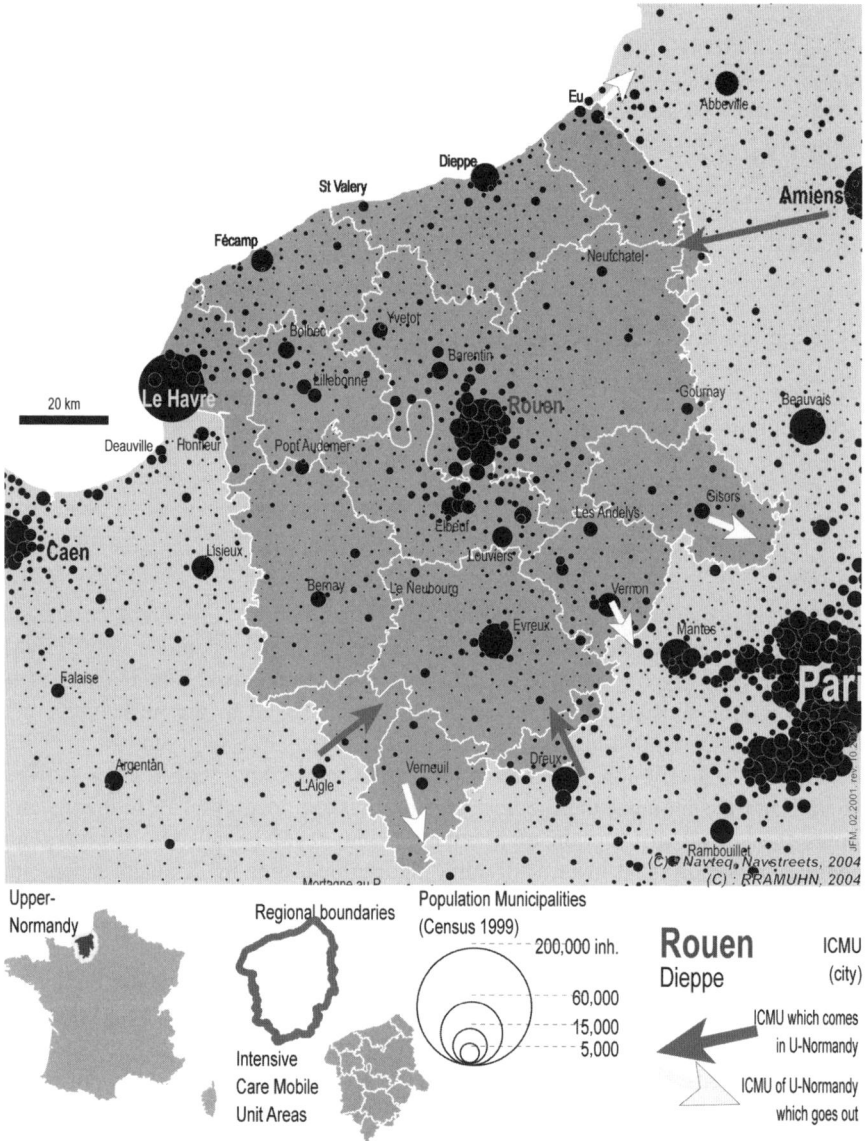

Figure 6.5. *ICMU area and spatial distribution of population*

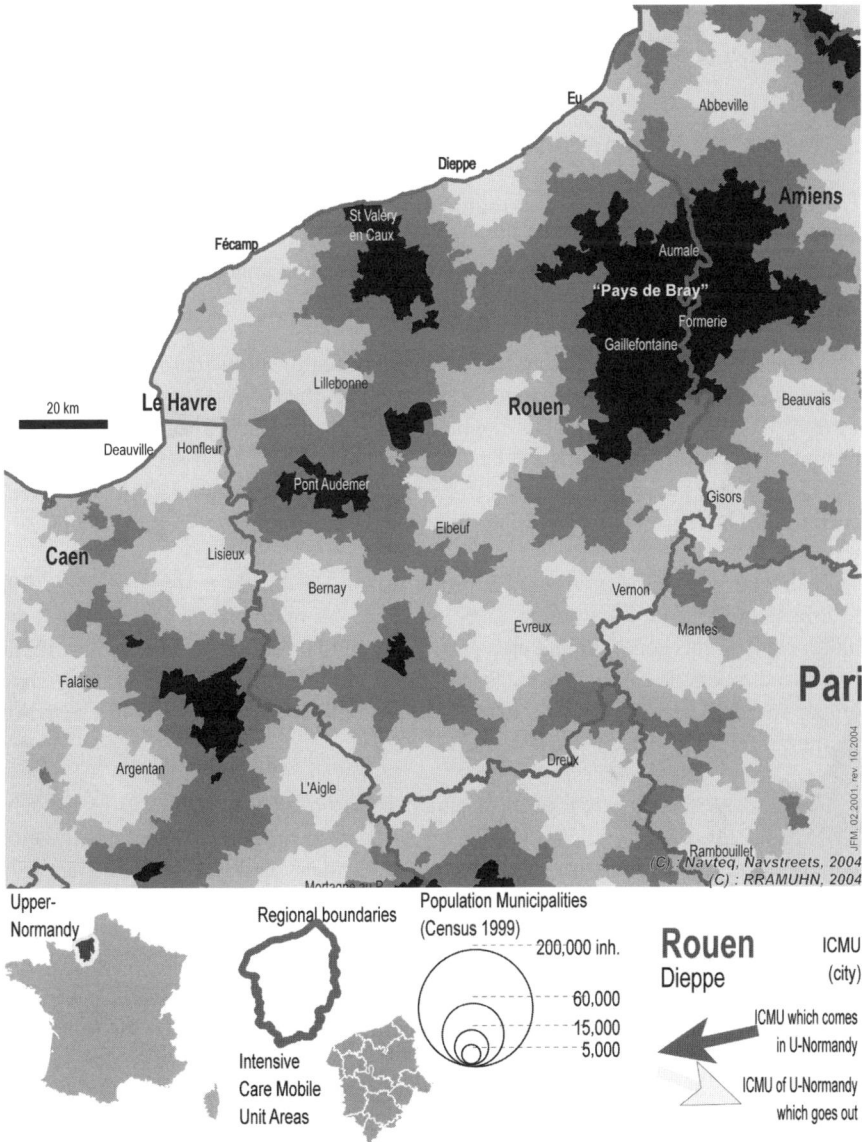

Figure 6.6. *Accessibility of intensive care mobile unit (ICMU) by road network (2001)*

The dark areas (Figure 6.6) are composed of municipalities which cannot be reached in less than half an hour (considered as a minimum time in terms of the public's medical safety). In Upper Normandy, this concerns mainly three rural areas:

Aumale, Gaillefontaine and Formerie, as well as the entire "Pays de Bray" and also Pont-Audemer (a middle size town) and "Saint Valéry en Caux".

6.5. The reallocation projects and their consequences

This approach could change regional health planning. The results are very significant with the road accessibility simulation. Authorities have to reconsider an optimization of the ICMU areas based on space-time and distance criteria. The boundaries change for a better population accessibility.

Following scientific and technical arguments, reallocation may be an objective for action. The proposal is to change the initial state, the official territorial allocation of the ICMU, for an optimized state, which could offer the best accessibility to every borough. The borough remains the basic geographical unit. Road accessibility is the only factor to be taken into account, except in the case of iso-accessibility between several ICMU, where the added criterion has been the number of medical teams.

Reallocation may be envisaged in three ways. The synthesis map presents first the kinematics on the whole area to be analyzed. A matrix shows the main flows able to reshape the geographical area (for example the population lost by an official ICMU to the benefit of a ICMU with a better accessibility). Finally a graph gives a comparison between total populations served in both states. Decisions are based on the interpretation of these three documents.

The map illustrates a reallocation which is based on the definition of the new ICMU intervention sectors. The ICMU have not been modified, with the exception of their operational boundaries. The creation of these new sectors (in gray with white limits in Figure 6.7) has been made on the basis of the boroughs grouped according to an optimized accessibility. The map also presents (in dotted lines) the current limits of the official ICMU, as they are known from the services of EMA. The interval between the official allocation and the new proposals gives the opportunity to analyze the scale of the reallocation, which is shown by the arrows.

The arrows are showing clearly, in the matter of accessibility, that an inter-regional cooperation (presently very weak) must be necessarily conceived, with exits out of the region (white arrows) and arrivals from outside the region (gray arrows). The reallocation may be estimated at 15% of the regional area, and also concerns the intra-regional territory.

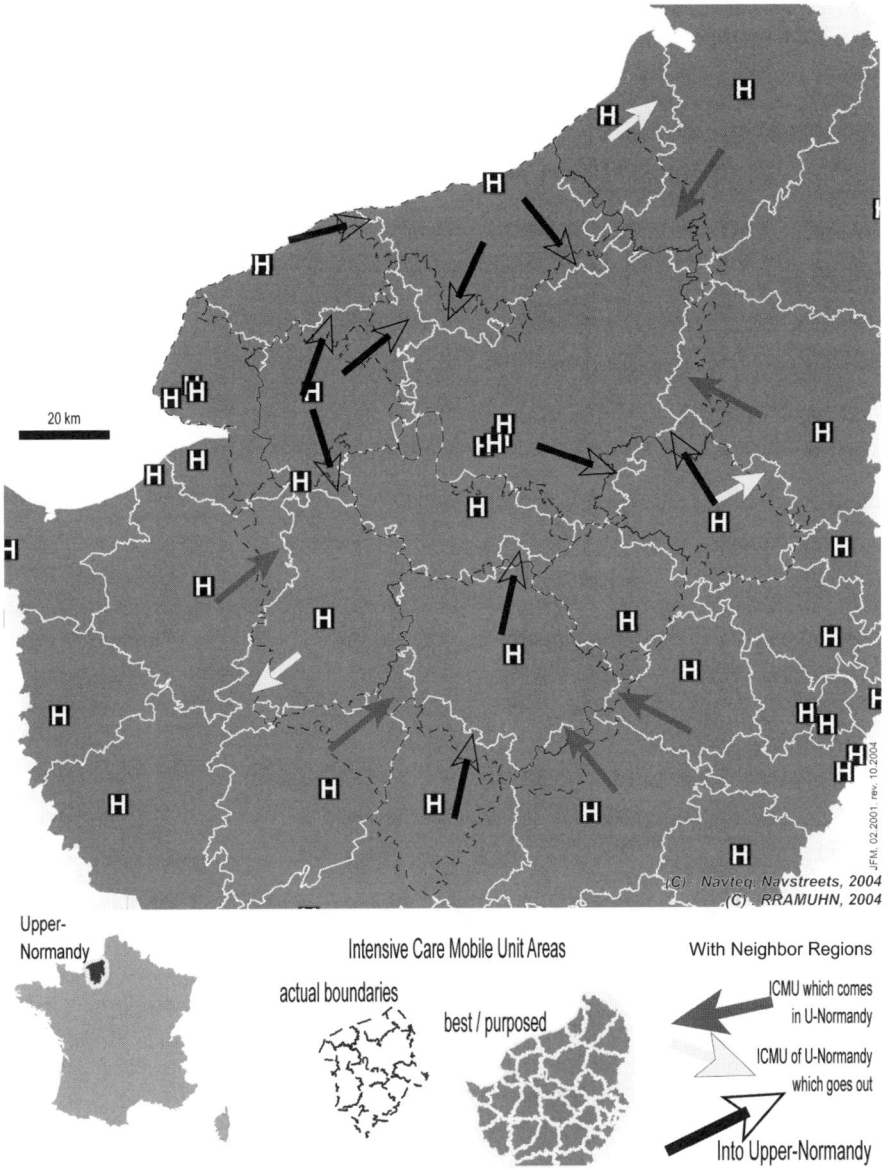

20 km

(C) Navteq. Navstreets, 2004
(C) RRAMUHN, 2004

JFM. 02 2001. rev. 10 2004

Upper-Normandy

Intensive Care Mobile Unit Areas

actual boundaries

best / purposed

With Neighbor Regions

ICMU which comes in U-Normandy

ICMU of U-Normandy which goes out

Into Upper-Normandy

Figure 6.7. *Changing boundaries to improve accessibility*

ICMU to → from ↓	ABB	BEA	DIE	ELB	GIS	LH	LIL	LIS	Others	Total exit
Amiens										971
Bernay							12,221	7,406	1,096	20,723
Dieppe									5,050	5,050
Dreux									1 572	1,572
Elbeuf									1,696	3,767
Eu	9,970									9,970
Evreux									7,898	8,650
Fécamp						2,011	6,134			8,145
Gisors									2,144	2,144
L'Aigle									468	656
Lillebonne										1,346
Rouen		11,664	12,532	11,388	2,095		34,285		290	73,614
Vernon					2,565				1,382	3,947
Others	1,749			1,056		1,843	1,723		1,118	
Total Arrivals	11,719	11,664	12,532	12,444	4,660	3,854	54,363	7 406	22,714	140,555

ABB (Abbeville), BEA (Beauvais), DIE (Dieppe), ELB (Elbeuf), GIS (Gisors), LH (Le Havre), LIL (Lillebonne), LIS (Lisieux).

Figure 6.8. *Redistributing population (before/after)*

The results of the reallocation may be expressed also by a redistribution matrix for the 1,420 municipalities of the region from the official ICMU (left column) to the optimized ICMU (headline). The kinematics which is measured here is the population, whose representation is limited to the moves of more than 2,000 inhabitants and to the ICMU receiving more than 4,000 new residents.

More than 140,000 inhabitants, which is about 8% of the 1,780,000 inhabitants of the whole region of Upper Normandy, could benefit from another ICMU, with a better access. This benefit is mainly due to the proposed reallocation of an important amount of population from the north-west of the ICMU of Rouen to the ICMUs of Lillebonne and Dieppe (near 50,000 people). The matrix proposes to send more than 25,000 residents of Upper Normandy to extra-regional cities (Abbeville, Beauvais, Lisieux).

The efficacy of an ICMU can be measured by the population reached within different time limits, the objective being to reach the greatest possible amount of population in a minimum time, and also to avoid leaving anyone over a given threshold.

A cumulative population curve of the accessibility function shows the efficacy of the servicing in a synthetic way for the totality of the ICMU and of the 1,420 boroughs of the region. The graph (Figure 6.9) allows a comparison of both accessibility potentials: the "official" situation (black curve) and the "proposed" situation (white curve). The optimization is shown by the space between the curves.

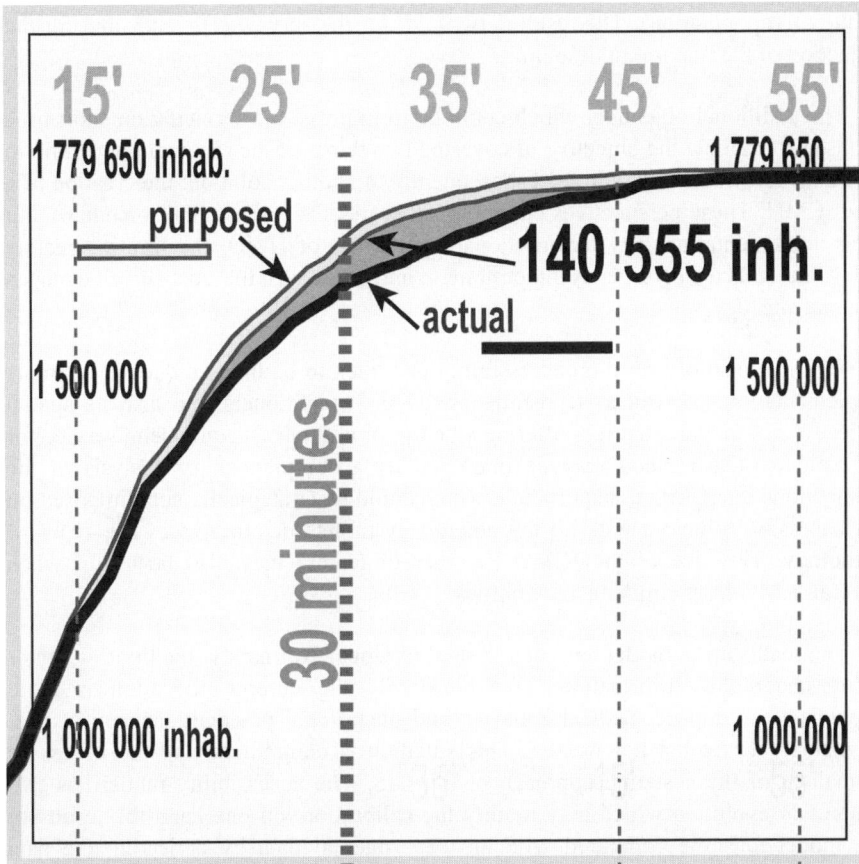

Figure 6.9. *Comparison of the servicing potential between both allocations (actual/proposed)*

The optimization is noticeable qualitatively for the long time limits (more than 40 minutes), and quantitatively for the medium time limits (20 to 40 minutes), but does not play a role for the shorter periods (less than 20 minutes). Qualitatively the longer time periods are reduced from the range (46-57 minutes) to the range (41-51 minutes). This data range only concerns 0.5% of the population, but is spatially delicate and politically sensitive. In the present proposal no one is left more than 51 minutes in the new allocation, which improves the maximum time limit by 6 minutes. Quantitatively a large population would see its distance to the ICMU reduced to 35 minutes, compared to 45 minutes actually (see the horizontal asymptote to the curve). The gap between the reference ICMU and the optimized ICMU is maximum at 32 minutes, where 68,000 inhabitants are better served (3.8% of the total population). The result is twofold, for the service efficiency and for the reduction of the maximum time limit.

The additional objective, which could be to improve the servicing on short time limits (for example the objective of covering two thirds of the regional population in less than 15 minutes), would lead subsequently to another solution: the creation of a new ICMU. These perspectives could use other models (such as the p-median), but their implementation may be questionable in terms of HMO, the limiting factors being the scarcity of urgency practitioners, and moreover the cost of relocating a hospital site.

This reallocation based on accessibility is a track to be favored due to its simple implementation, as well as its validation by the practitioners, but also its spatial efficacy, and the fact that it is offering a better, or at the least equivalent, service for all patients. The method may be improved by a complementary analysis of the destinations for ICMU transportation, which could be strategically superimposed on the accessibility layout with all the emergency targets determined by the types of pathology. The size of the ICMU (number of teams) may also be useful when several ICMU have similar accessibilities.

The reallocation model has been limited to Upper Normandy, the limiting factor being geographic information. For the ICMU the current allocation is more dependent on empiric medical behavior and operational practices than on official regulation: it is often not written. This situation is contradictory to the necessary formalism of the systemic approach of the GIS. The accessibility model has two sources of evolution which may modify the calibration: on one hand the results of the model must be compared with the experimental data by collecting the data concerning the interventions, and on the other hand every modification of the road network must lead to a renewal of the accessibility graphs. It is also possible to anticipate the impact of a new expressway on the servicing of the medical urgencies.

This allocation of the extra-hospital emergency service, as well as of the mesh of the hospital emergency units and of the ICMU, may lead to a chain of other re-allocations: the medical emergency transport system, the on duty practitioners sector, but also the phone operator area, which has the fundamental task of regulating and coordinating this set of operational means.

The regulation area (for the emergency phone calls) should be ideally an aggregation of the diverse intervention allocations, with coordinated means, to favor the operational against the administrative objectives, avoiding any overlapping of sectors, and clarifying decision sequences and territorial responsibilities.

6.6. Relocation of a medical clinic: simulation of a new accessibility

Here is a practical example of the relocation of a private medical and surgical clinic. The "Ormeaux medical and surgical clinic" is situated in the heart of on old district of Le Havre. It suffers greatly, not only from a lack of car parks, but also from poor road accessibility. Following various meetings of the clinic's administration council, it was decided to relocate the clinic to the "Vauban Dock" site which appeared to meet all requirements. Nevertheless, a further in depth analysis was decided to confirm their initial assessment. The method has been identical to the one used above for the Emergency medical assistance. In the present case, the territory is smaller and made up of the healthcare area called the "estuary" area (300 boroughs – Figure 6.10) and the town of Le Havre and its suburbs (2,230 blocks – Figure 6.11). The edges are specified with *a priori* speeds. With the referential preparation, we are able to calculate time-distances to both locations of the clinic (old/new) for all the geo-referenced points.

Starting from a mesh of inhabitant data and the access times between the different locations, the GIS software "Arc View®" and the extension "Spatial Analyst ®" is able to generate a contour map. The contours are drawn at 5 minutes for the global accessibility and 1 minute for the compared accessibility. The clinic is located in the main town. In this context, the outcomes are presented through two complementary scales: the "estuary" (more than 400,000 inhabitants at 1/400,000. It includes a part of Lower Normandy) and the downtown district and its close suburbs (250,000 inhabitants at 1/75,000). Then we can analyze with precision the information in a rural or dense urban environment as well.

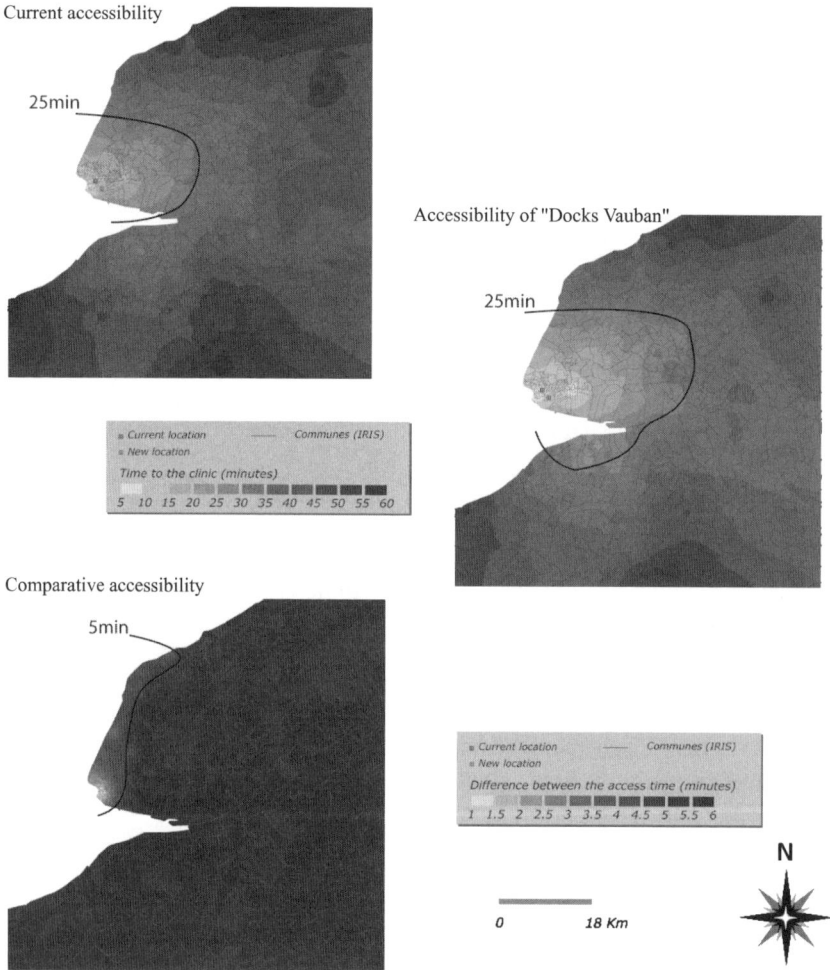

Figure 6.10. *Accessibility at the borough scale*

At the borough scale (Figure 6.10), a time improvement between the current location (downtown) and the new one ("Vauban Dock") is clearly observed: the average time saving is 5 minutes for the whole area, with the exception of the coastal region, due to the road network.

At the sector scale (Figure 6.11), the result is similar. The zoom into the town of Le Havre is a real revelation given our preconceived ideas. Indeed, centrality does not always mean accessibility: the advantage of the old location is limited to the downtown district, with a time saving which is not above 6 minutes. The new one shows a time saving towards the east and an expansion to the north.

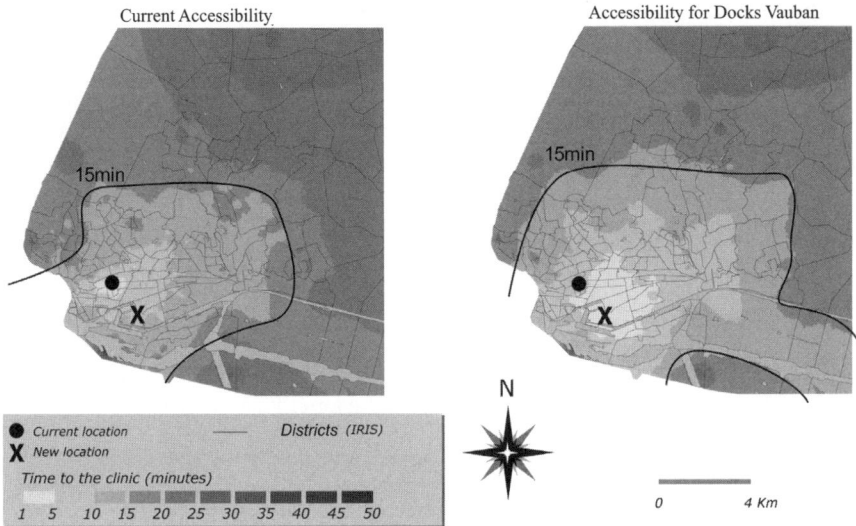

Figure 6.11. *Accessibility at the neighborhood scale*

In the urban area the structure of the road network and improved traffic flow give the advantage to the port even though some of the subdivisions are located between the 2 sites (2 km). Around the port, the network of road arteries improves the access to the new location while peripheral roads in the northern area improve the time by around 10 minutes. Today, the city dwellers need 15 minutes to go to the clinic. In the future, this will be cut by an average of 5 minutes.

The findings speak for themselves. A total of 94% of the inhabitants of Le Havre will need less time to go to the clinic. This study was necessary to help choose the new location, avoiding limited discussions and conflicts of interest.

In the early 21st century, health is more than ever at the center of debates. Indeed, our society may witness deep mutations in the coming decades. We are living longer and the cost of ageing is increasing. In addition to that, recent trends have shown a concentration of healthcare centers and the rarefaction of medical and nursing staff, as well as doctors (especially for anaesthesia and psychiatry). This situation amplifies the inter/intra-regional imbalance in healthcare supply, suggesting our system may have reached deadlock.

The solution seems to come from better management of the healthcare system. We must optimize efficiency by reconsidering the organization, offering new options and in short adapting ourselves. It seems that emerging techniques meet some of the issues such as telemedicine, home-care, interspeciality patient management, establishment of patient transportation standards and the computerization of patient data.

The geographical process brings a scientific approach to the decision, establishing patient location as the major concern. From this locational analysis, a set of chain reactions must be implemented: healthcare structures location, segmentation of the territory, choice of the services, better accessibility – a wide array of topics in which geographers must play a full part.

6.7. Bibliography

[HAI 98] HAINING R., "Spatial statistics and the analysis of health data", in GATRELL A.C., LOYTONEN M., *GIS and Health*, vol. 6, Taylor & Francis, London, UK, pp. 29-47, 1998.

[MAR 98] DUREUIL B., FALOURD J.C., GUERMOND Y., LECLERCQ O., MARY J.F., DE VARGAS F., VAGUET A., *Géographie des urgences en milieu rural: le cas de Neufchâtel en Bray (Seine-Maritime)*, 5th colloque Géographie et économie de la santé, Allocation des ressources – géographie des soins, pp.187-196, Paris, 1998 (http://www.irdes.fr/Publications/Bibliographies/bibresusom/1998/rap1242.htm).

[MAR 01] DRIEU C., MARY J.F., "Géographie appliquée", in *Gestions hospitalières* no. 407, pp.413-469, Paris, France, July 2001.

[MAR 03] DRIEU C., MARY J.F., "La valeur ajoutée des SIG dans l'urgence-santé à l'échelle régionale: expérience du Réseau régional de l'aide médicale urgente de Haute-Normandie", 15th conférence française ESRI, in *Géomatique expert* no.27, pp.26-38, Paris, 2003 (URL: http://www.esrifrance.fr/actu/SIG2003/Communication/homesante.htm).

[MAR 04] DRIEU C., MARY J.F., MEDEIROS R., MENARD C., "The added-value of geographic information system in emergency medical assistance", *10th Symposium in Medical Geography*, University of Manchester, UK, July 2003.

[PEE 01] PEETERS D., THOMAS I., "Location of public services: from theory to application", in SANDERS L., *Models in Spatial Analysis,* SANDERS L. (ed.), ISTE Ltd., pp.73-95, 2007.

[THI 11] THIESSEN A.H., ALTER J.C., "Precipitation averages for large areas", *Monthly Weather Review*, 39, pp.1082-1084, 1911.

[TON 03] LUCAS V., POLTON D., PORTAL S., TONNELIER F., TOUSSAINT J.M., *Une revue de méthodes et d'expériences d'analyse et de construction de territoires*, Appendix 2 in Working Group Report *Territoires et accès aux soins*, CREDES, Ministère de la Santé de la Famille et des Personnes handicapées, January 2003 (URL: http://www.sante.gouv.fr/htm/publication/).

[TOU 01] TOUSSAINT J.M., "Ouvertures de nouvelles structures de soins, la modélisation géographique comme solution", in *Gestions hospitalières* no. 407, pp. 408-410, Paris, France, July 2001.

[TOU 99] TOUSSAINT J.M., "Spatial analysis of the cardiac pacing centres", *11th Colloquium of Theoretical and Quantitative Geography*, Durham University, UK, 1999.

Chapter 7

Modeling Spatial Logics of Individual Behaviors: From Methodological Environmentalism to the Individual Resident Strategist

In 1913 Alfred Siegfried published his *Tableau politique de la France de l'Ouest* [SIE 13]. This pioneering geographer in political sciences confronts an astonishingly modern question throughout this founding work: the modeling of individual behaviors. He specifies in his introduction, "If, according to Goethe's word, even hell has its own laws, why shouldn't politics have its own as well?" However, this has been very little followed up by geographers, particularly by Vidal de la Blanche: Votes where the whim of the elector adds to the natural mobility of the crowds, where the opinion of the deputy becomes complicated with personal considerations, are they able to support solid conclusions?" [VID 14].

Vidal de la Blanche's report of the "*Tableau politique*" illustrates a recurring geographical skepticism: can we model individual behaviors, that integrate agent dynamics, by nature spatial or political? Moreover, the "laws" advanced by Siegfried will be widely criticized, for their determinism, whether dealing with the nature of soil, altitude, race, etc. The debate surrounding Siegfried's work recalls a much wider and ongoing debate: how does a geographical approach, that is to say by space, allow the proposal of systems of coherent explanations of phenomena where the dynamic manifests itself mainly on the individual scale?

Chapter written by Michel BUSSI.

In other words, how, from collective means ("ecological" approach), can we succeed in understanding the complexity of micro-decisions? how can models, whether graphic or mathematical, exceed the reference to means, or to a general interest, valid for all, but hardly applicable to each? For a long time, this question did not have a geographical answer, but on the contrary marked a clean boundary between researchers. It seems that it can be, for the most part, overcome.

7.1. Reconsidering spatial determinism: modeling versus local development

The organization (of the territory) provided the main frame of modeling applications in geography. Traditional geographical models are principally issued from economic theses and allow us to simulate economic implantations, demographic evolutions, flows, trading areas, etc. Most of these models are determinist. They stipulate that individuals possess standardized rational behaviors, most often linked to the principle of least effort, and are dependent on macro-economic logic or "spatial laws" (distance, cost, gravitation, etc.). These models appear well adapted to a centralized organization of territory, which largely resorts to structuring equipment and, as an inciter, the strict economic sphere.

However, these models adapt poorly to the current shift of organization towards the notion of local development, from government to that of governance, from zoning to contract, etc. The example of DATAR illustrates quite well the vagueness that is generated by passing from planning to decentralized participative development. In a recent reference document [DAT 00][1], it speaks "of a turn in territorial organization". Faced with a State limited to a role of anticipation and organization (notably institutional), it notes the rise in power of the notion of "factual territories" [FRE 99] and of the actors of development. Such an evolution modifies the paradigms of the geographical planning models, as the "lyrical" expression which hardly made caused a reaction among geographers shows: "territories are no longer the frames where things happen, but where things are invented". This passing from planning schemes to participative development can be interpreted differently. Some speak of "reversal" [LON 00], of "turning" [DAT 00],

---

1 DATAR (French administration for territorial planning and regional action) isolates in its action three very classic phases, simplified here to the extreme. Firstly that of the founding politics of 1960 to 1975: the State is "omniscient", "rational", "planning" and the "great builder"; space is abundant, cheap, and malleable; the republican principle of equality of territory commands the policies. Next, that of the politics of times of crises from 1975 to 1990: the civil society questions the hierarchal authority and the role of the State; it is followed by decentralization and the multiplication of the contract as a mode of relation between the State and local collectivities; nevertheless, the appeal to the State remains associated with that of a fire-fighter settling inequalities. Finally, the phase from 1990–1999 is called, by DATAR, "the return of territory to the center of the debate".

or even of "natural sloping" [WAC 00]. Beyond such a nuance, the important detail is that there now exists a consensus admitting the emergence of a new form of space management that obligates the rethinking of models of this organization.

What place does modeling have in these new logics of organizing territories by development? If we admit that local development is not based on a model but on experimentation and on the synergy of local abilities[2], having a short term project and not a theoretical clarification as our objective, we could conclude from it that "implied" researcher-actor geographers have better understood what is or what should "actually" be local development than researchers who study it "in theory" and from the outside. From the local development point of view, the right place would be from the perspective of the ants and not that of the entomologist. This is in part what E. Glon *et al.* attempt to explain [GLO 96] when they conclude that "one can affirm at the end of this study that local development serves geography before considering the inverse, that is to say a geography that would serve local development". They demand a local development that could be made of democratic practices and respect to others. This development would be sustainable, balanced, environmentally sound, non-segregated, and would invent new areas of abilities or new modes of governance. A geography serving local development, in the spirit of the authors, would imply on the contrary a geography that would search to impose pre-established spatial models. In a less geographical register, this is also what J.-P. Deffontaines and J.-P. Prod'homme affirm [DEF 00]: "the biggest threat would be to enclose local development in a cold and distant position. Local development is firstly the lives of concrete people, brainstorming, intertwined initiatives and ideas".

Unquestionably, such logics lead to distancing a theoretical and modeling geography from a "citizen and implied" thematic, and give priority to a very monographic geographical production, exhausting "the case study". "Local development" would become for geographers a relatively modern drawer to arrange the ensemble of studies having as a goal the place, or more generally a place observed from a systemic point of view (we will eventually speak of "locality"). Most special issues of geographical journals dealing with the "locale" or "place" do not escape this tendency. We could therefore think that it is hardly possible to model the "locale".

---

2 One of the most current definitions of local development is that of X. Greffe: "neither mode, neither model" [GRE 84].

However, the question is poorly posed. Sorbets [SOR 93] correctly recommends differentiating the "local object" from the "local question".[3] Quite often, the locale is not the object of the study itself, but an area of analysis, that takes place "in the locale". It is therefore current for scientists to prefer to define the locale as a question rather than as a goal, in the same way that anthropologists explain that they do not study the village, but in the village. The differentiation is however less distinct among geographers. They often claim, explicitly or implicitly, the "locale" as an object (working on a certain neighborhood, locality, or cultural area, etc.). The "locale as a question" therefore finds itself either confused with the "locale as a goal" (synthesis maps presented as graphic models for example), or marginalized (the widespread idea that the locale cannot be modeled; that it is, at best, the residue of the model).

In summary, the passing from planning to development, from macro-economic determinisms to micro-local initiatives, from centralization to contracted negotiation, would complicate the use of a purely "modeling" geography, less capable of simulating the complexity of social facts. Nevertheless, the reduction of the geography of local development to a monographic production remains disputable. If local development logics imply that the initiatives to begin are not necessarily reproducible from one place to another, the modes of participation, the diagnostic methods, and a certain number of spatial invariants allow a real scientific theorization. Many geographers have since gone down this road.

We can, on the methodological level, cite the applied use of the Fuzzyland model of territory evaluation by Rolland-May [ROL 99] [ROL 00] or the MTG laboratory studies on intra-urban analysis (applied to transport management, urbanism, town policy, etc.). On the theoretical level, we can cite the interpretation models of J. Levy on territory organization [LEV 00], or those of A. Chauvet [CHA 92] between "development and location", around the three themes of heritage, position and territory.

The ambiguous relationship between geography and local development, between practice and theorization, seems to be a surmountable problem. Nevertheless, we cannot conceal the fact that around local development a certain rift has been created at the heart of geography. Far too often this is claimed by certain geographers who do not thoroughly grasp the subject, while others distance themselves from such a

---

3 Nevertheless, Sorbets picks out with humor the limits of the scientific approach of a locale uniquely considered as a "question": "in one word, one could say that the locale is made just for the innocents (those who live without asking many questions) or for the cunning, whether they be elected, administrators, or researchers; the latter tend to speak too often on behalf of those that they are supposed to 'explain', and who will be affected by the heaviness of the ways of saying things. From there, who will risk to say but one part of this discrete world of local practice".

"popularized" subject.[4] Local developers struggle to communicate with theorists of the universal horizon. The heirs of the monographic school perceive quite well that local democracy opens new theoretical horizons for them, but they hesitate to rush in. Conversely, the followers of the quantitative school hesitate to admit that the transposition of economic models is no longer in phase with current organizational policies [GRA 97], although they should instead focus more than they do on the transposition of other models, and particularly the spatial adaptation of models dealing with "games of actors". At the very center of territorial organization, the geographers who are captives of the Saint-Simonian model, convenient for applying spatial analysis, poorly perceive where geography begins when planning is decided from "the bottom". At the same time, local development contributed to "dispossessing" geographers of their tools: they are no longer the only ones to master cartography, nor geographic information systems and other corollary complex spatial processing. These tools are henceforth current in local collectivities and are wisely used in the hands of high-level developers. This dispossession is however doubly beneficial for geography. On one hand it makes geographic methods otherwise reserved to a closed circle or specialized laboratories commonplace (map algebra, buffer utilization, automatic addressing, etc.). On the other hand it stimulates university geographers fond of quantitative processing to pass from the stage of efficient tool utilization to that of innovative simulation/modeling.

For lack of having imagined pertinent models and theoretical frameworks, this turning point in land and country planning is an opportunity that geographers are at least partially in the process of letting pass. In the 1980s, the external perspective placed on geography by local development specialists was hardly flattering. Greffe [GRE 84], in a reference manual, simultaneously denounces (excessively) the complicity of geographers in relation to power, their dispersion, their weak external readability, their confinement in description or in models exported from other disciplines, this being all the more regrettable since it recognizes in geographical science an inherent legitimacy in questions regarding territorial development: "the contribution of geographers to the debate on decentralization should be of primary importance. However, it happens that even in their own opinion, geography can provide fewer and fewer answers to today's problems compared with those of the past... geographers have lost their originality in adopting the regionalist debate or by lining up behind the emergence of new economic methodologies. They found themselves in an ambiguous gathering of oppositions to centralizing radicalism or in the description of the old or new polarizations of the territory".

---

4 The sphere of local development can be associated with debates on "sustainable development", "governance", etc.

The example of organizational/developmental evolution illustrates that in a society that is becoming more individual and decentralized, geographical models based on a determinist and "transcendent" vision expose their limits. However, the geographical scientific study of local development and ascendant logic is limited essentially to case studies, the opposite of modeling experiments. Our goal, therefore, is to provide some leads to reconcile modeling and the localized participative approach.

7.2. Ecological methodology

To tackle social questions, we often oppose approaches based on cartographic comparison and those based on survey. Therefore the term "ecological" is frequently used to describe the collective or geographical approach. Borrowed from natural sciences, it originally designates the study of the milieu where things live. By extension, it becomes the study of the territorial environment's impact on social behaviors, then, in a larger sense "the study of behaviors from given information in the frame of territorial units often called collective" [LAN 75]. For more information, we can refer to Rhein's long article [RHE 94] *segregation and its measurement*, which details the epistemology of ecological analysis in geography. The principle manuals of electoral sociology therefore oppose the "ecological" models with "psychological" models [LEW 01]. These same manuals remind us that originally the social explanation was more collective, then from the 1960s onward the hegemony of psychosocial explanations began, followed by a rediscovery in the 1980s the virtues of the ecological approach. This ordering of the social explanation, opposing the ecology-cartography-geography association to the sociology-psychology individual inquiries association seems relatively outmoded to me.[5]

7.2.1. *Individualism and ecology*

The justification for the use of opinion surveys to explain social behaviors is often made by the criticism of the ecological approach. Nevertheless, it is possible to

---

5 The criticisms of political scientists towards the ecological-cartographic approach of the vote provide an illustration: the principle criticism of the ecological approach is the shift that is happening from the elector towards the electorate, from a reality to a so-called fictive aggregation: "Electoral geography ignores individual voting to take an interest in groups that have voted, and which are defined by their collective determination. Would there be a sort of curse that a cartomatic apparatus does not construct other than a collection of petrified agents, removed from their two complementary dimensions, individualist and anthropologic?" [HAS 89]. Expressions such as *"la France qui vote"* are therefore denounced as abuses of language: it is not a space that votes but rather electors, whose opinions are differentiated, even if majorities emerge.

retort that the approach using surveys introduces exactly the same inverse bias, that is to say to totally forget the social aggregates: "A sample is not a real social grouping, but a collection of individuals. Such a method refuses to comprehend the systems of interactions that concretely characterize the existence of multiple groups (family, locality, enterprise, union, etc.) that individuals are a part of, and the impact of which is large on the dynamics of individual and collective attitudes and behaviors" [MIL 77].

What is more, the ecological approach is the only one that allows us to "exhaustively" deal with social data. A survey, even one concerning several tens of thousands of polled individuals, cannot report representative details of the inhabitants of a nation, a region or even an agglomeration.

Sociologists justify the paradox of their approach, treating social relations though the study of a series of decontextualized individuals, by disassociating the notion of "methodological" individualism (attribute of the researcher's approach) from that of "sociological" individualism (attribute of the studied subject). In theory, the two forms of individualism should have nothing in common, but this distinction remains debatable in principle: "there probably exists a relation between methodological individualism as a method and sociological individualism as a favorable climate for this method, although the individualist climate does not necessarily determine the choice of methodological individualism" [LEC 86].

The central concept of all explanatory attempts, the blurred relation of causality between an explanatory variable and explained variable is one of the major criticisms made against the ecological approach. "That two phenomena have the same distribution over the territory, does it necessarily signify that they are linked by a causal relation?" [LAN 74]. It is thus sometimes difficult to determine among two apparently linked phenomena, which of the two explains the other. More often, the causality is reflexive: the links are only indirect, by the effect of other hidden variables. But the survey, contrary to popular opinion, brings nothing to the relation of causality between the variables because those surveyed are often quite incapable of judging the reasons for their behavior.

In conclusion, we often criticize geographers for contenting themselves with resorting to maps, and by so doing over-evaluating the collective influence on behaviors using the ecological approach. To counter such an argument, we can ask why geographers did not think of resorting to the same "strategy" as sociologists, by distinguishing a "methodological ecology", which is a simple attribute of the researcher's approach, from a "sociological (and/or political) ecology", which is an attribute of the studied subject. Disassociating a "sociological ecology" from a "methodological ecology" would remind us that adopting a collective approach of behaviors (for example an approach of geographical modeling) does not presuppose an ascendant, community-based or determinist vision of society. Consequently, by confirming that geographic models come strictly from a "methodological ecology", we eventually render them compatible with a sociological individualism.

7.2.2. *What place does geography have in the systemic approach to societal phenomena?*

How can we integrate geography into global models explaining social behaviors? In other words, beyond the geographical models, the question asked could be that of the place of space in the models of social sciences. Levy [LEV 94] recognized that space could be a new explanatory variable of social behaviors that elsewhere other disciplines are in the process of rediscovering, notably political scientists. He argues that the geographical approach should not be limited to a new axis of multifactorial analysis of social behaviors: "Space can inform us if it constitutes a true model of reading, a way of understanding how the entire society, with all its dimensions and logics, generates its policy. It is the societal character of spatiality that must be mobilized – and not the only spatial projection of non-spatial phenomena." Such analysis is therefore concerned with not confining geography to a simple "factorial axis", but admitting that the strategies of actors possess an undeniable spatial dimension.

Nevertheless, this geographical claim entails a certain risk. The idea that all social phenomena possess a spatial dimension is widely accepted by non-geographers, but by being recognized as theoretically "everywhere", the space increasingly risks being scientifically "nowhere". The notion of "territoriality" is widely taken up by political scientists, sociologists or economists.[6] More than a recognition of a specificity of spatial models, space becomes the last resort with which we associate that which cannot be otherwise explained. In the classic models of social sciences, space explained the residual behaviors of sociological models, in the same capacity as history. The models have barely changed, but if turning to geography is becoming more frequent, it is simply because there are more residuals than before. We therefore stockpile these phenomena, incomprehensible for the moment, in the spatial receptacle while waiting to find their significance. We turn to space in the same way we used maps before opinion surveys: by default. Instead of the unflattering term of "space bin", political scientists, sociologists and economists prefer that of "melting pot". Space therefore becomes the melting pot in which opinions and attitudes mix together, until composing a "culture" that never has

---

6 Likewise, the new modernity accorded to Siegfried's work is witness to this return, for most political scientists, to the contextual approach, having exhausted the heavy explanatory variables of the vote or hardly having models to propose to explain geographic organization of new parts or the maintenance of old ones.

exactly "the same flavor" as before. Space is nothing more than a pot and does not take part in the mixing. Dargent's [DAR 99] conclusion is, that despite the use of the most sophisticated contextual inquiries (that of the inter-regional observatory of politics), we cannot explain the mystery of the permanence of regional behaviors (e.g. notable religious traits), and that this is symptomatic: "in fact, there are as many historical sociologies of politics as there are French regions that must be constructed if we want to take account of territorial specificities". Here is the definition of the territorial approach of regional cultures: "a historical sociology of politics", that is to say finally the ensemble of social sciences, and not so much geography! Dolez and Laurent [DOL 97], "the most geographic" of French political scientists, also write in the conclusion of their national analysis, "social history and the contextual analysis are irreplaceable to understanding why space always remains differentiated". As a last resort, in 1913, Siegfried evoked "the soul of the people", "the spirit of places" or the "ethnic mystery". We advance today, with more prudence, the term "culture"[7], but the fundamental principle seems the same to me. If the preceding examples are mainly taken from political scientists, the same issues of definition of the explanatory part of geography arise among economists or sociologists.

This comment should not be interpreted as a negation of the pertinence of all contextual study, but as a problem underlining that the value of geography as a science capable of producing models lies elsewhere. Geography would not be reduced to the "inspiration of place" that leads "an underground battle against major tendencies" [LEW 01]. Even when such an approach, presupposing the specificity of a location, sees itself as explanatory or modeling ("graphic" model for example), the fact that it is localized somewhere does not suffice to make a geographical work from a study of historical sociology.

An illustration of the difficulty interpreting geography as a science capable of producing models is supplied by the debate launched by the political geography review. It shows that despite the long tradition of modeling geography in Britain, the role of spatial explanation, here in political sciences, remains fragile. In this debate, in response to a text by Agnew on the effect of context on geography, the political scientist King [KIN 96] attempts to restrict geography to a curious "science of ignorance", opposed to a science capable of being the subject of modeling: "To take an extreme example, scientists understand some aspects of physics reasonably well, and because of this I don't think there are physicists writing papers on a geographical theory of the electron, coloring in detailed maps of Canada by the number of electrons per province. This is perverse of course, but precisely because many aspects of electrons are reasonably well understood. Wherever we find an

---

7 For example, the term "political culture" is widely used by political scientists: [OTA 97] [OTA 00] [PUT 93] [LAU 99].

electron, we understand its characteristics well enough so that it is exchangeable with any other electron on the planet, and presumably in the universe. In contrast, we need political geography because political scientists don't understand enough about politics".

Geography therefore only finds usefulness as a "science of evidence", to simple pedagogic virtues: "Political geography is so useful in large part because political scientists do not understand politics sufficiently (and because geography is perhaps the clearest way to understand what it is we do not know. Information can be organized in many other ways: we can list unexplained facts alphabetically, or by size, color, weight, our degree of uncertainly about them, or how important we think they are. Geography is useful because we really do know a lot about it, and because humans happen to feel very comfortable thinking geographically. Displaying data geographically helps because it connects a variable we wish to explain with numerous others coded on the same level of geography. Moreover, because most observers know the values of many of these variables without having to look them up, geographical displays are instantly recognizable and interpretable. Thus, geography is useful because of a standard pedagogy technique: it connects something we do not know to the information we do know...geographical variation yes, contextual effects no".

Geography would therefore be useful (and even essential) as a methodology, a descriptive science, and as a pedagogical issue since it is "meaningful" to all, but would launch itself into a dead end if it sought to isolate itself from any spatial effect. The comments of King come back to negate the modeling pertinence of social geography. King's position, like that of Hastings [HAS 98] towards the publication of *France qui vote*, even if being overtly caricatured, is interesting nevertheless in the sense of how it demonstrates a reluctance, sometimes admitted, including in Britain, to recognize the pertinence of spatial explanation. The "ecological" approach of social facts would be by nature determinist and therefore simplifying. For this reason, in order to integrate the spatial dimension to the heart of explanatory models of social sciences, the demand for a "methodological ecology" seems relevant to me.

7.2.3. The collective dimension of individual facts: the intra-urban example

The question asked *in fine* is therefore: is it possible to explain individual behaviors based on an ecological method? The previous examples were for the most part taken from political geography. Health geography provides another pertinent example. Here, I will make particular reference to a work on the consummation of psychotropic medications in collaboration with sociologists. The use of psychotropic medications is a personal, intimate, approach linked to an internal psychic equilibrium or disequilibrium. What is the meaning of cartography, even a cartography that is detailed and spread out over a geometric grid, for an average consumption? The traditional position of sociologists was to confine it to a role of spatial consumption typology serving to define the test neighborhoods where it was possible to launch individual studies – the only kind of study well fitted to a research of causalities – then explanatory models. Our position was more dialectal and consists of defending the idea that depression was equally linked to a rupture of equilibrium between the individual and their environment, something a strictly individual approach would be incapable of grasping. There will occur during and following the study a stimulating scientific debate [LEM 96] [LEM 96b]. The strongest correlation observed with respect to the cause of psychotropic medications were those particularly linked to the dilapidation of housing (we are dependent here on the poorness of INSEE[8] indicators in terms of precariousness). The temptation to make causality from this simple correlation could be great. Sociologists therefore denounce, with reason, the implied hygienist drift of such an approach (explaining a personal behavior by the harmful influence of a physical environment among the most stable).

These criticisms throw us back almost exactly to the criticisms of determinism launched against a certain quantitative geography. To avoid these "processes of intention", I am not persuaded that the collective approach must be renounced. On the contrary personal space integrates the perception of a close or distant environment, or neighborhood or multi-scalar, territorial or reticular, that only the comparison of "ecological" and contextual inquiries makes it possible to understand. The measure of differentiated averages on different zones does not seek to obtain the smallest possible standard deviation within each zone, in order to explain by correlation with other indicators the average individual behavior, but to understand which cumulative individual spatial attitudes succeed at these differentiated averages. The individual perception of these differentiated averages (and therefore of a socio-spatial differentiation, or even of a segregation) reveals itself as being one of the explanatory factors. This conception finds a natural application in the intra-urban approach, a specialty of the MTG laboratory. This approach rests on four principles:

1) The measure of socio-spatial differentiations at the intra-urban scale constitutes a research subject in itself due to its conceptual and methodological complexity. We can underline that "on the whole" urban society has a lesser specificity than rural society, since it is based on diversity and the complementary nature of customs and status. However, on the individual or intra-urban scale this is no longer true. Intra-urban spaces are more mono-functional and each citizen shares

---

8 *Institut National de la Statistique et des Études Économiques*: French National Institute for Statistics and Economic Studies.

their time, according to the location, between specific specialized activities, which justifies a study at this scale.

2) The processes of intra-urban socio-spatial differentiations cannot be understood unless we consider the city in its entirety, and not just a sub-space of it. Few geographers demand such an approach, with the exception of the work by the MTG laboratory[9] and more generally the applied works of town planning agencies.

3) The processes of intra-urban socio-spatial differentiations can only be approached by a systemic analysis, including objective inequalities and subjective representations of the space. This tends to limit the pertinence of sectional analyses in the intra-urban milieu, or rather obliges them to put the conclusions into perspective according to their position in a system of complex retroactive effects.

4) The processes of intra-urban socio-spatial differentiations are explained by a reflexive relation between individual behaviors and their collective consequence. In this case, the individual approaches by survey or "on ecological issues" by map are both insufficient to understand the process which is at work. Specializing in the collective scale should not persuade the researcher to try to find an average individual through minimal standard deviations whose behavior we could explain (we are therefore situated in "the ecological error"), but on the contrary to analyze how this "average", known or supposed by the inhabitants, influences their individual behaviors and contributes to evolving this average. On this note, I completely agree with Y. Grafmayer's definition of urban space [GRA 00]: "We should not conclude that space is a perfectly neutral recording surface, a sort of dual-equipment of social life. Even insofar as it is a product, space is an integral part of this social life. It constitutes less of a faithful replica than a particular register, that must be understood in its interdependence with the other registers." It is this dimension that we must try to integrate to urban models.

We could think that favoring such an approach comes back to ignoring the city as a space privileged with mobility, with maximal socio-spatial interaction due to the density/diversity couple, that is to say contrary to the urban nature of inhabitants of the "metapolis" that F. Asher describes to us [ASC 95]: "Their sociability takes root less in the neighborhood. Their daily lives unfurl simultaneously in the home and in public 'metapolitan' places while the neighborhood loses a part of its traditional functions". On the contrary, the position of the geographer must be to demonstrate that if the mobility of citizens increases globally, it implies on one hand that the range between spatial capitals of inhabitants increases at the same time, and on the other hand inhabitants can therefore mobilize more easily to benefit from the competitive advantages between intra-urban spaces. This generally causes an

---

9 The very weak mobilization of geographers against the renovation project of the French population census, condemning this type of approach for the city, is proof of this.

acceleration of the mechanisms of "urban selection" that the policies of institutional spaces, at least in France ("districts"), increase more than they correct. In a caricatured way, we could claim that in the "metapolis", neighbors tend to know each other less and less, but resemble each other more and more. Here, it concerns a theory that we could qualify as "post-individualist".

7.3. Towards a post-individualist behavior

The expansion of "individual spatial capitals" and the multiplication of the micro-decisions of actors (choices, strategies, itineraries, etc.) express themselves paradoxically by (very) spatially organized actions, according to the simple rules of contagions and confrontations...The individualist apprenticeship ends in spatial strategies generating forms of self-organization. These individual spatial strategies continue to confront strong territorial logics, which impose the practice of representative democracy, the perception and redistribution of resources, and the learning of shared values (school, culture, pastimes, etc.). It is the meeting between these spontaneous mobilities and such "formal social mixing" [BUS 02] that seems to me important to model today.

7.3.1. *Self-organization and segregation*

To explain this hypothesis of "post-individualist" behaviors, the example of urban segregations seems the most pertinent. According to Schelling [SCH 80] there are three types of segregations [GRA 00]. Firstly, that which results in an organized segregative intention. This segregation, principally political, can nevertheless be economical, associative, etc. Next, the second form of segregation can be the result of inequalities produced by social division: here it is concerned with an essentially economic process, on which the notion of prestige comes to the top, but which includes no intention to segregate, at least directly. These first two forms of segregation are very well studied, particularly through the generalization of NIMBY[10] practices, which according to the authors could be classed as type 1 [DAV 97] or type 2 [JOB 98]. These first two forms of segregation are widely the subject of geographical modeling, according to the Chicago school. The third form of segregation, the least studied, is that which Schelling develops: it is the "emerging collective result of the combination of individual discriminatory behaviors". By discriminatory, we must understand a perception of the other that influences as much the place of residence as the choice of equipment that we frequent or the route of travel that we choose. These attitudes determine the limits of desired or tolerated neighborhoods. The combined game of individual choices leads to logics of

---

10 "Not in my back yard".

segregation, which at the beginning were not wanted by any of the players. By modeling this "tyranny of little acts", Schelling demonstrated the importance of the domino effect: even if each individual has, at the start, only limited requirements of neighborhood, having 70% of neighbors of the same social class for example, after a few iterations, the simulation demonstrates that in order to satisfy each of their individual demands, we must strive towards a much more segregated system, extending far beyond our initial wishes. Spatial modeling of such cases is still rare in geography, even if Schelling's approach fundamentally draws many similarities with that of Levy, who presents the inhabitant as a fundamental player, and the habitat as a permanent spatial choice. We can nevertheless point out on the level of modeling the recent works of D. Badariotti [BAD 01], which simulated residential mobilities in the city of Bogota constructed from cellular automata, by testing different hypotheses of tolerance towards other social groups, particularly ethnic. Here as well, the simulation led to an increase of segregations, in opposition to the initial "rules".[11]

The analysis of the first two forms of segregation, which either refers to political processes, or economic processes, seems abundantly developed to me in urban sociology. They correspond to a "traditional" process of "descending" segregation. The third form on the contrary, corresponds to an ascending process. It expresses the result in a liberal frame of the cumulated effect of individual choices. Such an approach is similar to a geography of democracy that fully expresses the contradictions between freedom and equality, leading, for the urban areas, to the establishment of "negotiated spaces". This leads to us explicitly raising the question of Levy's density/diversity couple or Wirth's density/heterogeneity or more generally the fact that the city gathers together in one place differentiated populations [GRA 00]. It is this "non-existent distance" that renders the city more sensitive than other milieus to this "tyranny of little acts", but which in parallel aspires to the "moral density" of Durkheim. We enter into the fundamental urban paradox, which Roncayolo reminds us of [RON 90], "the city presents two complementary aspects: it is at the same time a place of differences that separate in a more or less visible way the social groups, the functions, the land use; it is also the place of regrouping and convergence that overcomes or erases, as much as possible, the effects of distance."

The advent of liberal societies implies therefore integrating to the models a form of "spatial contradiction", clearly identified by Claval [CLA 79]: "From the moment when all institutionalized hierarchal organization disappears, rich individuals lose the possibility of imposing their will on their poor neighbors and the means to avoid

---

11 During the 2004 year, a team of geographers from the MTG team began a reflection on the application of the Schelling model on the agglomeration of Rouen (Michel Rolland, Michel Bussi, Patrice Langlois, Eric Daudé).

the nuisances that they generate. Nevertheless, their is no longer a direct method to protect oneself from the inconveniences of cohabitation; there is one possible strategy: spatial segregation...as long as the agglomerations are subjected to one political authority, the chances of seeing the social situation expressed as a geography of neighborhood inequality are modest: poor groups carry enough weight in municipalities to defend themselves. They lose all means where the urban space is fragmented into districts interdependent from one another. We see here one of the contradictions of liberal societies taking shape (and it concerns a spatial contradiction): to fight against the alienation caused by the growing dimension of bureaucracies, we find the tendency to reinforce the autonomy of political social units; this brings the citizens closer to those who govern them, facilitates the direct expression of needs and allows them to have a decisive influence on the decisions of large organizations. But the more the recognized competencies of elementary territorial units grow the more the risks of seeing them used in a strategy of local differentiation increase."

It seems to me that complex systems and models of self-organization perfectly allow testing by way of modeling (new) social organizations.[12] Yet we must not forget that it is a "liberal modeling of space" that is taking place. Well beyond a methodological evolution, it is the paradigm of geographic modeling itself that finds itself modified. The geographer is no longer the specialist on models and organization and control of territory. He becomes one of the theorists of the complex effects of a liberal society. This acknowledgment tends at the same time to modify the societal usefulness of the "geographer who models": he could pass from a status of "expert distanced to the ivory tower of the elites", to that of interdependent analyst attentive to movements, actions and aspirations of the civil society.

7.3.2. *Space/individualism: two interpretations*

Nevertheless, the debate on the sense of the production of differentiated spaces in a society marked more by individualism can be subject to two interpretations, ultimately in opposition with each other: "cultural and identity" explanations and "self-organization" explanations.

The first interpretation states that faced with the decline of behaviors of class and ideologies, we are witness to the growing importance of questions of identity, particularly on infra-national scales. The Europe-wide rise in of nationalisms, regionalisms and localisms would reflect this. These tendencies would reverse the process of geographical homogenization of behaviors. It is this thesis, which does not limit itself to a simple *"culturalism"*, that concludes Leca's work on

---

12 These in particular are developed in Chapter 13.

individualism [LEC 86]. Pizzorno [PIZ 86] claims that "there is a value that democracy can realize: it is not the freedom of political choice (we have demonstrated that this is an illusion) but the freedom to participate in the processes of collective identification". According to him, the main function of democracy is to keep collective impulses under control (territorial, religious) within political rules. This "collective" freedom would have been accorded by the nation-states at the very moment when these traditional identities disbanded and when the States no longer had the means to control individualist demands. If today, "the collective self-control produced by the mechanisms of political representation" seem superfluous, at least in western democracies, the new values of democracy remain by nature rooted in an "identifying" landscape, whether it be to demand new collective identities, or to refuse all imposed identities (an attitude that Pizzorno qualifies nevertheless as "private identification").

Levy [LEV 94] also recognizes that the individualization of society does not beget a homogenization, but rather recompositions: "what is fascinating for the geographer is that these transformations do not make space disappear, but make a new one appear" [LEV 99]. However, contrary to the preceding approach, he refutes resorting to holistic logic (or identity) to explain these new spatial organizations. According to him, the spatial reorganization of behaviors can be explained first and foremost by the emergence of the inhabitant-agent, controlling (or not) their spatial capital. It is in this way he claims that the adherence to the "terrain" is stronger than religious, patrimonial or class adherence. We could therefore speak of the "post-individualist-agent". The socio-spatial differentiations would find themselves principally explained by forms of self-organization, stemming from the strategies of citizen-inhabitants.

7.4. From neighborhood effect to the theory of the citizen-resident-strategist

For a long time, the weight of the collective environment on the individual was summed up by the neighborhood effect. Tocqueville [TOC 39] insists well before Cox [COX 69] on the importance of this effect on opinion, and that it is at the same time one of the explanatory factors for the permanence of behaviors: "At all times that the conditions are legal, general opinion weighs heavily on the mind of each individual; it envelops them, directs them, and oppresses them. As all people resemble each other more, each feels weaker and weaker when compared to everyone. Not only do they doubt their strengths, but they come to doubt their rights, and they are ready to recognize that they are wrong when the greater number claims it. The majority does not need to force, it convinces...This favors marvelously the stability of beliefs".

The ecological measure of the domino effect expresses itself by the acknowledgement that the behavior of an area is not, generally, simply in accordance with its social structure in light of national behaviors measured by individual surveys, but exaggerated in favor of the majority opinion of the social structure. If the neighborhood effect is the subject of few writings in France, it is the complete opposite in Great Britain[13]. Some research does exist in France. For example, the voter registration cards in 2002 were the subject of a simulation project, though the elaboration of a model of diffusion by proximity from neighborhood centers: the map of the French National Front, in particular, can be simulated quite precisely with simple neighborhood effect rules [BUS 03].

However, the explanation behind the process leading to this effect remains a source of debate. The effect of personal conversations was developed by the Columbia school and Cox [COX 69]. Nevertheless, the role of such conversations and local information are today minimized due to their limited influence with regards to other types of information, national media in particular. So, rather than limiting the neighborhood effect to a simple circulation of information, Taylor and Johnston [TAY 79] prefer to place it in the vaster field of "political socialization". They briefly evoke in their conclusion a theory that they term "self-reproduction": "self-selection processes lead people to live among and act like those whom – on objective grounds such as census occupational classifications – they differ from". It pushes the individuals of a social class to search to develop strategies for living in an environment where the majority of the inhabitants are from a different and/or higher social class.

This "self-selection" allows us to reconsider the whole of the neighborhood effect project. From its starting premise, the measure of neighborhood effect poses a fundamental problem. It consists of observing whether or not two individuals belonging to the same social class, in two different geographical contexts, possess similar behaviors. Still, differences can exist within a social class, and the geographical context could influence the position within this class, most often measured at the place of settlement. Going but a little further, we can claim that the settlement location emanates from an economic constraint and/or a free decision. In this case, to take a simple example, workers from bourgeois neighborhoods will be richer, will have accumulated more property than those from working class neighborhoods and/or will make the deliberate choice to live in a place where they will be the minority, which implies that they accept to merge with the values of the

---

13 Taylor and Johnston devote nearly 50 pages to the "neighborhood effect" in their manual. They rely at the same time on ecological data and surveys from the UK and the USA to show that within a space of aggregation, the force of the electoral results in favor of the majority social structure is very much the general rule. These reports could have been verified in turn in France, notably through the social structure model [BUS 98] [GIR 00].

majority. In both cases, the neighborhood effect is not proceeded by a more or less constrained or organized domination of the majority towards the minority, but an initial congruence between personal values, an economic status and the political perception of a location. We are no longer content with explaining a social behavior by a place of residence (determinism), but we recognize in the system that the choice of housing location is, in part, explained by social behavior (self-organization).

Girault [GIR 00] comes to the same conclusion, simultaneously studying urban ecological data on a fine scale, qualitative data of spatial political strategies and contextualized survey data. However, this exploratory inversion of the ecological approach remains absent from the principle explanatory models of behavior, shown for example in political sociology manuals. We could therefore put forward the idea that this "self-reproduction" or theory of citizen-inhabitant-strategist is only realistically perceivable in an urban and peripheral context. The urbanization of society, just like the growth of mobility, as much residential as daily, tends to give a growing importance to this theory. It should integrate itself completely in the current modeling of the structuring of urban segregations. This theory of self-reproduction is not however that new, since it partly takes its inspiration from the model of "voting with feet" developed by Tiebout since 1956: the localization of actors in space depends on the relation between the quality of local services and the local fiscal cost of these services. This model is principally applied in urban spaces where there are differentiated municipal policies (Tiebout's model was mainly tested in Los Angeles) [TIE 56].

The preceding developments explain why the rise in power of the individual-actor do not imply the "end of territories", and therefore an excess of spatial modeling, but on the contrary a rise in *contractualism* that supposes a collective agreement in a location, at the crossing of the ascendant and of the descendant. In a sense that is no longer represented, the map can therefore be presented as the new "social contract" [GUI 00]. Sometimes, the territory continues to impose itself on the individual, for example when it deals with accessing public equipment (scholar or sanitary for example), tax collecting or voting. It is logically these topographical constraints that explain, in part, the spatial strategies of individuals controlling their mobility. With the rising amplitude of spatial capitals between individuals, the differentiated strategies of bypassing this "formal social mixing" strengthen, and with them segregative logics. To curb them, the State, the district or the agglomeration have not found to date any other solution than to attempt to impose a formal social mixing.

My intention is to open the way for new forms of modeling, turned more towards the individual-agents and simulations of their possible actions, by integrating reflexive effects of the collective (whatever its scale) on the individual. So, the debate between qualitative and quantitative geographers could and should be bypassed. This could have been decisive and fundamental for the modernization of the discipline.[14] Today we can hardly claim that we should choose between investing time in quantitative methods of spatial analysis or in field work with agents. Technical progresses (and in the education of students) allow us to conjointly accomplish both. To take just one example, we now have access through GIS, by simple "macro-commands", to most methods of spatial analysis that were new just 15 years ago, even if such progresses does not prevent us from imagining new methods. For example, Ron Johnston in a recent article for the *Political Geography* review constructed a particularly new methodology: around the place of settlement where a major individual survey took place, the author found a "buffer" allowing him to know the socio-economic environment of each person surveyed. This allowed him to build bridges between two traditionally opposed approaches [JOH 04].

If we witness a major evolution for geography in the coming years, I hope that it is one that allows us to continue ranking in two opposing categories those who measure and those who estimate, those who work with aggregates and those who work with individuals, those who isolate residues and those who establish typologies, those who believe in "pure geography" and those who "tackle social issues", those who work on the tyranny of spatial laws and those who work on the chaos of agent freedom. We must, I think, hold both ends of the ball of string if we hope to untangle it, or at least evaluate with pertinence the whole organization of the system. The modeling of the "micro" scale of local individual actors allows this. Such an evolution could allow researchers to avoid claiming, whatever the sophistication of their reasoning that the explanation of a phenomenon is found in the "culture" of the inhabitants. There is in this attitude, to say the least, an explanatory idleness. The work of the researcher is exactly, whatever the methods, to take apart the system and look to explain why the actors from a certain place behave differently than elsewhere.

Individual conviction is often incapable of perceiving collective and/or spatial logics that structure this very conviction.[15] For this reason, the recognition of the individual as an agent justifies that the only possible scientific horizon is the individual approach, since we come to take refuge behind their "personality" in the same way that we take refuge behind group "culture". It is towards a plurality of

14 I consider myself as an heir having greatly benefited from the actions of the pioneers of this revolution.
15 It is possible to do the test of stopping to read this paragraph for a few moments and trying to remember for whom you voted during the last elections, and then look to rank the series of causes that explain this vote. There is little chance that this exercise will lead each person to a simple, unique answer, but on the contrary a great chance that it will assign the causalities in series, in which it is nearly impossible to know which aspects reflect our individual freedom and which those of external factors.

methods and models that we must concentrate; geographers can privilege, without shutting themselves in, the methods they can control, to know the ecological approach and spatial modeling.

Theories of complexity and self-organization can be privileged. As we mentioned earlier, this principally concerns liberal models that for the moment often marginalize Keynesian corrective models. It is possible to study these "self-organized" logics to expose drifts, notably in terms of socio-spatial segregation. A less contemporarily common method in geography, although fascinating for the future, is to integrate hypotheses stemming from the theory of cooperative games into spatial models; a technique widely used in other disciplines, but still very little in geography, notably in French geography.

Economic sciences and sociology of organizations have popularized these theories of cooperative games (Nash's famous equilibrium). The major objective of these theories remains the modeling or the simulation of individual (the agents) behaviors, which according to rules adopts a rational attitude aiming to optimize gains. However, the particularity of the "games" insists on the uncertain character of the result, and the possibility for the agents to adopt differentiated strategies. The principle contribution of games theories is to have shown the interest of cooperation in a strategy of maximization of gains for an individual ("tit for tat").[16]

We must now determine the relation between these theories of cooperative games and geography. They can appear weak since they break away from determinist models that are often spatial. However, these game theories integrate directly the retroactive effects between individual strategies and collective effects. In this sense, they help us to understand how the sum of individual acts leads effectively to differentiated spatial organizations, whether dealing with segregation [SCH 80] or polarization [VEL 97]. Yet the theory of cooperative games integrates a supplementary spatial dimension: several authors have shown that the cooperation between agents is more probable, more stable and therefore more efficient when

---

16 As can be seen from the works of Axelrod who underlines the efficiency of the "cooperation-reciprocity-pardon" [AXE 84].

they operated in the frame of similar and stable territories [AXE 97][17]. In other words, these theories of cooperative games give an interest back to the territory, not through a descending and planned organization, but through the self-organization of individual acts. The example of democracy, which can be presented as the most refined of cooperative games, is significant for geography. The institution over the past few decades of free democracy as a major mode of political organization (as well as economical) throughout the world does not produce an "aspatial" world, any more than a geopolitical model based on the impermeability of borders, the distance to resources and the central position of fortified towns does.

7.5. Bibliography

[ASC 95] ASCHER F., *Metapolis ou l'Avenir des villes*, Odile Jacob, 345 p., 1995.

[AXE 84] AXELROD R., *The Evolution of Cooperation*. Basic Book, N.Y. *Donnant-Donnant- Théorie du comportement coopératif*, Odile Jacob, Paris 1992, 1984.

[AXE 97] AXELROD R., *The Complexity of Cooperation: Agent-based Models of Competition and Collaboration*, Princeton University Press, 1997.

[BAD 01] BADARIOTTI D., "Modelising the urban residential mobility Som Bogota: a cellular automata and a multi-agent prototype", *12th Colloque européen de géographie quantitative et théorique*, Saint-Valery-en-Caux, 2001.

[BUS 98] BUSSI M., *Eléments de géographie électorale. L'exemple de la France de l'Ouest*, Presses Universitaires de Rouen, 1998.

[BUS 02] BUSSI M., "Les faux-semblants de la mixité sociale urbaine, in repenser les politiques de mixité sociale", *Pouvoirs locaux – les cahiers de la décentralisation*, no. 54, September 2002.

[BUS 03] BUSSI M., LANGLOIS P., "The organisation of the electoral behaviours: a spatial and socio-economic model", *13th European Colloquium on Theoretical and Quantitative Geography*, Lucca, Italy, 2003.

[CHA 92] CHAUVET A., *Approche géographique du développement local, in Actions et recherches sociales,* "Le développement local", no. 1, p. 31-41, 1992.

---

17 "The basic assumption is that the opportunity for interaction and convergence is proportional to the number of features that two neighbors already share. Stable cultural differences emerge as regions develop in which everyone shares the same culture, but have nothing in common with the culture of neighboring regions…In the near future, electronic communications will allow us to develop patterns of interaction that are chosen rather than imposed by geography. If individuals are linked together at random, we could expect substantial convergence over time. In the more likely case that interactions will be based on self-selection, people will tend to interact with others who are already quite similar to them on relevant dimensions. An implication of the model is that such self-selection could result in an even stronger tendency toward both 'local' convergence and global polarization."

[CLA 79] CLAVAL P., *Espace et pouvoir*, PUF, 224 p., 1979.

[COX 69] COX K.R., "The voting decision in a spatial context", *Progress in Geography*, no. 1, 1969.

[DAR 99] DARGENT C, "La notion de culture politique régionale est-elle pertinente aujourd'hui?", in *Le vote incertain, les élections régionales de 1998*, P. Perrineau and D. Reynié (eds.), 1999.

[DAT 00] DATAR., (2000), Aménager la France en 2020 – Mettre les territoires en mouvement, La Documentation Française, p.87.

[DAV 97] DAVIES M., (1997), *City of Quartz*, 1997.

[DOL 77] DOLEZ B., LAURENT A., "Trente ans d'élections présidentielles françaises: les dynamiques territoriales", *Revue Internationale de Politique Comparée*, vol. 3, no. 3, 1997.

[DEF 00] DEFFONTAINES J.P., PROD'HOMME J.P., *Territoires et acteurs du développement local – de nouveaux lieux de démocratie*, Editions de l'Aube, 2000.

[FRE 99] FREMONT A., *La region, espace vécu,* Flammarion, Paris, 1999.

[GIR 00] GIRAULT F., Le vote comme expression territoriale des citadins. Contribution à l'étude des ségrégations urbaines, PhD thesis, University of Rouen, under the direction of Y. Guermond, 2000.

[GLO 96] GLON E., CODRON V., GONIN P, GREGORIS M.T., RENARD J.P., "Le développement local au service de la géographie?", in *Bulletin de l'association de géographes français*, "la géographie au service du développement local", 1996.

[GRA 00] GRAFMAYER Y., *Sociologie urbaine*, Nathan University, p.128, 2000.

[GRA 97] GRASLAND C., *Contribution à l'analyse géographique des maillages territoriaux*, research report, University of Paris 1, 1997.

[GRE 84] GREFFE X., "Territoires en France. Les enjeux économiques de la decentralization", *Economica*, Paris, 1984.

[GUI 00] GUIGOU J.L., *La mutation silencieuse des territoires - 1990-1999*, Ministère de l'aménagement du territoire et de l'environnement, DATAR, 2000.

[HAS 89] HASTINGS M, "Les démiurges de l'introspection cartographique", *Politix*, no. 5, 1989.

[JOB 98] JOBRET A., "L'aménagement en politique ou ce que le syndrome NIMBY nous dit de l'intérêt general", *Politix, définir l'intérêt général*, no. 42, 1998.

[JOH 04] JOHNSTON R, JONES K, SARKER R, PROPPER C, BURGESS S, BOLSTER A, "Party support and the neighbourhood effect: spatial polarisation of the British electorate 1991-2001", *Political Geography*, no. 23, 2004.

[KIN 96] KING G., "Why context should not count", in *Political Geography*, vol. 15, no. 2, 1996.

[LAN 74] LANCELOT A. "Sociologie électorale", in *Encyclopédie Universalis*, 1974.

[LAU 99] LAURENT A, BRECHON P, PERRINEAU P., *Les cultures politiques des français*, Presses de Sciences Po, 1999.

[LEC 86] LECA J. (ed.), *Sur l'individualisme*, Paris, Presses de Sciences Po, 1986.

[LEM 96a] LE MOIGNE P., "La faute au faubourg? Le recours aux médicaments psychotropes en milieu urbain", *Les Annales de la Recherche Urbaine*, no. 73, 1996.

[LEM 96b] LE MOIGNE P., "Les territoires de consommation des médicaments psychotropes", *Courrier du CNRS*, no. 82, 1996.

[LEV 94] LEVY J., *L'espace légitime. Sur la dimension spatiale de la fonction politique*, Presses de Science Po, 1994.

[LEV 99] LEVY J., *L'espace et le politique enfin réconciliés?*, in *Géographie et liberté, mélanges en hommage à Paul Claval,* (ed.) J.R. PITTE and A.L. SANGUIN, l'Harmattan, 1999.

[LEV 00] LEVY J., "Aménagement, fin et suite: l'Etat, l'Europe, la société et leurs géographie", in WACHTER S (ed.), *Repenser le territoire – Un dictionnaire critique*, Editions de l'Aube, 2000.

[LEW 01] LEWIS-BECK M., "Modèles d'explication du vote" in *Dictionnaire du vote*, (ed.) P. PERRINEAU and D. REYNIE, 2001.

[LON 00] LONGHI C, SPINDLER J., *Le développement local*, Librairie Générale de Droit, 2000.

[MIC 77] MICHELAT G., SIMON M., *Classe, religion et comportement politique*, Presses de Science Po, 1977.

[OTA 97] OTAYEK R., "Démocratie, culture politique, sociétés plurales: une approche comparative à partir de situations africaines", *Revue Française de Science Politique*, December, 1997.

[OTA 00] OTAYEK R., *Identité et démocratie dans un monde global*, Presses de Sciences Po, 2000.

[PIZ 86] PIZZORNO A., "Sur la rationnalité du choix démocratique", in LECA J., (ed.), *Sur l'individualisme*, Paris, Presses de Science Po, 1986.

[PUT 93] PUTNAM R., *Making Democracy Work: Civic Traditions in Modern Italy*, Princeton University Press, 1993.

[RHE 94] RHEIN C., "La ségrégation et ses mesures" in *La ségrégation dans la ville*, BRUN J., RHEIN C., (ed.), 1994.

[ROL 99] ROLLAND-MAY C., "Fuzziland, modèle de détermination et d'évaluation de territoire de cohérence. Application au pays et au département de Moselle", in NORIOS, *Pays et développement local*, no. 181, vol. 46, 1999-1, p. 39-82.

[ROL 00] ROLLAND-MAY C., *Evaluation des territoires – concepts, modèle, méthodes*, Hermes Science, 2000.

[RON 90] RONCAYOLO M., *La ville et ses territoires*, Gallimard, 1990.

[SCH 80] SCHELLING T., *Micromotives and Macrobehavior*, Paris, PUF, 1980.

[SIE 13] SIEGFRIED A., *Tableau politique de la France de l'Ouest*, A. Colin, 1913.

[SOR 93] SORBETS C., "Le mot et la chose", in Mabileau A (ed.), *A la recherche du local, actes du colloque le local dans tous ses Etats*, L'Harmattan, p. 29-40, 1993.

[TAY 79] TAYLOR J., JOHNSTON R.J., *Geography of Elections*, Harmondsworth, Penguin Books, Series "Geography and Environmental Studies", 1979.

[TIE 56] TIEBOUT C.M., "A pure theory of local expenditures", *Journal of Political Economy*, vol. 64, no. 5, 1956.

[TOC 39] TOCQUEVILLE A de., *De la démocratie en Amérique*, libr. de Médicis, 1839.

[VEL 97] VELTZ P., *Mondialisation villes et territoires*, PUF, 1997.

[VID 14] VIDAL DE LA BLACHE P., "Tableau politique de la France de l'Ouest", *Annales de Géographie*, 1914.

[WAC 00] WACHTER S., "L'agenda de l'aménagement du territoire entre prospective et rétrospective (1980, 1990, 2000)", in WACHTER S (ed.), *Repenser le territoire – Un dictionnaire critique*, Editions de l'Aube, 2000.

Chapter 8

Temporalities and Modeling of Regional Dynamics: The Case of the European Union

The expectations and uses of modeling in geography have evolved over time. The polysemy of the word model, the variety of scientific uses pertaining to this notion and the diversity of ways in which models are put together has seen much change in scientific practices relating to this term. In this chapter we will argue from the perspective of systemic modeling. This on the one hand means accepting the theoretical posture of Claude Bernard's maxim: "*systems are not in man's nature, but in man's soul*", while on the other hand accepting that this approach is distinct from Cartesian analytical precepts. Once this frame has been stated, we can move on to the definition of modeling processes according to J.L. Le Moigne [LEM 95]: "*modeling is about identifying and formulating problems and attempting to solve them through simulation.*" While the idea of simulating reality to better analyze or predict the future is not new, the conditions in which these simulations are realized and the objectives assigned to these simulations have changed considerably. Simulation[1] no longer aims to determine, by prolonging current trends, what reality will be, or what reality would be if initial conditions were different, by changing certain parameters.

From an application for normative ends or optimization (finding the optimal location or *ex post* analyzing to compare what exists to a norm produced by a model), a new way would be to propose some likely scenarios, depending on the organizational logic specific to certain variables.

Chapter written by Bernard ELISSALDE.
1 For the section on "simulation" in this chapter, see Chapters 12 and 13.

Ways of thinking tend to be unpredictable and improbable, not by intellectual gratuity, but rather because this or that trajectory, once seen as unimaginable according to determinist models, could one day become reality. Modeling today is also making the presence of the temporal dimension of reasoning explicit. Introducing time is in fact about anticipating the near or distant future through the elaboration of plausible scenarios of evolution. Simulation and prediction do not attempt to find a unique solution to a problem but, through multiple responses, look to integrate uncertainty and propose a range of credible options in a world which is constantly changing. Modeling in geography applies therefore to a reality which is not static but evolutionary and to a reality which is not made up of isolated and fixed units but according to interactive relationships. Despite belonging to an interdependent body like the European Union (EU), each one of the spatial units follows a trajectory specific to them, where differences in reaction time are observable. The pace of transformation varies strongly from one region to the next. All of these differing time rates will together produce a regional European mosaic making a mechanic relationship improbable between the directives of the Union's regional policy (structural funds and cohesion funds) and the initially imagined rebalancing. Beyond the traditional debate on the effectiveness of the European Commission's regional policy, we will consider European spatial policies as a system of governance at multiple levels (*multilevel governance*) [MAR 92, HOO 96] and non-linear evolution, producing a partially self-organized system. The EU paradox has emerged from the fact that although each actor, organization and institution is programmed to realize fixed objectives (gradually changing over time), the final result corresponds neither to the dominant objectives of a single actor, nor to the smallest denominator common to them all.

8.1. Integrating time and temporalities into spatial models

8.1.1. *A renewed approach to time*

It was not among the objectives of this chapter to outline the relationships between spatial modeling and the question of time in geography [ELI 00]. A large number of spatial models explicitly or implicitly integrate the temporal factor. In view of moving beyond purely statistical and chrono-determined spatial models, recent research has altered the thinking on temporal categories in geography [PUM 98, DUR 01]. Contrary to a long-held belief in social sciences, where structural research entailed the elimination of each temporal category, this research revealed that spatial structures had a historicity [GRA 96]. We have been able to show [ELI 99] that spatial structures, as resilient objects, were animated by varying temporal rhythms, whose trajectories do not necessarily follow the same path through time.

Paradoxically it is at the point where the present, the ephemeral or the "end of history" seems to triumph that geographers become involved in the multiplicity of possible outcomes: not to transform geography into a forecasting discipline, but to guide decisions being taken today. Concerning ourselves with the question of whether *"geography will be able to invent the future?"* (1998), Pumain proposes to reverse this perspective, relating time and simulation. *"By allowing an exploration of a range of possible outcomes"*, he attempts to elaborate an evolutionary theory of spatial entities which should enable us to study *"in a nomothetic way, changes in geographic structures"* by looking at processes which *"predict and become part of the geographic field"*.

This field of research makes it possible to integrate social time, upon which spatial analysis works, and to go beyond notions of uniqueness and non-reproduction of events in their historical approach, for accepting that geographical phenomena could be shaped as sequential events, according to the steps of time adapted to the studied objects.

This perspective has opened a field of research which made it possible to relate the interaction between innovations, even technological revolutions, and transformations in the organization of the European space. As a number of authors have shown [JUI 72, JAN 69; BRE 99, OLL 00], because of the increasing rates of transfer over time, there is a shortening of relative distances and a widening of initial interactions. Then we notice a clear enlargement in how spatial cells operate in the spatial system, beginning with parishes or boroughs, then districts, and finally regions. The changing spatial size of the reference units and parameters which facilitate the movement of spatial interactions should therefore be integrated into the models' operation.

Using the spatio-temporal contraction, Janelle [JAN 69] has tried to evaluate the pattern through which two places "merge". He relates the gains in transport time between two dates on a given journey, and the number of years taken to obtain the gain (following technological innovation or changes to infrastructure). Behind this idea of functional distance, accounting for the contraction of space through the effects of increasing rates of transfer, two temporal consequences have occurred. The first has been the well-known transformation over a period of a few years of the accessibility of cities, which can be measured by calculations plotted on graphs. The second enables us to observe over a longer period, how changes in accessibility affect the urban hierarchy and the relative positions of one city in relation to another.

8.1.2. *Temporalities and complex systems*

The ways in which systemic procedures apprehend time has greatly evolved since the Forrester method was applied to the theory of self-organization developed by the Brussels School.

Barraqué has criticized the Forrester method's "system dynamics", condemning its lack of flexibility and for making the possibilities of evolution too rigid. The idea of determining the evolution of the system's variables at each time interval would have meant focusing on the strategies of each actor in the short term and omitting each regulator in the long term. The addition of choice in the short term may lead to catastrophic scenarios because their impact is only perceptible in the very long term.

When looking at the first systemic models, a contradiction was possible between fixed rules and the absence of space reserved for unpredictable events. This locking would confer a strong rigidity against externally imposed transformations and would have weakened their heuristic power. [BAR 02] rightfully asserts that the components would be nothing more than "puppets of a systemic destiny" and contribute to the elimination of history.

The determinist nature of these first systemic applications, in the field of global evolution as well as in the relationship between elements, has been challenged for its lack of aptitude and failure to take into account the unpredictable nature of human behavior. Learning from this first generation, we try today to integrate uncertainty into our models. Simply because uncertainty does not come from improbable chance or from unchecked disturbances but can emerge from a number of processes even in fully determinist systems (determinist chaos theory) [DAN 03; KIE 97].

Work around the application of chaos theory in social sciences [KIE 97] and on self-organized spatial systems relies on the inclusion of chaos theory through its capacity to integrate the temporal variable not in its retrospective dimension, but through the unpredictability of human behavior: "*chaos theory appears to provide a means for understanding and examining many of the uncertainties, non-linearities and unpredictable aspects of social systems' behavior.*" There are two aspects worth highlighting in this procedure. On the one hand, it invalidates the traditional criticism of inaptitude, on account of determinism, for the theories coming from natural sciences in their application to systems integrating human behavior. On the other hand, it opens up possibilities for the social sciences, with the exception of history, for using temporal series (economics, sociology, geography) in their attempts to identify how non-linear behavior and changes appear at given times.

A chaotic or self-organized system is different from a mechanical system because at a certain level or geographic scale, its situation does not correspond to the initial objectives for which it was programmed. It continually evolves according to interactions between actors and between the different levels of operation.

This type of system does not incline towards a unique and stable state; its complex behavior has several attractors, and several directions towards which it could incline. The theory of determinist chaos tries therefore to explain discontinued and complex behaviors. In this theoretical frame "bifurcations are the product of the internal system processes being investigated. The self-regulations and non-linearities create bifurcations by increasing the internal fluctuations" [DAU 03].

Through its association with time, this theory has the advantage of accounting for the changes and transformations of a structure or for the emergence of a new configuration. In the relationship between temporalities and geography, this point refers to the appearance and identification of novelties, in an unperformed way through a configuration or spatial organization.

The link with time enables us to date these alternations of linear and non-linear dynamics. The fact that in the self-organized system, the evolution is non-linear, means that the processes affecting each component used to build the macroscopic structure changes over time. Besides the variations in the respective weight of the variables, the process can modify and readjust the structure of the totality when a few thresholds are reached. The self-organized systems manifest a degree of unpredictability in their behavior, and how they evolve is still not known.

"The instability of dynamic systems and the fluctuations which characterize them, mean that it is impossible to prepare initial conditions which would produce similar outcomes. By this approach, unpredictability is put forward *a priori* as a theory. However, analyzing the dynamic behavior of systems, their sensitivity to the theoretical value of parameters, makes it possible to explore a limited number of future outcomes and configurations towards which the system is likely to incline, due to assumptions on the evaluation of the parameters...The order observed or in other words the self-organization of the system, emerges from the continuous fluctuations in the reactions between elements. The observed configuration is but one of the many possible configurations starting from the interplay of these interactions. It is the result of bifurcations appearing in the lifecycle, which have left an indelible mark" [PUM 89].

This quote is used to support the idea of there being "a limited number of possible outcomes". This means that the unpredictability of complex systems does not emerge in a situation of absolute uncertainty where any outcome may occur. On the contrary, with any spatial system, only a few scenarios are possible at a given moment, and the whole modeling activity consists of recreating the processes through which time transforms the range of possible outcomes into a single past [LES 85].

Although past schemes allowed nothing other than the prolongation of current trends, the theory of complex systems does account for the possible occurrence of bifurcations. These systems are characterized by the alternate periods of linear evolution during which the system undergoes a relative continuity, and others non-linear patterns. In the first phase some longer range correlations in time and space may appear among the system's components, contributing to the creation of strong, identifiable spatial structures. In the second phase, the changing relationship between some components may engage in a process which modifies the entire structure.

Every external impulse and internal inflection is nevertheless not a source of bifurcation. Self-organization always refers to processes which engender or promote the system organization and therefore foster a tendency towards temporal stability. (see spatial system durability in D. Pumain and S. Van der Leeuw) [ARC 98].

Resilience seems to be one of the major characteristics of spatially self-organized systems, such as cities or other strong spatial structures like the "European dorsal", for example.

As a way of establishing a relationship between modeling and system dynamics, [DAU 03] has made an important amendment to attempts made to assimilate the complexity of a spatial system to the observation of all temporal irregularities. The concept of bifurcation *"has nothing to do with a rupture in a temporal curve…during a bifurcation the system modifies its trajectory and heads towards a new attraction, it is effectively changing state"*. We will later look at the application of this amendment by examining the relationship between changes to the EU's regional policy and the dynamics of regional inequalities.

8.1.3. *A necessary introduction of polytemporality into modeling*

Thanks to the contributions of philosophers like Herder in the 18th century and the more recent written accounts of Elias, we know that there are as many times as social objects, because of the co-evolution of the different components of the system or model and the problems of different rates and speeds of evolution. However, a problem emerges in geography when space, or parts of that space, run at transformation speeds belonging to a long term category (population systems, spatial structures).

Spatial system dynamics offers several kinds of temporalities. One empirical example of conjunction between differentiated temporalities, and of unforeseen consequences on the organization of European space, is the flow of finance, goods and people from the UK to continental Europe and particularly western France. At the origin there is an ongoing attraction of Britons towards the south of England and the Thames basin. As a result of the meridian tropism there has been a rise in land value in the affected parts of the British Isles, which also coincides with the fact that the region is known for tourism and recreation. The price rise in these parts of the UK provokes, in turn, a movement, assisted by European integration and construction, towards Normandy and Brittany of an ever increasing number of Britons (for holiday homes and retirement purposes), and this is not even limited to people living close to the channel. A last temporality is linked to British policy with respect to the EU (Euro, Schengen, etc.), which is added on to the first two, and which acts, according to the period, as either a break or an accelerator to continental integration.

Like other complex systems, natural systems show some relationship to time, marked by the irreversibility of the phenomena. This irreversible and continually evolving time means that even if spatial structures remain stable for a period, spatial systems continue evolving, despite ruptures, and frequent reproductions of cyclical systems. We see scales or time intervals covering expansive periods whose occurrence and consequences frequently surpass human generations. The example of debates around the consequences of climate change illustrates quite well two conceptions of the relationship between modeling and time. Even if climate specialists largely agree to accept the principle of terrestrial climate change, they will disagree on its consequences.[2]

On the one hand there is a "cyclical" approach in which current events feature in a long history of terrestrial temperature variation together with predictable consequences, because they are reproducing what has happened already (ice melting, rising sea levels).

On the other hand there is the approach centered on atmospheric composition, which shows that, besides the global warming observed, the simulations executed with modifications to atmospheric gasses will induce, across an unknown time horizon, the complete overhaul of atmospheric circulation and thus a change of terrestrial climate towards a new and unknown state.

Applying a model to this conjunction of a reiterative time and temporally diachronic and irreversible transformation is understandably difficult in social

---

2 For more information on these debates, see the website of the French Festival of Geography: Acts of the Fig – Fig 2003 Water and Geography; http:// fig-st-die.education.fr.

systems [MOR 99]. There is the joint representation of doing and becoming, an organization by regulation and reproduction of the system, which is not fully identical. The most difficult problem in representing and moreover modeling this polytemporality comes from the continuous fluctuation of all the components (variables, spatial units) of a system and of all the different sizes.

The model by Allen has been used by the PARIS team as part of a study into the dynamics of intra-urban spaces[3]. This model is composed of non-linear differential equations which describe the variation by unit of time of the different variables (employment, population) in different zones of an urban area.

To simulate the evolution of each zone, the model considers its potential (based on the external demand and the number of activities practiced) and attractiveness. Through the aid of a mathematical formalization, this model will allow us to simulate the phenomena of concentration and saturation of activities and people, including threshold effects and also the possibilities of bifurcation.

The aim is to simulate, by taking into account the initial conditions in each zone, the evolutionary trends of the different parts of an urban area. This model has attracted the interest of a number of researchers for its use as an experiment in dynamic modeling, integrating the temporal dimension across different "time periods". It also incorporates interdependence in the evolution of different zones, since the potential variation of employment or population in a given zone includes, at each stage, the relative position of a spatial unit in relation to all other zones in the urban area.

8.2. Introduction of complexity theory in the interpretation of regional inequalities in Europe

Can we conceive of the regions of the EU as a complex system? This conception of European space is based on possible analogies with other European themes, and particularly on the classical analysis, conducted by some political researchers into the ways in which the EU is changing and the ways in which negotiations between Member States are being organized (C. Lequesne and A. Smith). The political decisions of European summits, like the policies implemented by the Commission, could not be enacted as a pre-planned scenario. Looking beyond the supposed existence of a European project to the results and potential hurdles which are likely to occur, we must consider them as neither the triumph of a single dominant actor, nor as the minimal result corresponding to the smallest common denominator among Member States. Some incidental strategic alliances allow certain governmental

---

3 For more information on the presentation of this research see [PUM 89].

actors to influence the decision-making process whatever the size of the country. Beyond the strict institutional definition, the real activity of community policy is to foster "multi-level governance" [MAR 92, HOO 96], including complex relations between multiple actors (community institutions, territorial groupings at various levels, professional bodies, various social actors, etc) rather than definitively hierarchical power relations.

Further, by looking at the CAP we find support for the idea of a complex European system. The first mechanisms implemented to assist the European Agricultural Guidance and Guarantee Fund in 1962, were intended to reduce the agricultural deficit of the common market. This policy attracted so much positive feedback, that by the 1980s it was causing overproduction. Hence, in 1992, multiple reforms and adjustments of the agricultural policy were introduced, which have succeeded in challenging a significant part of the production-led principles and initial protectionist mechanisms. In sum, integration and deepening of European policy is related to a mixture of incremental adaptations guided by the economic situation, and by major ruptures at other periods.

For these reasons we can look to R. Geyer [GEY 03], in order to explore how the EU may be said to contain a number of specific characteristics of a complex system:

– it is made up of a large number of elements, which together form a whole with relative coherence (commercial integration of member countries, community policy);

– there are multiple interactions between these elements, occurring at different levels, according to variable perimeters, generating feedback loops;

– it is opening up to a larger environment (e.g. PSEM, ACP countries, WTO, etc.) but still possesses a certain robustness during times of difficulty (financial crises, international relations, conflicts), originating from this environment;

– the dynamic of these different structures is both linear and non-linear.

To appreciate the complexity of the EU's dynamics in this way forces us to consider the exchange between different decision making levels and the different temporalities which engender them. For as long as it seems certain that in the short term, base structures (institutions, positions of each Member State, etc) will not radically change, the more it seems unlikely that we may predict the results of certain community policies in the mid-term as well as the overall evolution of the EU in the long term (with or without additional enlargements). This observation is only one aspect of the paradoxical functioning of the EU. This uncertainty cannot be compared to an empire in decline, or more traditionally to the biggest or smallest success of a public policy. This inability to predict arises from both the incremental nature of European construction and from the growing number of potential interactions between the elements which together make up the system.

In a study on competition between the EU regions, it would be wrong to assume that the territories are constantly competing with each other. Not only do territorial solidarities exist within each state but the EU provides structural funds to eligible regions, particularly through the intermediary of the European Social Fund and the European Investment Bank. A study of the relative positions of European regions in a competitive framework must take into account both the variations of financial support over time, and also the reforms undertaken for eligible zones in the 2000-2006 period.

In the case of regional policy and relations between European regions, the above mentioned specificities oblige us to consider the consequences of change in the modalities of public intervention and the diversity of actors. Following the creation of the ERDF in 1975 the rules of redistribution and the allocated amounts have evolved and the funds have not gone to the same regions throughout the period. The reforms of 1984, then 1988, adopted for the requirements of social and economic cohesion, have altered the size of the allocated budget: twice between 1994-1999 in relation to the 1989-1993 period. Cohesion funds for four countries (Spain, Portugal, Greece and Ireland) were introduced in 1994 to compensate the budgetary efforts of these countries in view of convergence towards the criteria for participating in the EMU. These "cohesion countries" received aid which accounted for a substantial amount of their GDP; they have been added to the structural funds according to a logic which sought to reinforce an overall policy of cohesion among Member States.

State	1989-93 % of GDP	1994-99 % of GDP
Portugal	3.07	3.98
Greece	2.65	3.67
Ireland	2.66	2.82
Spain	0.75	1.74
Italy	0.27	0.42
UK	0.13	0.25
France	0.14	0.22
Germany	0.13	0.21
Denmark	0.08	0.11
EU average	0.29	0.45

Table 8.1. *Importance of structural funds in the State's GDP*
Source: European Commission

A number of works have evaluated the impact of community policy on reducing regional disparities in the EU [RIO 02, CHA 03]. They agree that the general philosophy of these aids, especially since the Delors report, is to assist the competitive positions of the poorest regions in order to help them meet the competitive standards needed to function within an integrated market. Given that this support has, over the course of the last two decades, been principally used in the development of transport infrastructure, this has significantly altered the potential mobility and the real flow of goods and people across Europe's regions (Elissade, Langlois, 2004).

Is it possible to identify a clear relationship between regional policy and progress among the laggard regions, or are the reductions or widening disparities exclusively due to EU policy? Besides evaluating the positive effects of GD REGIO structural funds, it is necessary to recall a number of budgetary realities. In the EU budget, which is at a level below national budgets[4], the structural funds have never exceeded 0.20% or 0.35% of the total GDP of the EU. We therefore have a right to question the redistributive capacity of EU funds when they are separated from the budgetary shares of Member States. The modifications of the rules managing the interactions between regions, and between some regions and the remaining part of the EU, must be taken into account. The need to nuance the overall diagnostic on the rates of convergence or on the maintenance of disparities is explained not only by the fact that the aid system has considerably changed since the 1988 reforms (ranking the "objectives") but also by the fact that the economic situation of each country and region has never stopped changing. The self-organized nature of unequal regional dynamics must be addressed again in view of the overall results. The evaluation of regional development policies cannot be exclusively based on correlations between aid and GDP/capita variations, it needs to include the question of the reasons for the differentiated receptivity of funds by one region, compared to another, as well as the share of other factors in regional dynamics: national policies or fiscal status, as exemplified through Ireland's spectacular rise.

As we address the issue of redistribution policy, an important distinction must be made according to the origins of the programs, between national initiatives which account for the majority of the funds, and community initiatives (INTEREG, EQUAL, LEADER, URBAN) which as their acronyms suggest, are intended to serve strictly delimited objectives. Taking into account the other institutional actors such as large European metropolises and the networks created by territorial groupings like CPMR (conference of peripheral maritime regions) which together

---

4 The EU's budget represents only 1.27% of the EU's GDP, against 15% of the American Federal Budget and 20% for the majority of large European countries (M. Basle in G. Baudelle and C. Guy, *Le projet européen,* PUR, 2004).

aim to influence the EC's decisions, it is difficult to determine the weighting of these various influences.

8.2.1. *The European Union: regional convergences or divergences?*

Successive studies conducted by the EU or INSEE[5] all point towards the same diagnostic: wealth created in Europe is highly concentrated. Furthermore, the employment market in Europe is characterized by major regional imbalances. One-third of European countries produce two-third's of the EU's GDP, while a quarter of Europe's population live in regions where the GDP per capita is lower than 75% of the EU average. Moreover, in six of the seven extreme-peripheral regions, GDP per capita is only half of the European average. Regions with prominent geographic features: mountains, peripheral and maritime locations, including islands, suffer from both structural problems as well as isolation from the main urban centers which together hinder their competitiveness.

Conversely, metropolitan regions are among the most important contributors of GDP. The Île de France (Paris region) with 5% of the EU's GDP and 3% of its population, is situated in the first rank, ahead of Lombardy (Milan). Among Europe's capital regions, Berlin is an exceptional case, holding 8th position in Germany, falling behind the metropoles of Munich, Düsseldorf, Frankfurt, etc., which partially confirms its status in Europe as within Germany, as the new capital of a reunified state. The evolution of "metropolization"[6], concerning functions as well as overproduction in relation to national and EU wealth, has never declined in two decades. This overall polarization of EU space supports a number of theoretical hypotheses on the positive exchange between the effects of concentration (positive externalities) and growth.

Thinking about trends of convergence and disparities brings us back to evaluations of redistribution policy. For a number of years studies have tried to evaluate the consequences of deepening EU economic integration on regions, as well as the effectiveness of EU structural funds.[7] All agree to identify phases of catch-up in the so-called cohesion countries (Ireland, Spain, Portugal, Greece) from 1991 and the regions eligible to receive structural funds, notably Ireland, which reached the European average in 1997. There is an alternation in these phases of catch-up and less definable stages. Trends of convergence are apparent at the regional level, with an observable increase in regional GDP between 1980 and 1995, which is inversely proportional to the levels before funding. This positive diagnosis is however tempered by the existence and deepening of regional disparities within

---

5 "Statistics in brief" no. 13/2001 and INSEE First no. 810, October, 2001.

6 J. Barrot, B. Elissade, G. Roques, Europe/Europes, ed. Vuibert, 1997.

7 Extracts from the CEPII magazine of July 1997 and March 1999.

countries, as has been seen with Spain, where the poorest regions are advancing more slowly than Catalonia. Taking the EU regions as a whole, the rate of growth among the richest regions remains higher than the EU average. According to Maurel [MAU 99], EU economic integration is expressed spatially by an unequally distributed and polarized growth:

"The eventual catch-up and convergence of national economies does not ignore the issue of geographic division of economic activities between European sub-regions…yet regions and states must not be conflated. European integration could promote a regional rather than national specialization. The gradual removal of national borders could potentially provide more support to an increasing regional specialization, which until now has been restricted by national policies. Regional specialization would then rely more on the geographic and economic situation of the region within the European Union".

The gap between the constantly reiterated objectives of solidarity and cohesion, and reality is highlighted in the text of the second report on economic cohesion. The text deplores the "very centralized" state of the Union, remarking that "the gap observed at the level of the south-west periphery of the Union, including the level of education, is not sufficient to ignore the center-periphery model, which should, on the contrary, become reinforced with the addition of central European countries".

Inequalities have also emerged from an unequal sector distribution of employment and infrastructure, particularly in transportation. Although the ratio of the industrial sector is not overtly discriminatory, the primary sector contains gaps of between 1 and 7, between the centers and the peripheries.

Thus, the spatial dynamics of the EU takes the shape of a self-organized system, where the reactivity of all the spatial units would act in an autonomous way with respect to the redistributive logic of the Commission, which seeks immediately positive results from the distribution of regional aids.

8.2.2. *Which interpretive models?*

The relevance of current spatial models (e.g. center/periphery) and of the methods applied for interpreting the European spatial dynamics, is an important focus of our work. Following the comment that the economic model upon which the EU is based is responsible for spatial inequalities, numerous studies have been carried out to identify the processes at the origin of those inequalities.

Following the preliminary steps taken by Krugman, the new geographic economy proposes to observe, amongst other things, the effects of economies of agglomeration on regional development and how they can oppose the intended objectives of convergence among regions.

The Krugman and Venables [KRU 95] model, concerning the inequalities between different countries, demonstrates firstly how opening borders generates a concentration of activity in larger centers. It shows that, in a second stage, the differences of productivity and wages should inevitably lead to outsourcing towards regions offering low wages. Does this model apply to the current processes operating within the EU? If there is really, as we shall see, a concentration of the higher GDP/capita in the central regions, the so-called cohesion countries (Spain, Portugal, Ireland) have received industrial investment from the rest of the EU. Does this phenomenon share certain characteristics with "product cycle" type diffusion?

Besides empirical evidence, a number of studies have tried to link changes in the relative position of regions to economic changes, particularly with the transition from Fordism to the regime of flexible and generalized accumulation. Rodriguez-Pose (*Economic Geography*, 1994) questions the way in which post-Fordist growth and the restructuring which it entails, have influenced regional territorial organization. The logic behind the rationale is based on economic transformations being the principle or even exclusive regulators of spatial organization.

His methodology is based on indicators of value-added GDP and employment in the 113 regions of the EU during the 1980-89 period, studied with linear regressions and correlations between econometric models of growth and independent variables.

The results suggest that nation states were, during that period, still relevant units for studying spatial restructuring. Calculations carried out on the regional average annual GDP growth indicate that it is more likely distributed according to national contexts than dispersed at random, thus reinforcing the dependence hypothesis referring to national economic situations. The results of this work indicate that throughout the 1980s, restructuring has not produced a trend towards a dual European space, and that the key sector as far as growth rates are concerned is still industry. The apparent resilience of spatial systems can be interpreted to mean that in many areas the characteristics of "Fordist" spatial organization is still evident and barely unaltered.

Other researchers like Ghio and Van Huffel[8] question the capacity of an infrastructural policy based on reduced transport costs as a solution to urban concentration or as a way of dispersing activities. They base this on the premise that

---

8 Ghio and Van Huffel (RERU, 2001, II).

in developing new infrastructure, the state has a role in reducing transport costs and invites firms to relocate. Yet, as in the case of regional GDP, the authors show that the results are unequal in terms of the spatial concentration of activities.

The economic studies that have been carried out provide crucial information on the dynamics and eventual motors of regional development. They do however fail to conceive of European territories as areas of interaction, through contiguity or networks. Through neighborhood effects, the mechanisms of trickle-down will be guaranteed. Boiscuvier [BOI 00] in her PhD entitled "Integration, convergence and cohesion of countries in the European economic space", has attempted to design a regional growth model which includes, after calibration, the variables of initial wealth, unemployment, spending on research, and the degrees of openness, industrialization and productivity. The model enables us to categorize the European regions into three groups: the 41 poorest regions, 34 intermediary and the 19 wealthiest regions. Despite how well her model appears to reflect reality, it omits the phenomenon of spatial interaction in regional growth. Each regional unit develops faster or slower according to the evolution of internal variables, which itself omits the neighborhood or contagion effects.

The models by Martin and Ottaviano (1999) and Baldwin, Martin and Ottaviano (2001) on the role played by local externalities in the creation of growth inequalities are based on the relationship between two regions and two factors. This simplification of spatial interactions, particularly the potential relations of contiguity, limits the number of possible evolutions to binary situations and prevents us from simulating complex reactions across a large number of spatial units.

8.2.3. *The evolution of regional inequalities in Europe*

Beyond the European Commission identifying disparities, through structural policies, by analyzing the relative positions of areas under the auspices of EU macroeconomic and budgetary values, we can evaluate the overall effectiveness of the policies adopted and provide prognostics on the shape and nature of territories integration into the Union.

Given the statistical data missing for a number of countries in the information compiled by the REGIO database (EUROSTAT), and the need to develop a complete series of measures, we have chosen a sample of 129 to 161 regions from the NUTS2 level on which to carry out our different calculations. Aside from the relative stability over time of the hierarchies and the relative positions of the different European regions, we can see a general widening of trajectories.

In order to respond to the question of regional convergences/divergences, and the question of when peripheral regions will catch-up, we have carried out a number of calculations over different dates, measuring the annual GDP per capita and examining the regions which do or do not receive structural funding. The results demonstrate that there is no trajectory common to all regions and that there is a clear diversity in rates of evolution. These two observations lead us to question the existence of a European economic climate at regional level.

When we plot annual GDP per capita variations against purchasing power on to a Cartesian diagram at two successive periods (1991-95 and 1995-99) we can see a major diversity of evolutions in each NUTS2 grouping with regard to the European average.

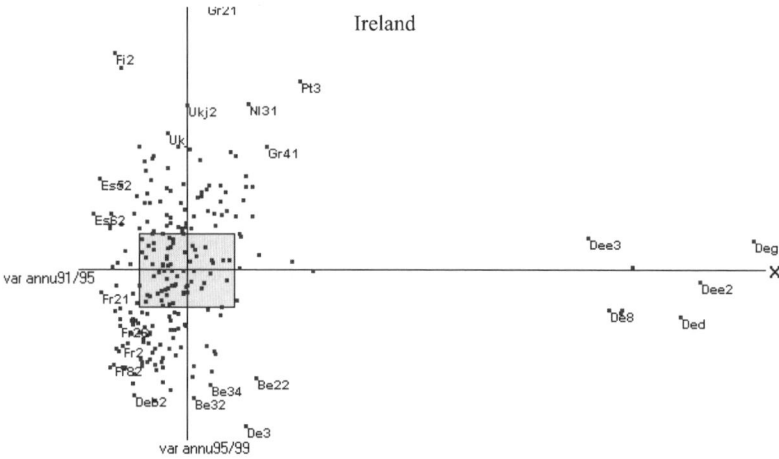

Figure 8.1. *Annual variations in GDP per capita (1995-99 compared to 1991-95)*

Not only do evolution rates and regional trajectories diverge from the European average, but there is also no fixed hierarchy in growth rates of the different periods for a given region, and of the different regions for a given country. A number of scenarios emerge from the diagram. Looking at overproduction during the 1990s in comparison to the EU average, we find Ireland and the new federal German states (states from the former GDR) and a number of radically opposed examples like Epirus (Greece) and the south of England (Sussex and Surrey). On the other side are regions with levels which are still below the average such as the Bourgogne or the Champagne-Ardennes.

Ireland is a case which needs further consideration. With a GDP growth throughout the 1990s of 5-7%, as a result of welcoming foreign businesses looking to enter the European market at a reduced cost, Ireland is different from all other European countries. The fact that this so called "region", according to the NUTS criteria, is an independent country, has clearly added to its positive evolution. Ireland has implemented a number of prerogatives, mainly fiscal, which other regions are not controlling. The economic voluntarism which has been used to attract foreign investors has been consolidated by the creation of the governmental agency IDA (Industrial Development Agency) which issues subsidies to new firms that help generate employment. Other well known examples include the "Third Italy" (Benko, 1990) and the neo-Marshallian industrial districts which are scattered among a number of regions within the Union (Caetano, 1995). The development of these districts is both endogenous and autonomous, which further highlights how important the funds are but also reveals how they often prove insufficient. These emerging industrial situations also highlight the loose relative positions of regions in the mid-term.

This idea of there being a lack of compensation in the relationship between growth rates and European subsidies requires closer inspection. Evidence of the non-automatism of the spatial distribution of the benefits of growth is supported by the correlations between successive periods. A number of linear regressions of annual GDP variation (1986-91 against 1980-85 and 1992-96 against 1986-91) have produced weak correlation coefficients of 0.10 and 0.09 for the adjustments recorded between the different periods. The different tests of significance underline the independence of the variables for the calculation periods.

In order to complete the annual variation data, which will eventually be affected by the differential of inflation between the countries (as was seen particularly during the 1980s), we have selected a sample of 155 countries, comparing gross GDP levels in 1986 and 1996. The tables below are based on descriptive statistical data and on a linear regression between 1986 and 1996 data.

	1986	1996
Mean	10051	16867
Standard deviation	2928	4447
Variation index	0.29	0.26
Max/min ratio	5.2	4.4.
Interquartile coef	39.6%	32.2%

Table 8.2. *Comparison of regional GDP in 1986 and 1996*

The clear falling margins between regions must be influenced by the results of linear regression and the most extreme residual values (positive and negative).

The 10 most positive standardized residuals		The 10 most negative standardized residuals	
Ile de France	4.86	Flemish Brabant	-1.14
Brussels	4.27	Groningen	-1.14
Bremen	2.55	Aegeus	-1.29
Luxembourg	2.49	Andalusia	-1.37
Baden Wurtemberg	2.05	Castile la Mancha	-1.45
Hamburg	2.03	Murcia	-1.55
Anvers	2.02	Epirus	-1.65
Catalonia	1.7	Languedoc Roussillon	-1.67
Ireland	1.3	Limburg	-1.97
Denmark	1.19	Picardy	-3.44

Table 8.3. *Regional GDP: the extreme residuals*

It is worth pointing out that the highest differences of GDP between 1986 and 1996 separate, on the one hand, metropolitan regions and the countries which have undergone high growth leading to the most positive growth margins and, on the other hand, the less prosperous regions (Epirus, Andalusia) and the regions in transition (e.g. Picardie), which account for the most negative margins below the EU average. Using living standards (GDP/capita) as a comparison with other chronological variables, we can confirm the existence of regions which are in a state of permanent wealth and conversely regions in permanent deprivation. With the exception of Ireland, there is not one piece of evidence shown, using this variable, to suggest that there has been a catch-up during the decade.

The graph below is being used to evaluate the more or less homogenizing rates of EU dynamics in the long-term.

For GDP/capita we have compared a temporal series which includes both the standard deviation and the mean, weighting each value of our regional sample by the mean at each date. This enabled us to reduce the bias coming from the comparison of a very disparate statistical series between an EU of 9 and an EU of 15 members. The curves represent the evolutions from 1975 to 2000 of the GDP per capita in the EU. Whatever the indicator of dispersion chosen (one or two standard deviations above or below the mean), they still produce an irregular representation of regional contrast variations in divergence or convergence.

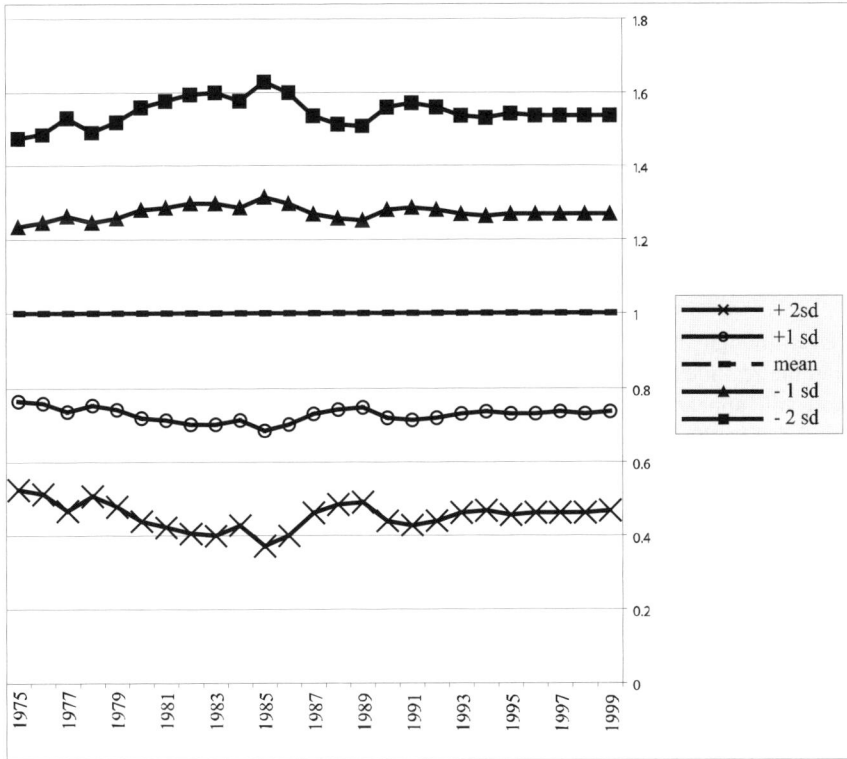

Figure 8.2. *Convergence or divergence in relation to the average GDP/ capita value of the European regions*

Beyond the sequential nature of regional fluctuations, the question is to find out the reasons, within the same period, for occasionally wayward trends observed between different groups of regions in relation to the mean.

In order to correct the threshold effects, linked to disparities between the "wealthiest" and "poorest" regions, around the limit of 75% of the mean of the European GDP, we wanted to find out if similar trends were separating subgroups among the European regions. The suggestion of a discriminate growth is confirmed when we look at the evolution of indexed GDP (100 = EU average) for the regions receiving ERDF funding, as the table below, taken from the report on cohesion in the EU, attests.

	1995	1996	1997	1998	1999
Standard deviation	10.66	10.82	11.44	11.51	11.68
Mean	**69**	**70**	**70**	**70**	**71**
Variation index	0.154	0.155	0.163	0.164	0.165

Table 8.4. *GDP per capita statistics for the regions eligible for Objective 1
Objective 1 regions – 1994 to 1996 (Abruzzo included).
Source: European Commission*

Even though the mean balances out during the 1994-1999 period (falling well within the 75 index threshold), the disparities between laggard regions (see variation index) still appear to be deepening, despite benefiting from the same policy of support during those five years.

Whatever the sample taken and the reference period in question, the results challenge the existence of a regular and general trend towards falling regional disparities in the EU.

8.2.4. *Evaluating the issue of possible catch-up and convergence*

Among a number of available indices of convergence and divergence, Sala-y-Martin (1995) established a method to enable us to carry out an appraisal of the evolution in time of regional inequalities around notions of catch-up and convergence. The author distinguishes between sigma and beta convergence, the first resulting from falling standard deviation and variation index values, the second from a negative correlation between GDP/capita growth in a given period and the initial values of this variable, which is otherwise known as "absolute convergence".

In order to adjust the findings at successive dates to meaningful points in time, a number of studies have attempted to link the regional dynamics of economic cycles. They have done this by showing an overall correlation between periods of economic growth and falling regional inequalities and, conversely, between the slower phases and processes of regional divergence [BUZ 94].

The findings of the results on EU regions demonstrate, through the negative correlations, a beta convergence, which suggests a trend towards catch-up among laggard regions. However from one period to the next, the strength of this trend diminishes and thus pervious studies which have applied this index have often highlighted the need to backup these results with the results of sigma convergence.

	1991-1995 period	1995-1999 period	1991-1999 period without the new Länder
Regression	y=.13.21 logx+57.91	y=30.21logx+17.71	y=.3.91logx+19.7
Correlation coefficient	r = -0.60 (-0.34)	r = -0.29	r = -0.44
Determination coefficient	$R^2 = 0.36$ (= 0.11)	$R^2 = 0.08$	$R^2 = 0.19$

(in parenthesis: without the new Länder for the 1991-1995 period)

Table 8.5. *Indicators of convergence and divergence in GDP/capita for the European regions in the 1991-95 and 1995-99 periods*

The difference between these two periods may to a certain extent be explained by a change of the orientation of growth. It seems nevertheless difficult to assimilate these inflexions in the rate of changes to bifurcations in the future of regions. The regions of the ex-GDR have undertaken, for well-known reasons, an upheaval, linked to an "external event" whose impact receded after 1995. The integration of the new German states provoked a significant destabilizing of the clusters, increasing the value of the negative correlation. By removing the new German states, the same regression leads to values closer to the following period (r = -0.34 and R^2= 0.11). The inversion of trends from one period to the next, explaining the negative correlations, and therefore meaning a different process of redistribution of the benefits of growth, is not evident and is in no way a tidal wave. Throughout the 1990s, "absolute convergence", through inverting evolution rates, is much clearer (r = -0.44), because the period of observation has been extended

Whatever the period in question, a number of territories like Ireland and Luxembourg (independent countries, yet still added to NUTS2, for the purposes of EUROSTAT's classifications) have extremely positive residuals in relation to the mean. In the second period (1995/1999) they are accompanied by regions in the Netherlands and the UK, as well as regions in Greece and Spain, which suggests that the highest regional contrasts in GDP/capita growth in relation to the starting point, concerns regions other than just those belonging to the so-called cohesion countries.

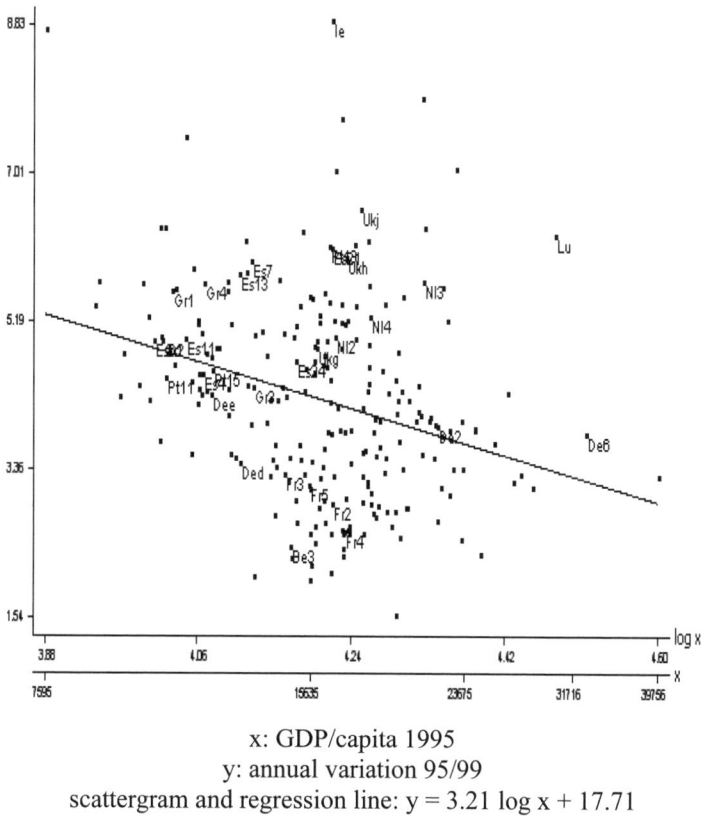

x: GDP/capita 1995
y: annual variation 95/99
scattergram and regression line: y = 3.21 log x + 17.71

Figure 8.3. *Annual GDP per capita variation and
the situation at the start of the 1995-1999 period*

Furthermore, by plotting the spatial units on to the axis we would be devaluing the predictions made for the likely mechanisms affecting European rebalancing. Supported by the table of residual values, we find that the different regions tend to group together according to the country (e.g. French regions: Fr, or Dutch: Ndl, Spain: Esp, Greek: Gr, etc.), suggesting that their evolutions can be attributed more to their economic situations and national policies than to responses to policy initiatives from the Commission. With the aim to re-evaluate national policies, we may find the principle of additional European aid allocated only to an equivalent to national income. The same European programs of national initiatives help finance extremely diverse projects across regions. The impact these projects have on development (transport infrastructure, education, production) is extremely variable and not all regions within the same country will display signs of relative convergence in the given period.

Charleux's PhD thesis [CHA 03] has brought a great deal of information on the regional disparities within the Union during the 1988-2000 period. His evaluations have revealed there to be a noticeable absence of convergence, which can be explained by the overall independence between allocations of structural funds and regional growth. This has led us to question the European policy of granting regional aid to disadvantaged regions. Having established a typology of trajectories for the major regional categories (four classes corresponding to the unequal level of GDP/capita, productivity, female employment, etc.) during a period of 12 years as the graph above shows, Charleux demonstrates that "in 161 regions, only 38 (less than a quarter) have changed category over a period of 12 years…these regions are generally found in the heart of the Union: western Germany, Belgium, the southern half of France and the north-east quarter of Spain (p.152)".

Among the 38 regions which changed category between 1988 and 2000, only 10 have benefited from objective 1 aid, in which two, Belgium and French Hainaut, have experienced negative trajectories. Only eight (less than one-seventh of the regions eligible for objective 1) have therefore seen a sufficient socio-economic improvement to enable a change of category (see p.156).

Similarly, studies carried out by Maurel [MAU 99] have tried to identify factors, supported by empirical data, of both convergence between countries and regional divergence occurring simultaneously within countries.

Maurel's team has used the absolute and conditional convergence model of Salah-y-Martin to test the hypothesis of "absolute" catch-up among the most disadvantaged regions. The team then included complementary explicative variables (e.g: infrastructural allocations) in order to assess their impact on growth and regional disparities. These calculations were carried out by separately comparing tests of convergence at the state and regional levels. His diagnosis is clear: "The negative correlation in European regions between growth rates and initial levels of GDP, can be almost exclusively attributed to national differences in revenue, which means that once inter-country differences in growth rates and initial levels of GDP have been incorporated, there will be no regional convergence in Europe" ([MAU 99], p.55).

Nevertheless, the chosen methodology is based on an evaluation of the weight of different public investments in regional growth, separating the impact of each allocation in turn. In attempting to isolate the respective importance of each variable we reduce the mechanisms of regional development to an additive approach of factors, and we fail to identify the eventual emergent phenomena which come from the crossed interactions between several variables. The interactions between close spatial units are not even taken into account when testing a hypothesis of growth "contagion".

8.2.5. *Hypothesis of the neighborhood effect*

It has become necessary to check the hypothesis on the influence of spatial interactions and neighborhood effect on the trajectories. In terms of temporalities, we do it because it is necessary to outline the fluctuations of each regional unit's relative position alongside the trajectory of the regional block it belongs to. The regional block is plotted alongside the trajectories of all EU regions in a partially regulated system of competition. Aside from the Pascalian aphorism on relations between parts and the whole, the difficulty is in knowing how to incorporate both the fluctuations of each region, and the factors which disrupt all of the regions in the model of the European regional system.

In the spatial dimension, the difficulty is knowing how to simultaneously measure the dynamics of spatial entities and the effects linked to the relative disposition (contiguity, barriers) of these same entities. The majority of economic studies omit spatial interaction phenomena linked to the effects of proximity, neighborhood and contiguity. The regional growth models above analyze the region as an autonomous unit, even though external changes are one of the variables measured by the model.

Given that the distance-factor of migrations and exchanges is very large, what happens in adjacent regions is often overlooked, while a number of studies have shown that economic proximity, combined with territorial proximity, will together multiply the number of interactions. In his thesis on the integration of exchanges in the EU, Robert (2000) shows that the geographical shapes assumed by integration are closely linked to the process of weakening internal European borders. However even when neighborhood effects are no longer affected by commercial or political borders, the ubiquity of cross-border exchanges is still not a reality.

In a study officially commissioned by EUROSTAT, Heylen and his team (2001), using data from 1996 on the volume of goods transported, show that the majority of goods transported in the EU moved within national and, more importantly, regional borders. According to the study, only 16% of all goods transported, on average, were destined for other countries. The main inter-regional exchanges of goods generally tended to follow the phenomena of contiguity. The same may be said for the international movement of goods. Hence the most populated region of Germany (North-Rhine Westphalia) is the number one European region for the exchange of heavy goods, while the five largest international exchanges are all directed towards the bordering regions (Wallonia, southern and western Netherlands and lower Saxony). The region of North-Rhine Westphalia transports a very small amount of goods by road to Portugal, the interior of Spain, Ireland and Greece, which is either because of distance or because the goods are transported by alternative means. We are therefore witnessing an evolution in the spatial structure according to a gravitational-type scheme. The volume of goods being exported falls in relation to distance.

The explanation proposed by Robert (2000) is that by virtue of sharing one or more characteristics through proximity or neighborhood, the number of goods being exchanged will increase. From this hypothesis arises the question of whether spatial interactions would operate across large regional groupings?

A number of empirical findings, evaluating the degree of self-correlation between European regions, measured in GDP per capita, lend support to this explanation. In order for this to happen we have calculated the coefficient of Moran's self-correlation for a sample of European regions, grouping them into successive rings of 100 km. We have successively applied this approach from 1994 to 1999, plotting the data on to the curves of the graph below.

The existence of territorial cores, corresponding either to regional groupings or to national spaces with similar development characteristics or trajectories, corroborates the idea of there being a large stability of relative positions and regional similarities at the European level. These findings lead us to develop the hypothesis of co-evolutions by neighboring and contiguity, even though the development of networks would intuitively lead us to envisage a relative uncertainty in the ascription of growth.

Given that there is a state of spatial competition between the activities and territorial units, it has become crucial to study the phenomena of contagion, mimicry and power struggles linked to the neighborhood effects. By opting exclusively for the indicators of spatial self-correlation to measure the co-evolutions of two contiguous units, omitting interactions linked to the numerous exchanges between two units, there is a risk of it not being fully evaluated.

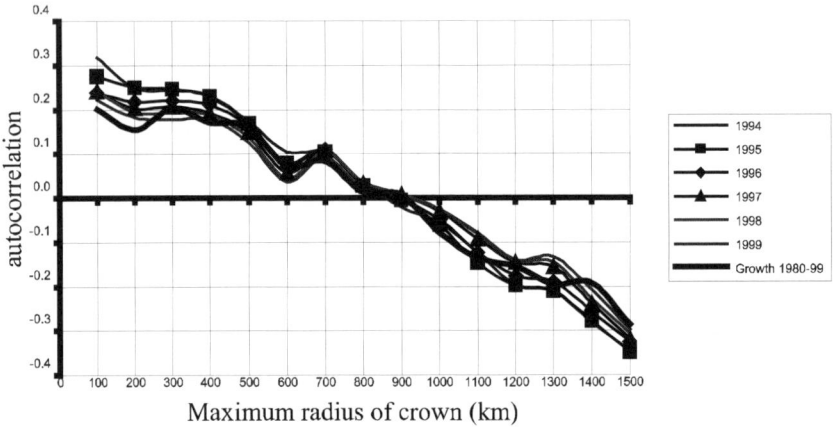

Maximum radius of crown (km)

Figure 8.4. *The spatial self-correlation of European regions (GDP/capita), measured by the Moran indicator along buffers of 100 km*

This is where the ability to create a model for the inter-regional reactions, using tools of systemic modeling in a cellular automata or multi-agents system, would be useful. This is what Patrice Langlois develops in Chapter 12.

8.3. Conclusion

The aims of this chapter, to articulate the different temporal rates at work in European regional dynamics, have been opposed by the problems caused by integrating the multiplicity of time into modeling. How can we dissociate the temporalities of each region in their interactions with the overall dynamics of the European regions? How can we incorporate the dramatic splits linked to the changing regimes of assistance for laggard regions and integrate the changing system configurations, levied through the introduction of new Member States into the Union?

Several temporalities animating the evolution of European regions have nevertheless emerged:

– in the first place a large stability of relative positions. Even though the indicator of GDP/capita growth gives the impression of there being differentiated variations in each region according to the period, the variations coexist alongside very stable situations within the hierarchy and overall organization of the European regions. This global configuration of the regions appears particularly resilient. No matter what policies the European Commission adopts nor what benefits these policies have, even for the wealthiest regions, the relative positions evolve slowly while the processes at work within the European economy favor or hinder certain territorial profiles;

– individually envisaged, each region exhibits an absence of regularity from one period to the next in rates of GDP/capita variation, linked or not to national and international economic situations. The irregularity reflects neither the processes of convergence nor the growing disparities. The factors affecting each region's rate of evolution has become an area of debate between the respective influences of endogenous factors and contagion factors (the latter is discussed in more detail in another chapter of this work);

– finally, a temporality linked to the differentiated reactions of regional entities in response to European policy. The number of years it will take countries to potentially meet the European average is a good indicator of the complexity of processes at work in the European system. It enables us to compare the example of Ireland with other countries or regions labeled as "laggards". In 1988 Ireland had a marginally higher GDP per capita[9] than Portugal and lower than Spain. In 1997 it surpassed the 100 index (then average of what was the European 15), while Spain and Portugal did not reach that level until 2001. This example highlights the limits of European regional policy in its effectiveness at tackling the major trends when this policy is not affected by factors endogenous to the region or country. Looking ahead to the future, this example will undoubtedly serve as a lesson for future regional policies, which from 2006 has included the regions of 25 Member States.

Whether we are looking at remarkable temporal stability or even the unpredictability of the consequences of regional policy on each region concerned, all of these indicators suggest that it is possible to assimilate the system of European regions into a self-organized system. While we are far from saying that the hierarchy of the regions necessarily corresponds to a logic of uneven development, we cannot deny the more or less random nature of the regional reactions to budgetary impulses, their rates of catch-up, and the counter-productiveness of realizing certain infrastructures.

There is still uncertainty over how much time is needed to establish and evaluate policies. Firstly, if the idea of there being differentiated speeds in the effects of policies which are intended to address disparities is confirmed, which the so-called "cohesion" countries are affected by, what will be the impact of the new regional differential, created by the eastern enlargement, on the most laggardly meridian or peripheral regions? Secondly, the weaknesses of regional rebalancing must confront the numerous rallying calls accompanying each period of programming. This in turn

---

9 Figures published in the second report on economic and social cohesion (European Commission, January 2001).

raises the question of response time needed for changes to structural policy. Do they act like simple inflections or as reorientations in which the effects are seen only in the long term? In this second hypothesis, the introduction of the "territorial cohesion" idea (Treaty of Amsterdam, 1997; and ESPD, 1999) constitutes a turning point in European regional policy, however it can only be evaluated over several periods of planning. The notion of territorial cohesion therefore entails surpassing the traditional objective of regional policy which sought to reduce disparities between countries and regions by complementary action at the European and national levels. This goal of more cohesion refers to a project involving all the public and private actors and all the territorial levels, which means that it can only take place in the long term.

8.4. Bibliography

[ARC 98] ARCHAEOMEDES, 1998, "Des oppida aux métropoles", *Anthropos*.

[BAL 01] BALDWIN M. and OTTAVIANO, 2001, "Global economic divergence trade and industrialization", *Journal of Economic Growth*, no. 6, 5-37.

[BAR 02] BARRAQUE B, 2002, "Modélisation et gestion de l'environnement", in P. Nouvel (ed.) *Enquête sur le concept de modèle*, Editions PUF.

[BEN 90] BENKO G, 1990, *La dynamique spatiale de l'économie contemporaine,* Editions de l'Espace Européen.

[BOI 00] BOISCUVIER E, 2000, Intégration, convergence et cohésion de l'espace économique européen, PhD thesis, Aix-Marseille University.

[BRE 99] BRETAGNOLLE A, 1999, Les systèmes de villes dans l'espace-temps, PhD thesis, University of Paris 1.

[CHA 03] CHARLEUX L, 2003, La politique régionale de l'Union européenne: des régions à l'espace?, PhD thesis, J. Fourier, Grenoble University.

[CAE 95] CAETANO C, 1995, "Les nouveaux districts industriels dans le développement récent du Portugal", *Revue de Géographie de Lyon*, vol 70, 1/1995.

[DAU 03] DAUPHINE A, 2003, *Les théories de la complexité chez les géographes*, Anthropos.

[ELI 99] ELISSALDE B, 1999, "Les temporalités des structures spatiales en Bretagne" *Travaux de l'Institut de Géographie de Reims* no. 101-104.

[ELI 00] ELISSALDE B, 2000, "Temps et changement spatial en géographie", *L'espace géographique*, 2000, no. 3.

[GEY 03] GEYER R, 2003, "European integration, complexity and the revision of theory", *Journal of Common Market Studies*, vol. 41, 1, March 2003.

[GRA 96] GRATALOUP C., 1996, *Lieux d'histoire*, Editions RECLUS.

[HEY 01] HEYLEN C *et al*, 2001, *Transports de marchandises par la route au niveau régional*, Eurostat.

[HOO 96] HOOGHE L (ed.), 1996, *Cohesion policy and European Integration, building multi-level governance*, Oxford University Press.

[JAN 69] JANELLE D.G, 1969, "Spatial reorganization, a model and concept", *Annals of the Association of the American Geographers*, vol. 59.

[JUI 72] JUILLARD E. 1972, "Espace et temps dans l'évolution des cadres régionaux", *Etudes de géographie offertes à Pierre Gourou*, Paris, 1972.

[KIE 97] KIEL L.D and ELLIOTT E, 1997, *Chaos Theory in the Social Sciences*, University of Michigan Press.

[KRU 95] KRUGMAN P and VENABLES A.J. 1995, "Globalization and the inequality of nations", *The Quarterly Journal of Economics*, vol. 110, no. 4.

[LEM 95] LE MOIGNE J.-L., 1995, *La modélisation des systèmes complexes*, p. 15, Dunod.

[LEM 99] LE MOIGNE J.-L., MORIN. E, 1999, *L'intelligence de la complexité*, L'Harmattan.

[LES 85] LESOURNE J., 1985, *A la recherche d'une théorie de l'auto-organisation*, Economie Appliquée, p. 559-567.

[MAR 92] MARKS G. 1992, "Structural policy in the European Community", in SBRAGIA A. (ed.): *Europolitics*, Brookings Institute.

[MAR 99] MARTIN and OTTAVIANO, 1999, "Industry location in a model of endogeneous growth", *European Economic Review*, (43).2, p. 281-302.

[MAU 99] MAUREL F., 1999, *Scénario pour une nouvelle géographie économique de l'Europe*, Commissariat général du Plan/editions Economica.

[OLI 00] OLLIVRO J., 2000, *L'Homme à toutes vitesses, de la lenteur homogène à la rapidité différenciée*, Presses universitaires de Rennes, coll. espaces et sociétés.

[PUM 89] PUMAIN D., SANDERS L., SAINT JULIEN T. (ed.), 1989, "Villes et auto-organisation", *Economica*.

[RIO 02] RIOU S., 2002, "Géographie, croissance et politique de cohésion en Europe", *Revue française d'économie*, no. 3, vol. XVII.

Chapter 9

Modeling the Watershed as a Complex Spatial System: A Review

The influence of morphology is important at different levels in the hydrological behavior of calibrated watersheds. The combination of the shape of the basin and the organization of the thalweg networks controls the concentration of water flow paths. According to the lithologic and tectonic conditions and geomorphologic history, the basins and their systems have different shapes. Among the most classical shapes, Lambert (1996) and Salomon (1997) quote the oak type, the poplar type as well as the "parasol pine type". On the one hand, when the setting up is done under homogenous lithological conditions, systems take a dendritic shape, as the branches spread out randomly in all spatial directions [LAM 96]. On the other hand, the development of waterways can be strongly constrained by regional tectonics. The trellis (also called "bayonet") systems characterize the folded regions and the subsidence basins are often drained by centripetal systems [BRA 97].

Many researchers have developed their own methods to quantitatively characterize basin shapes and system organizations in order to better define their influence on hydrological behavior. However, these results have never really been usable in hydrological modeling and this is due to the different reasons that will be set out in this chapter.

For the last few years, the methods resulting from the complex systems theory offer new perspectives with more synthetic and dynamic approaches of the morphologic organization.

Chapter written by Daniel DELAHAYE.

9.1. Shape indices for measuring various forms of a watershed

Measuring the compactness of a basin is possible by comparing its shape with the shape of a perfect circle or a perfect rectangle. The most famous of such indices is the Gravelius index relating the perimeter of the watershed to the circle circumference having an equal surface:

$$Kc = 0.28 \frac{P}{\sqrt{A}}$$

[9.1]

P: basin perimeter
A: basin surface.

The closer the index is to 1, the more likely the basin will have a perfect shape according to a circle form (GRA 1914).

The "equivalent rectangle" method, created by Gravelius (1914) and performed by Peguy (1942) is based on a purely geometric transformation in which case the basin is compared to a rectangle having the same perimeter and the same surface values:

$$P = 2(l_1 + l_2) = \frac{Kc\sqrt{A}}{0.28}$$

[9.2]

L^1 and L^2: the sides of the rectangle
P: the perimeter
A: the surface
L^1 and L^2=A.

$$l_1 = \frac{Kc\sqrt{A}}{1.12}\left[1 + \sqrt{1 - \left(\frac{1.12}{Kc}\right)^2}\right] \quad \text{and} \quad l_2 = \frac{Kc\sqrt{A}}{1.12}\left[1 - \sqrt{1 - \left(\frac{1.12}{Kc}\right)^2}\right]$$

All these indices measure the distribution of distances to the final outlet (compactness) and thus, relatively evaluate the time of concentration within the basins. It is possible to choose other shape indices, but their significance is also limited and their interpretation very subtle. On the other hand, without the integration of the hydrographical network within the indices, these shape indices have minor signification.

9.2. Organization of the networks

9.2.1. *Genesis of hydrographical networks*

Rules of development of a hydrographical network are not linear in space and time so it is difficult to show such genesis. Indeed, the size of the waterways is not proportional to the flow volume and the network dynamics is strongly due to extreme events [NEW 90]. However, many theories are scattered throughout the hydrographical literature. The statistic models of evolution are generally the most quoted by researchers. Three distinct theories have been proposed to explain the development of a hydrographical network.

Chronologically the first theory was that by Glock [ROD 97]:

– setting up by quickly digging a general skeleton;

– lengthening towards the upstream water;

– evolution of the tributaries by bifurcations;

– finally, simplification of the network by the disappearance of some tributaries.

The second theory was developed by Horton [HOR 45]:

– concentration of diffusive flow on the hillslopes and creation of rills parallel to the slope when the shear stress exceeds the soil resistance;

– migration of the confluences from downstream to upstream by a retrogressive erosion process.

Finally the last theory comes from the works of Schumm [ROD 97]. It is based upon the hypothesis that the hydrographical network occurs from the dissection wave progressing from a downstream wave towards the upstream boundaries.

The choice between the three development laws is very difficult. The scientific community agrees with the coexistence of these three forms of evolution in nature according to local conditions (climatic, geological, tectonic or topographical), but also according to different scales. Horton's theory is confirmed by numerous recent works [MON 92] showing how a channel is developed from a run-off concentrated in the downstream of a source area and how a channel is developed by retrogressive erosion. This development has been largely demonstrated by all the works undertaken over recent years on the topic of soil erosion. Thus, the genesis of a hydrographical network is linked to a complex interaction between, on the one hand, the topography which determines the size of the source area and the length of the runoff, and, on the other hand, the nature of the land use and the edaphical characteristics which condition the run-off amounts. Observations in nature effectively demonstrate that a variation of the drainage networks and of the soil erosion forms are due to the impact of changes in land use. However, all these studies undertaken on a large scale are made in a homogenous environment and they do not integrate all variables that condition the setting up of a hydrographical network (geology, hydrology, etc.).

After this first generation, other models were developed such as the "evolutionary" models based upon a more in depth study of the channels dispersion in space. A spatial model called the "invasion percolation model of a drainage network", developed by Stark [STA 91], was built on the rules that govern the percolation phenomenon. In this case, the hydrographical network develops according to the local lithological constraints, to the topography and to the characteristics of the nature of the land use. This model is richer than the previous models since it strives towards the integration of factors that are more numerous and closer to the reality of the watershed system .

Other models favor a physical vision of the phenomenon, especially in Rinaldo's works on auto-organized networks [RIN 92] [RIG 93]. The concept is different from all previous models since the author does not attempt to directly describe a given network, but rather measure the gap between this one and a theoretical network where the energy losses in the different reaches are minimal (Optimal Channel Networks). This theory was developed by Per Bak [PER 87].

The main objective of this study is not to analyze the genesis conditions of networks, but to better highlight the impacts of the network organization on the hydrological responses. Works in this field are numerous. However, even if all these models reveal important information, no model is able to reproduce the real organization of the observed networks in nature.

9.2.2. Researching network laws

Following the development of the above theories, researchers have tried to extract laws which govern the organization of hydrographical networks. There is a strong relationship between the basin surface (A) and the length of the main waterway (L) that is defined under Hack's Law [MON 92] [HAU 01].

$$L = 1.4 \, A^h \tag{9.3}$$

where the exponent h varies from 0.5 to 0.6.

The length L does not take into account the sinuosity of the waterway.

This network index absolutely does not give any information about the hydrographical network itself which can have multiple structures. Relating the total length of the waterways to the surface of the basin, the drainage index (Dd) calculates the density of a network. This index is efficient in order to compare watersheds, developed in different environments. The incidence of the climatic, lithological or orographic parameters became well described [BRA 97]. The drainage index also provides results on the relationship between slopes and hydrographical networks. Montgomery and Dietrich showed that the length of the source area boundaries L_s is equal to the opposite of the drainage density [MON 92]:

$$L_s = \frac{1}{D_d}$$

[9.4]

The drainage density (Dd) informs us about the transition point between the runoff dispersion phase and the concentration phase. In the same way, the average length of a slope L_v can be expressed according to the drainage density:

$$L_v = \frac{1}{2} D_d$$

[9.5]

In spite of the interests supported by these indices, they remain inefficient in describing the spatial organization of the hydrographical network since they do not integrate its topology.

The hydrographical network is a branched out object organized into a hierarchy in the watershed. In order to better describe its organization, several authors have set up various methods to draw hierarchies, the first being the ordering. Whether these indexations are derived from Horton, Shreve or Strahler, all have a topological grounding. Measuring or counting the river reaches are good resources and help in the setting up of a number of waterway ordering rules. Classifications also help to tackle the quantification of the networks structural and functional organization. The number and average length of the sections of that same order follow the geometric laws classically called Horton's ratios.

The first law is called the confluence ratio [HOR 45]:

$$R_c = \frac{N_i}{N_{i+1}}$$

[9.6]

N^1 being the number of order i sections.

The second law characterizes the length ratio:

$$R_1 = \frac{L_{i+1}}{L_i}$$

[9.7]

L_i being the average length of order i sections.

The average slope and the drainage area of the same order sections follow the same laws which help develop two other ratios [SMA 68]:

– the slope ratio:

$$R_p = \frac{P_i}{P_{i+1}}$$

[9.8]

P being the average slope of order i sections;

– and finally, the area ratio:

$$R_a = \frac{A_i}{A_{i+1}}$$

[9.9]

A representing the average area of the basins of order i.

The values of the R_C ratio are close to 4. This means that an average of 4 streams of the same order 1 are necessary to form a stream of an order 2 and thus 4 streams of order 2 to form a stream of order 3 [BRA 97]. The ratios R_1, R_a and R_p respectively vary around the following values: 2, 5 and 2.

Such ratios are reliable under homogenous environmental conditions. The control of the network by tectonic factors or strong lithological variations can change these rules. These statistical "laws", above all, help to characterize the average network structure but do not provide tools to compare the structure of two different basins with various slopes and forms. However, these geometric series underline the networks' internal similarities with indices which take the shape of power laws according to the order that plays the role of scale factor in this case. This purely statistical or mathematical approach, from the English-speaking school, has often been criticized by several authors [BIR 22; LAM 96]. Of course, such criticisms are well justified, but the Horton ratios describing this "natural tendency to the flow concentration" give us something to think about, as far as hydrology scale transfer is concerned, a fact that is of great interest to the geomorphologist community.

9.2.3. *Towards a law concerning reach distribution*

The Horton's ratios analyze the statistical distribution of the average sections but do not characterize their spatial distribution. Some researchers have studied the distribution of the reach lengths in order to better understand the network organization [CRA 95]. These results show that some rules seem to exist for the small reaches but the distribution of the confluences generally occurs randomly.

Shreve had considered this random distribution and had proposed the "topologically distinct channel networks" (TDCN) to quantify the number of possible configurations that a hydrographical network can take according to a specific number of sources (level 1). Table 9.1 shows all the complexity of the shapes that a network can take for the same number of sources and shows how combinations grow in a exponential manner for a low increase of the order 1 number.

Number of sources	Number of combinations	Number of sources	Number of combinations
1	1	10	4862
2	1	20	$1.767 . 10^9$
3	2	30	$1.002 . 10^{15}$
4	5	40	$6.804 . 10^{20}$
5	14	50	$5.095 . 10^{26}$
6	42	100	$2.275 . 10^{56}$
7	132	200	$1.29 . 10^{116}$
8	429	500	$1.35 . 10^{296}$
9	1430	1000	$5.12 . 10^{596}$

Table 9.1. *Number of possible combinations in the organization of the reaches according to the number of sources [CHO 72]*

The complexity is great, but the lay-out for the network in the basin is still fundamental in order to evaluate the transit time of the water between the sources' area boundaries and the final outlet. Various studies have tried to approach this travel time by studying the distribution function of the contributive areas, according to the distance covered by the water to the outlet. This distribution is often called "area function" or "distance area function" [BLÖ 95] [ROD 97]. The latter is classically approached through the width function by Shreve, which gives the number of reaches in the basin according to the distance to cover [KIR 76]. This width function, dating from 1969, is the first step towards a spatial analysis of the functional organization of a hydrographical network. Remaining purely graphic, this function gives a quick and synthetic idea of the watershed shape and of network density. Another quality is that such a function takes into account the real trajectory of the flows and not only the Euclidian distance of the outlet. Nevertheless, the "width function" remains a graphic tool that does not help to obtain a measure of the internal organization of the basin.

To properly approach such an organization it is possible to reconsider Hack's Law. Shreve, in 1974, demonstrated that the value h used in the power function decreases with the size of the basins [SHR 74]. The coefficient approaches 0.6 for large basins and 0.5 for the small ones. Mandelbrot attributes a characteristic to the fractal nature of the thalwegs, and shows that the coefficient h helps to calculate this dimension [MAN 68]. In the same manner, the Horton ratios which translate, through geometric growth of the concentration, the internal homothety of the thalwegs networks, identify the fractal nature of these shapes [HAU 01].

According to such results, it was reasonable to consider fractal geometry in order to study the organization of hydrographical networks. This approach seems particularly interesting because it goes beyond the statistical distribution in analyzing the structure of the complex systems. Fractality is the property of the objects in which every element is a scaled-down image of the entity. As the Horton ratios are stable throughout the scales, it is possible to imagine a fractality of the hydrographic networks.

The works concerning the fractal nature of the streams are numerous and all conclude that the networks are fractal [CUD 2000]. Among various works, some tried to measure the fractal nature of the networks from Horton's ratios, thus La Barbera and Rosso (1987) propose the following ratio [LAB 87] [CUD 2000]:

$$D = \frac{\ln R_c}{\ln R_l}$$

[9.10]

$D = 1$ if $R_c <= R_l$

This relationship has since been used by other researchers [BEE 93]. Liu names D the topological dimension of the watershed [LIU 92]. When the network drains the entire basin, the fractal or topological dimension approaches 2. This value would translate the maturity of a drainage network [LAB 89].

Other fractal dimensions are based on the area ratio [LAB 90].

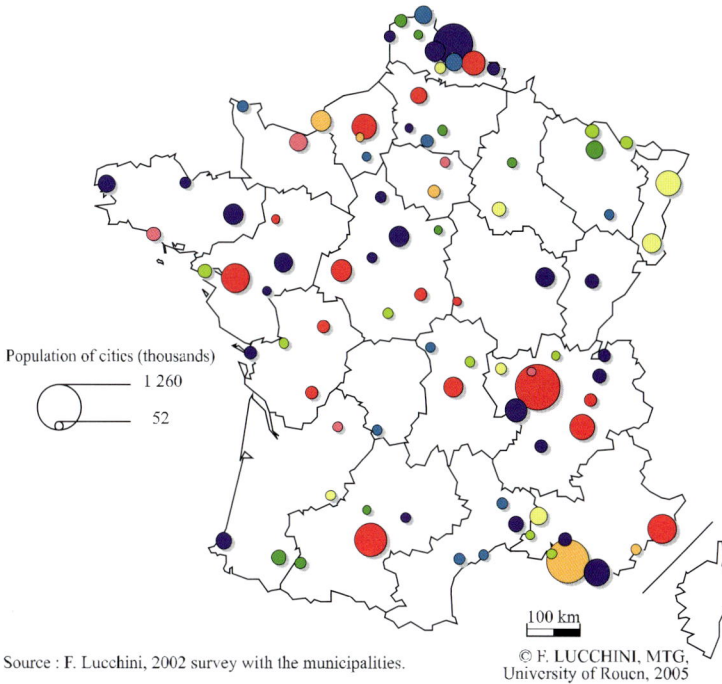

	Number of cities	Significance
■ (red)	16	Multiplicity of cultural choices of partnerships, and major cultural expenditures
■ (pink)	5	Less intensive cultural profusion and choice of local identity
■ (orange)	5	Specialization towards the internal life of the city with the help of major human resources
■ (pale yellow)	7	Less intensive specialization of city liveliness
■ (dark green)	9	Towards a financial and decisional autarchy, weak human and financial resources
■ (light green)	9	Greater human resources and less autarchy
■ (blue)	11	Small cities with fewer human resources, economic orientation and external cultural influence
■ (navy)	25	Choice of original domains and numerous cross-subsidisation

Figure 3.3. *Differentiated urban behaviors for the 110 surveyed municipalities – results of factor analysis*

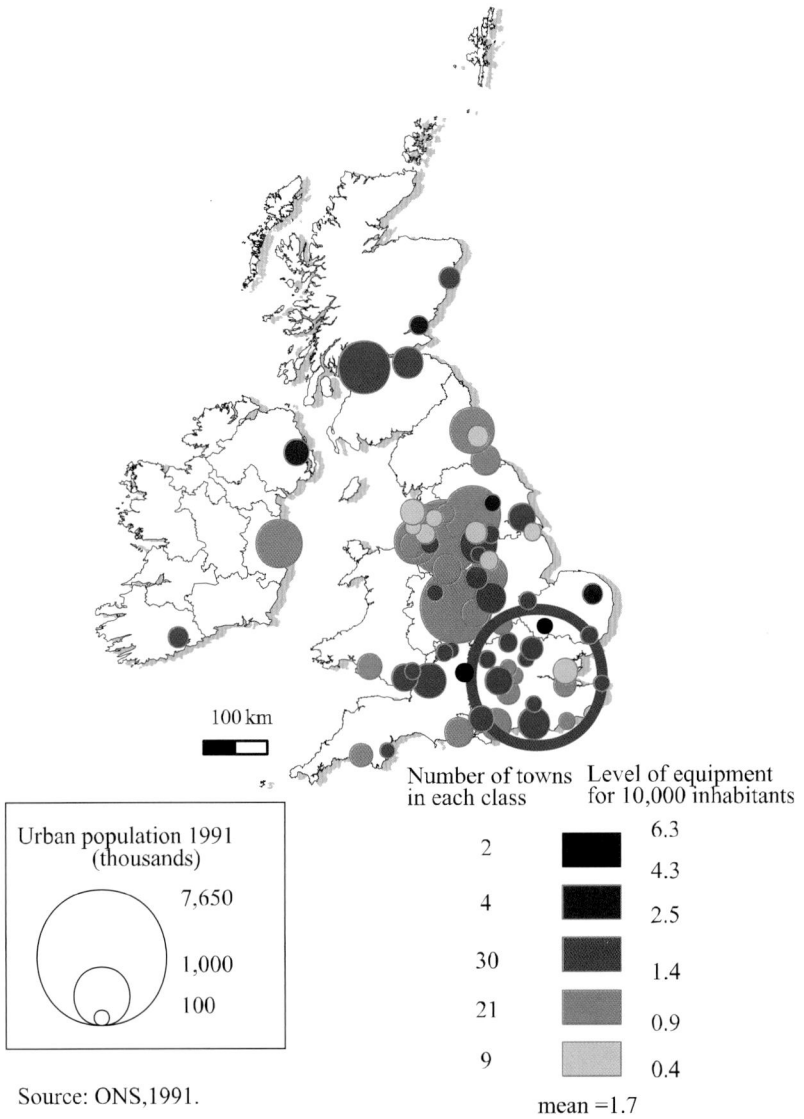

100 km

Urban population 1991
(thousands)

7,650

1,000

100

Source: ONS,1991.

Number of towns
in each class

2

4

30

21

9

Level of equipment
for 10,000 inhabitants

6.3
4.3
2.5
1.4
0.9
0.4

mean =1.7

F. Lucchini, MTG, University of Rouen, 2005.

Figure 3.11. *Cultural service per inhabitant in the UK and Ireland, 2002*

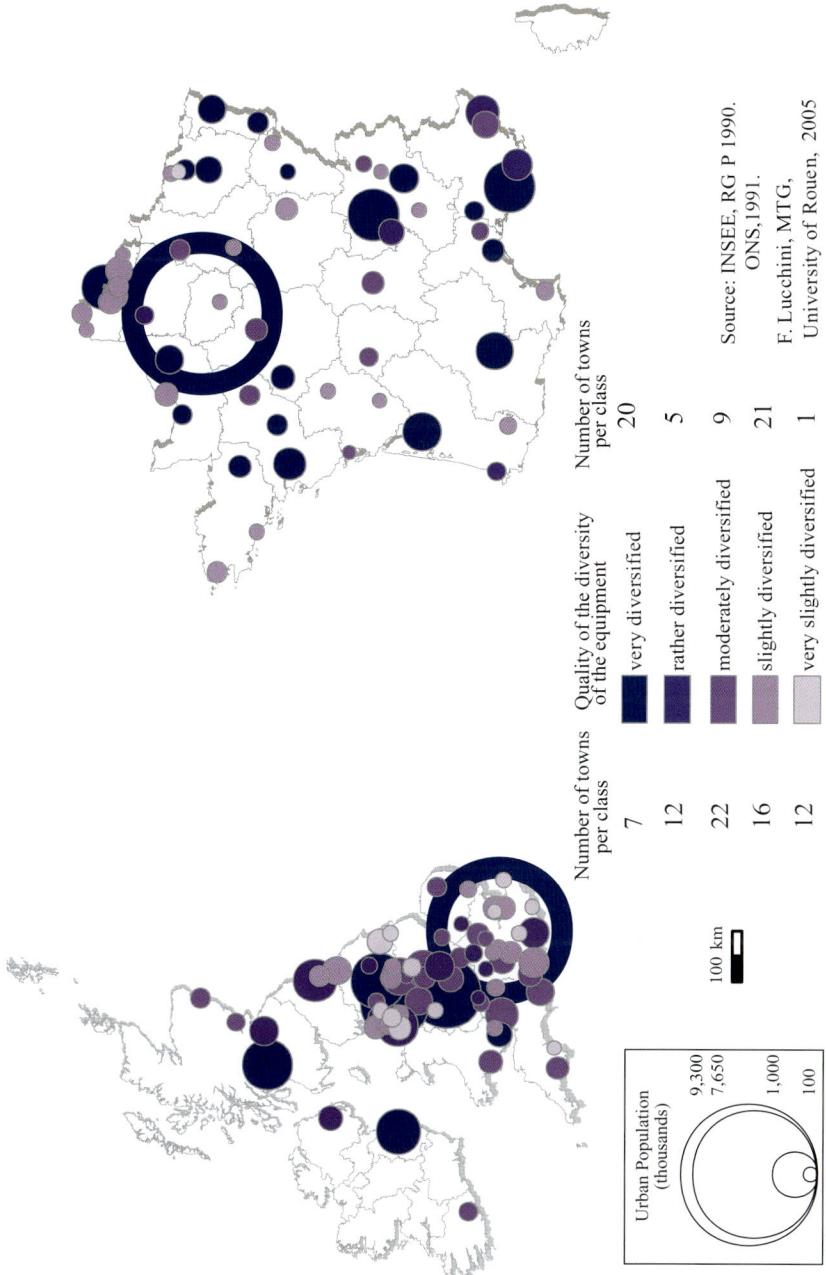

Quality of the diversity of the equipment

	Number of towns per class
very diversified	20
rather diversified	5
moderately diversified	9
slightly diversified	21
very slightly diversified	1

	Number of towns per class
	7
	12
	22
	16
	12

Urban Population (thousands)

9,300
7,650
1,000
100

100 km

Source: INSEE, RG P 1990.
ONS,1991.

F. Lucchini, MTG,
University of Rouen, 2005

Figure 3.12. *Cultural diversity of British and French cities, 2005*

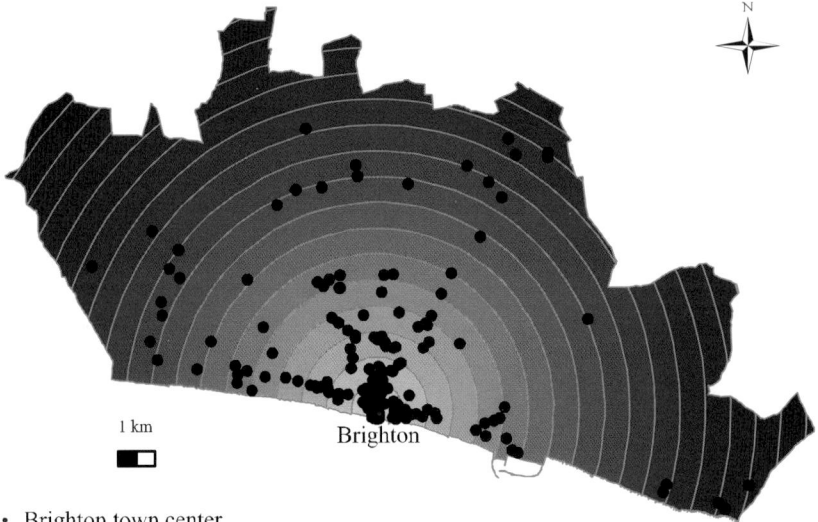

Urban area of Brighton and Hove
2004

1 km

Brighton

· Brighton town center
● Cultural equipment

Distance to the town center:

Concentric rings every 500 meters, starting from Brighton town center

F. Lucchini, MTG, University of Rouen, 2005

Figure 3.13. *Cultural equipment and distance to the town center of Brighton-Hove*

Urban area of Rouen
2004

- Town center of Rouen
. Cultural equipment

Cultural equipment and distance
to the town center of Rouen

Density per square km

$$y = 48.141x^{-1.7097}$$
$$R^2 = 0.7526$$

Distance to the town center (km)

0   4   Km

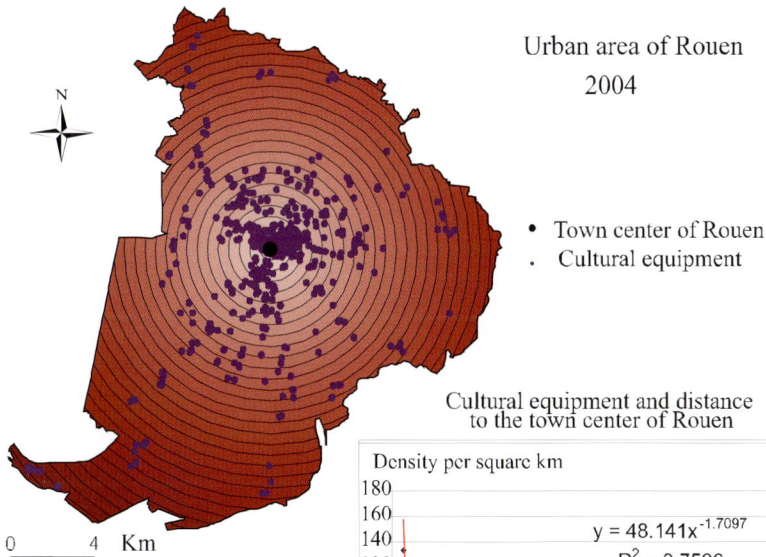

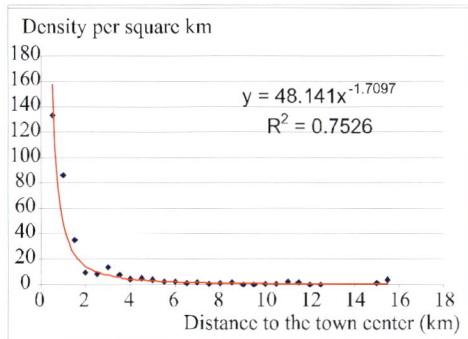

F. Lucchini, MTG,
University of Rouen, 2005

Identification of the ring	Surface of the ring km²	Distance to town center km	Number of services	Density per square km
1	0.78	0.5	104	133.09
2	2.34	1	202	86.17
3	3.91	1.5	137	35.06
4	5.47	2	52	9.51
5	7.03	2.5	58	8.25
6	8.59	3	119	13.84
7	10.16	3.5	78	7.68
8	11.72	4	50	4.27
9	13.28	4.5	65	4.89
10	14.85	5	58	3.91
11	16.18	5.5	39	2.41
12	16.97	6	41	2.42
13	17.85	6.5	22	1.23
14	19.11	7	30	1.57
15	19.69	7.5	14	0.71
16	18.61	8	19	1.02
17	17.72	8.5	33	1.86
18	16.72	9	3	0.18
19	14.59	9.5	4	0.27
20	13.44	10	9	0.67
21	11.92	10.5	7	0.59
22	8.52	11	20	2.35
23	6.49	11.5	14	2.16
24	4.51	12	1	0.22
25	3.55	12.5	1	0.28
30	1.97	15	3	1.52
31	1.54	15.5	6	3.89

Figure 3.14. *Cultural equipment and distance to the town center of Rouen, 2004*

Urban area of Rouen 2005

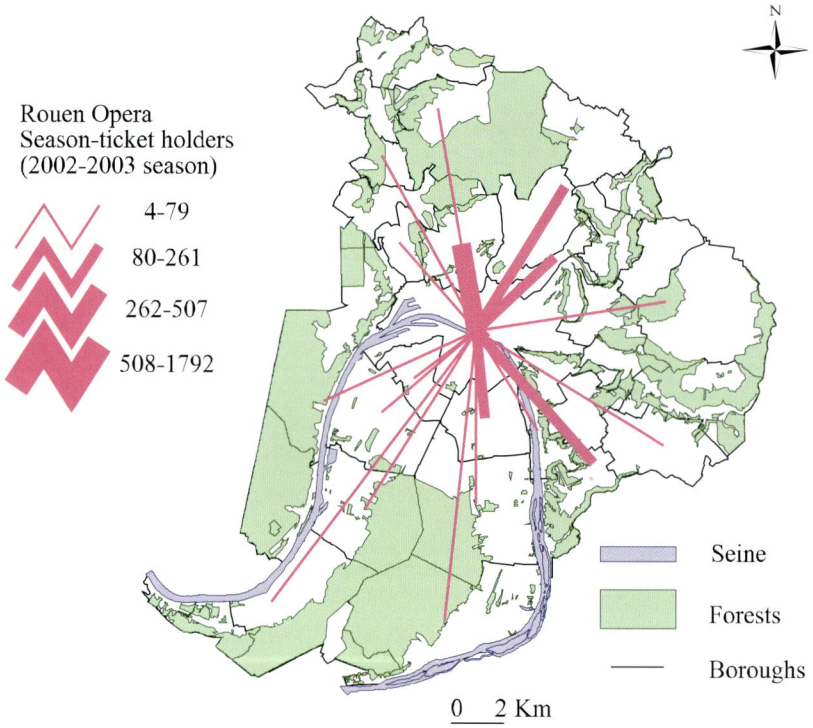

Rouen Opera
Season-ticket holders
(2002-2003 season)

4-79

80-261

262-507

508-1792

Seine

Forests

Boroughs

0   2 Km

F. Lucchini, MTG, University of Rouen, 2005

Figure 3.15. *Area of influence of the Rouen Opera, France, 2005*

Urban area of Rouen

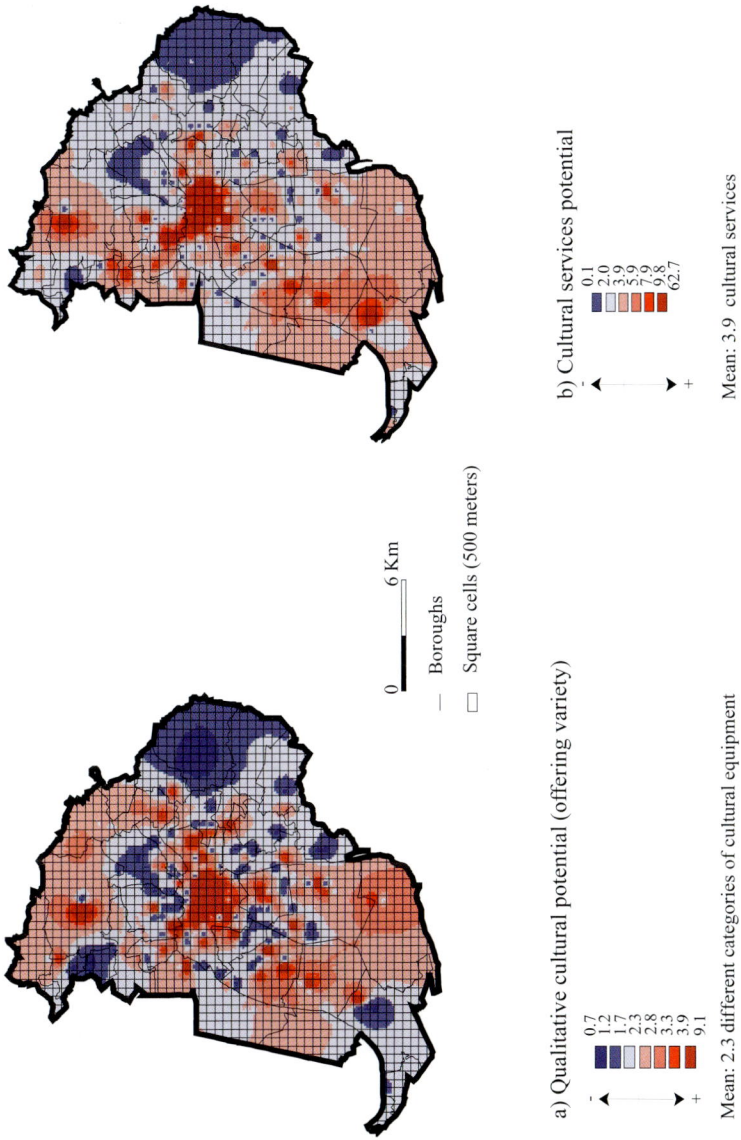

0 ———— 6 Km

— Boroughs

☐ Square cells (500 meters)

a) Qualitative cultural potential (offering variety)

0.7
1.2
1.7
2.3
2.8
3.3
3.9
9.1

– ◄————

————► +

Mean: 2.3 different categories of cultural equipment

b) Cultural services potential

0.1
2.0
3.9
5.9
7.9
9.8
62.7

– ◄————

————► +

Mean: 3.9 cultural services

F. Lucchini, MTG, University of Rouen, 2005.

Figure 3.16. *Two measures of the cultural potential in Rouen*

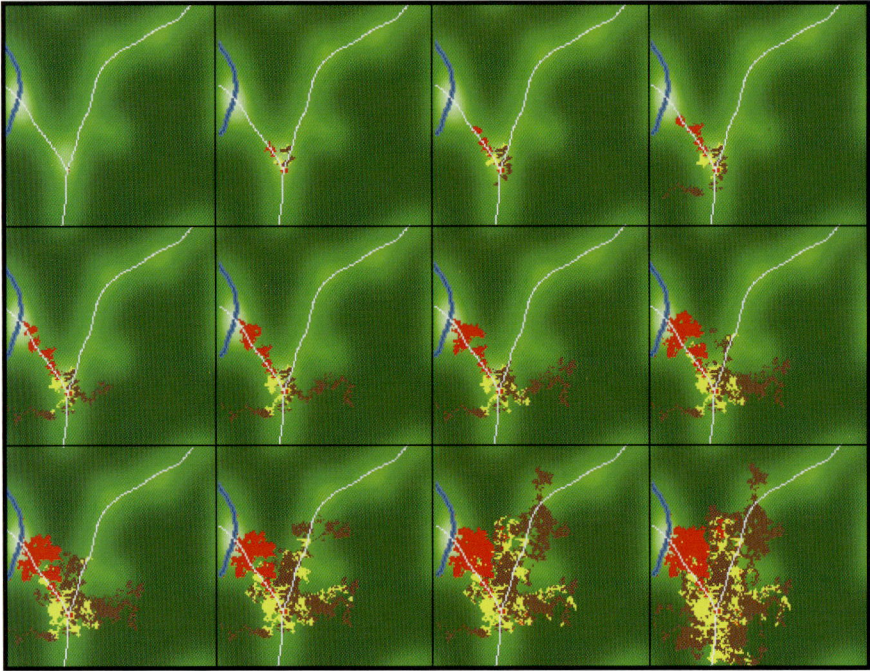

Figure 13.3. *Initial environment – land-use (variations of green according to altitude), main road (white) and river (blue) – and simulations: extension of habitat (brown), of commerce (yellow) and of industry (red)*

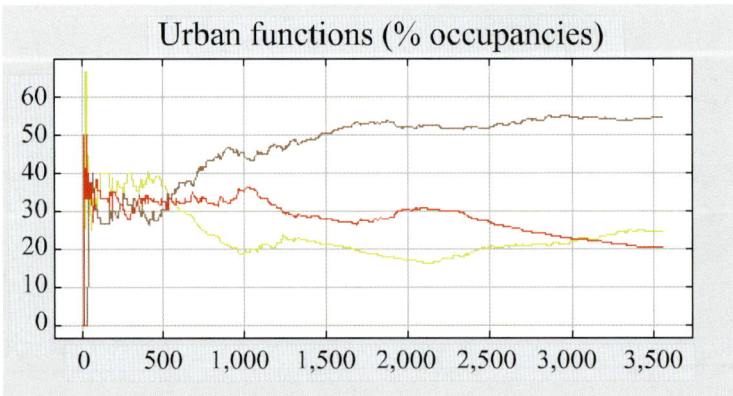

Figure 13.4. *Simulation of an urban structure, according to different strategies*

$$d = 2 \frac{\ln R_l}{R_a}$$

[9.11]

This ratio translates the lengthening of the drains according to the basin area. Finally, these first two calculations make it possible to arrive to the following synthetic fractal dimension:

$$D_t = dD$$

[9.12]

$$D_t = 2 \text{ if } R_c >= R_a$$

[9.13]

This dimension leads to both the network spreading into the space and its hierarchical organization according to the size of the basin.

Liu proposes two other fractal dimensions, one is a dimension of dispersion and the other one is a spectral dimension [LIU 92]. The dimension of the dispersion D_w translates the transfer functionality through the network:

$$D_w = D_t(1 + \frac{1}{D})$$

[9.14]

The spectral dimension D_s qualifies the network connectivity:

$$D_s = \frac{2D}{1 + D}$$

[9.15]

Testing numerous basins, Liu proposes the following average values [CUD 00]:

$D_t \approx 1.82$
$D \approx 1.55$
$d \approx 1.2$.

The dimension d is calculated from the length and area ratios corresponding to the variations of Hack's coefficient of h with, as demonstrated by Mandelbrot, d = 2h. The value of h varies between 0.5 and 0.6, and it is normal to find a fractal dimension close to 1.2. Hack's exponent translates the lengthening of the basin but also the sinuosity of the reaches [CUD 00].

9.3. Synthesis concerning the shape and organization indices

A small part of the studies conducted in those fields has been presented here. It was possible to multiply the examples by developing all research on the auto-organized systems [PER 87] or discussing results obtained by the various types of deterministic and statistical fractal analysis. However, the intention was, above all, to present the status of knowledge by putting forth the tools already available to answer our issues. More widely, it was equally important to show choices made by the researchers to better understand the general philosophy of this work and its contribution to geography and geomorphology.

All these approaches, aimed at the modeling of the network structuring, are interesting because they make it possible, starting from the power laws, defined and validated on numerous basins, to measure the deviations from the laws, and thus to find anomalies serving as a basis for more naturalistic approaches. It is in that sense that they are extremely fundamental.

Results also show the relevance of this type of approach taken in order to study the network structuring and some type of spatial organization. However, such an analysis helps to measure the statistical characterizations of the networks and not potential dynamics. In a concrete way, if we can define the lengths of the reaches, their ordering and their diffusion into space, the analysis is still ineffective to predict the effect of this structuring on the functioning of the basin at to the outlet. The Horton ratios are based on a discrete and statistical analysis of the networks and at any time do not allow the introduction of the transfer dynamics that could occur within the networks. The fractal dimensions give richer spatial information, but then again they do not help to appreciate the organization of the sub-watersheds that produces the watershed's general behavior. The width function is without a doubt the most dynamic approach but the curve that is obtained is difficult to use in order to define the influence of the network architecture on the hydrological behavior to the release.

Finally, even the "network" is a notion not well defined in those studies. The instability of Horton's ratios for small basins, underlined by the fractal analysis, reveals such a methodological problem. At what moment does a hillslope system change to a concentrated thalweg system?

All those approaches are not operable for translating the basin dynamics and moreover are not used in the hydrographical models. To be convinced, just look at the indigence of the morphometrical parameters used in the classic empirical formulae for the calculation of the basin concentration time:

$$Tc = 0.02.L^{0.77}.S^{-0.385} \quad \text{(Kirplich)}$$

$$Tc = 7.62 \left[\frac{A}{S} \right]^{0.5} \qquad \text{(Ventura)}$$

$$Tc = 6 \left[\frac{A}{L} \right]^{\frac{1}{3}} / S^{0.5} \qquad \text{(Passini)}$$

Tc: concentration time in mn
L: length of the longest hydraulic way in m (Kirplich) in km (Passini)
S: average slope of the hydraulic way in m/m
A: surface of the basin in km²

	Characteristics of the watershed
Surface of the watershed (in ha)	400
Distance of the hydraulic path (in m)	4950
Difference in height (in m)	59
Thalweg's average slope (in %)	1.2
	Rates of Tc in mn
Kerby/Haryhaway's formula (in mm)	11'50
Kirplich's formula (in mn)	76'50
Passini's formula (in mn)	51'10
Turraza's formula (in mm)	33'30
Ven Te Chow's formula (in mn)	75'
Ventura's formula (in mn)	140'20

Table 9.2. *Variation of time of concentration (Tc) for the same watershed according to different empirical formulae*

All these formulae give very different results which make the objective definition of concentration time difficult [PAP 1986]. Table 9.2 shows the results obtained with six different formulae for the same calibrated watershed. Values range from 12 mn to 140 mn!

Beyond the dispersion of the results, the construction of these formulae show the very weak integration of the morphological dimension which is limited to the length of the longest hydraulical path and to the average slope of this path. This shows all the interest of a more synthetic and dynamic measure of the morphological component.

Researching universal laws of the development of the hydrographical networks is probably without perspectives, while the extraction of invariant and multiscalar characteristics common to all networks helps to focus our attention afterwards on local parameters (geomorphological, tectonical, etc.) revealed once the system is cleared of the "background noise". While respecting the concentration laws, the networks can present different forms; however, the results of these studies provide tools in order to measure a part of these variations, thus allowing very rich comparative approaches.

These results also prove that the structure of the watersheds present an organized complexity in the way that it is defined in the complex system theory [HEU 98].

9.4. From morphometry to complex systems

Methods derived from the theory of complex systems have opened up new perspectives, notably the use of cellular automata as models of physical systems. As promising as it is, this new paradigm has seldom been exploited by geomorphologists. However, the automata networks constitute a very relevant method for translating the hydrological influence of the organization of a watershed in a dynamic and iterative manner. A specific cellular automata has been created by the Rouen team and directed by Patrice Langlois [LAN 2002].

9.4.1. *Methodological framework*

For a better explanation to the novice reader, the following framework is a brief methodological reminder; in addition, the set of results has already been presented in other papers [DEL 02] [DEL 01] [LAN 02]. The main difficulty was linking on one hand topographic variables such as elevation and its derivatives, and on the other hand, hydraulic variables such as water flows and direction [DEP 92; CRA 01; MIT 01]. The classic concept of cellular automaton (CA) has been generalized [LAN 97] in order to be able to model both the variable structure of the elements of a ground surface through surface (slopes elements), linear (portions of thalwegs) and punctual cells (local minima or steepest slopes) and on the other hand, the cellular routing scheme of the automaton which can no longer be guided only and uniformly by the neighboring topology of the cellular network. Here, it is performed by the morphologic links structuring the surface (outflow links between various elements, an overflowing link between the sub-basins) and in the same time the connectivity between these cells. Flow pathways cannot be guided uniformly by the rules of vicinity of the cellular network. Nevertheless, they depend, in this model, on the morphological links structuring the TIN surface.

This approach is based on a generalized cellular automaton in which cells have different forms and variable size (point, line, surface), and whose flow pathways make it possible to translate the real morphological structure of surface and not only its topology [DOU 05].

9.4.1.1. *Input data: the digital terrain model (DTM)*

Use of a digital terrain model (DTM) is of paramount importance to integrate on a small scale the basins' forms. The triangulation irregular network (TIN) is realized according to a DTM with a 50 m grid interval. To manage or to simulate spatial phenomena with DTMs, different methods can be used. The "finite element method" used to build the mesh enables a continuous interpolation of the surface between all the DTM's points [PAL 1998].

Thus, we can represent at all points P(x,y) and in a continuous manner its altitude z_p, and its normal vector $\overrightarrow{Up}$ at the surface. This procedure enables us to calculate all sizes related to the local form of the terrain (slope, exposure, vector flow, surfaces, volumes, outflows, discharges, etc.). The choice of the mesh is crucial in this case. A square mesh seems to be more adapted to available original data (hypsometric pictures, grid data), but it presents the inconvenience of not being interpolated by a planar portion (Figure 9.1).

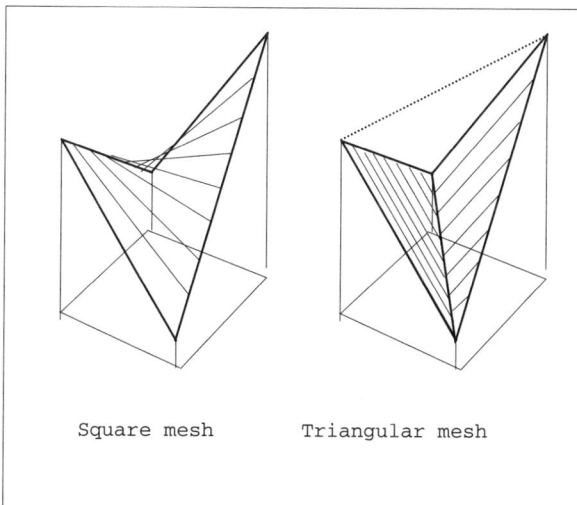

Square mesh   Triangular mesh

Figure 9.1. *Mesh choice for the structure of the automaton [LAN 02]*

To resolve this problem, each square has been cut into two triangles choosing one of the diagonals. This choice is not neutral, as the diagonals do not intersect each other at the same height. To favor the runoff, the diagonal which does not risk blocking the bed of a thalweg, meaning the one at minimal height at the level of the intersection, has been chosen.

The space gridded at the beginning is then decomposed into a triangular mesh. The mesh is regular (right-angled triangles) if it is built from a DTM, but it can also be elaborated from random points by the Delaunay triangulation (irregular triangulation). This allows a unique model, simplifying the programming and which adapts itself to all types of data. Moreover, by their intrinsic linearity, the triangular elements offer the most simple finished element model. This is useful when the model runs on a mesh defined by a great number of cells ranging from a few thousand to many hundreds of thousands of cells.

9.4.1.2. *The cellular routing scheme*

To optimize access to geometric information, data must be carefully structured in such a way as not to have to research a particular piece of information (altitude of a vertex, morphology of an edge, slope of a triangle, etc.) while scrubbing the database introducing the least possible amount of redundancies. The choice has been a "RZ-topological-graph" model [LAN 94] applied to a triangulation, that is to say formed by three principal entities: vertex, edge, triangle (Figure 9.2). This scheme enables the direct access of geometric information from the edges or the triangles without duplicating the information.

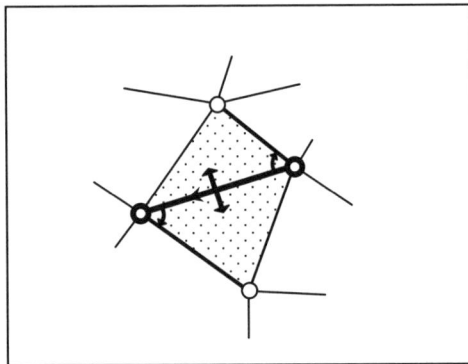

Figure 9.2. *Topologic links associated with an edge [LAN 02]*

9.4.1.3. *Local morphological characteristics carried by the edges*

A good morphological knowledge of the surface structure is needed for the construction of the upstream-downstream links between the elements. Firstly, this depends on the knowledge of the structuring role of the elements from the topological graph in this surface. Here, the edges play a central role and will carry the local morphological structure from the whole of the surface.

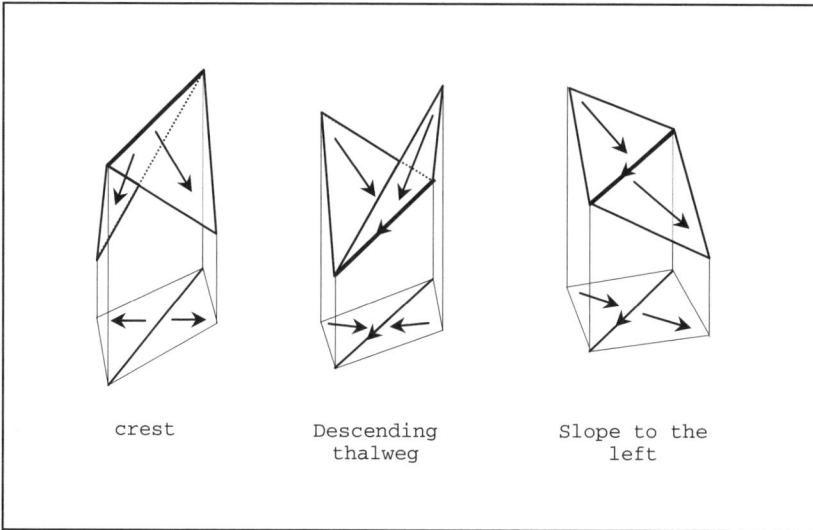

Figure 9.3. *Local morphology of an edge [LAN 02]*

In fact, for each edge connected to two vertices and two triangles, the relative altitudes and the relative slopes give the edge a morphological attribute. By comparing the altitudes of both vertices, we can tell if the edge is descending, ascending or flat. So, an edge that has a final vertex lower than its initial vertex is descending. If its two adjacent triangles present the same slopes (in the example, descending to the left), this is a descending thalweg (Figure 9.3). We now have a theoretical typology at $3^3=27$ possible combinations.

9.4.1.4. *The upstream-downstream links graph or the outflow graph*

Characterization of the thalwegs and the constituted linear network is not simple in that there is no continuity in the local morphology. When a thalweg enlarges, it is often composed of a large number of triangles which compose the base, in which the water circulates in a surface manner (locally, slopes), transversally to the edges. Local characterization of the arches is no longer sufficient to build a network. We have to globally and quantitatively consider all the upstream-downstream links between the different elements (poles, arches, triangles) of the topological graph.

The result is a graph of the upstream-downstream links, or outflow graph, which is similar to an oriented dual of the topographical graph, but in fact more complex, insofar as the dual graph links only the slopes, when this links vertices, edges and slopes between themselves.

A triangle can be connected upstream to three adjacent triangles at the most, but can also receive surfaces deriving from one or more thalwegs, coming from one or more of its vertices, which correspond to a link with the connected edges to its vertices. Likewise, it can be connected downstream at adjacent triangles. Particularly, when a triangle is connected downstream at two other adjacent triangles, we have to calculate the transition coefficients to define the proportion that will flow into each of the two triangles. These coefficients are calculated from the triangle's largest slope's line, which defines the proportion of the surface that naturally flows towards either one of the two downstream triangles. The problems linked to the organization of the outflow graph (a discharge of the basins, outflow on the flat parts) are numerous [DEL 01] [LAN 02].

9.4.1.5. *Cells and automaton*

The automaton's cells are elements of the topological graph, that is to say the triangles, edges and vertices that play a role in the outflow. The links between cells are those defined by the outflow graph, previously defined. Each cell is characterized by a definite number of attributes and properties, which can be summarized as follows:

– TabObjIn: table of entries, formed by the list of cells connected to the upstream of the cell;

– TabDataIn: table of entries' values in the cell coming from upstream;

– TabObjOut: table of outlets, formed from the list of connected objects downstream from the cell;

– TabCoeffOut: table of transition coefficients (in which the sum is worth 1) that serves to calculate the values of outlets towards the downstream;

– Etat, EtatPrec: state of the cell at time t and at the previous time t-1;

– DataExtIn,DataExtOut: external entry data and external output data;

– Motor: functional attribute which can affect a specific algorithm to each cell.

The motor calculates the state of the cell from its previous state, from the entry values and external entry data. It can also calculate or update all the external entries during the output.

Cells are built from an object hierarchy that we will not explain here, but in which triangles are inserted (TCellTrg), edges that serve to modelize the thalwegs (TCellArc), and particular vertices that manage the local minima or deepest slopes within the basins. Each one of the derived classes contains specific pieces of information which allows the knowledge not only of the geometry and topology of the element to which it is attached, but also the material content of the cell (surface, type of land use, infiltration capacity, precipitation, quantity of accumulated or in transit water).

9.4.1.6. *The cellular routing scheme (CRS)*

After an initializing phase, the automaton's mechanism is synchronous and iterative. Each iteration is divided into two phases: a communication phase between the cells and an evaluation phase from each cell's new state.

The communication phase consists of assigning its entry records to each cell (TabDataIn) from the upstream cell links (TabObjIn) that correspond to the cells' output links (TabObjOut). The transmitted values which have been calculated from the upstream cells' previous balanced state by the coefficient of the output of this link (Figure 9.4.). The upper boxes (V) refer to surfaces inputs in cells per iteration; the lower box (S) summarizes the cumulative sum of surfaces drained in each cells at the end of simulation. And values out of the box indicate surface flows between the two iterations t_n and t_{n+1}.

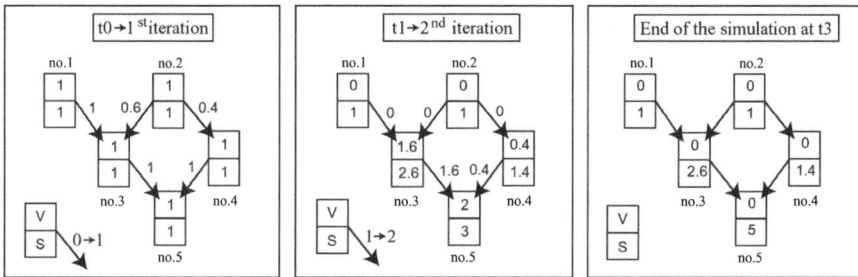

Figure 9.4. *Automaton example calculating the upstream surfaces. The upper square (V) refers to the entries in the cell at each iteration and the lower square (S) refers to the cumulated transit in each cell [LAN 02; DOU 06]*

The evaluation phase recalls each cell's motor in order to calculate the new state of this from its previous state and the value of entry registries communicated in the first phase. This iterative process is controlled by two ways. It can either become void as a result of the convergence of the accomplishing system towards an attractor or by the passing of a number of iterations fixed by the user. It is important to note that the step of discretization of time, configurable by the user, must consider the size of the cells (depending on the scale of the DTM) and the general dynamic of the basin (average slope).

Before reaching an operational stage, it is necessary to validate the mechanism of the system through an automaton application test, which is the calculation of each triangle, of the surface situated in upstream of this triangle. The cellular routing scheme, necessary for this calculation, is very simple: at each automaton iteration, we must cumulate the surfaces "input locations" (that is to say the cells immediately upstream). When the system converges, each triangle contains the sum of the areas of all the cells upstream (Figure 9.4).

In the initialization phase, the v_i state of each triangular cell is affected by the surface of its proper cell (the arches and the basins are initialized at 0).

In the communication phase, the entry records (w_{1j}, w_{2j}...,w_{nj}) of the j cell are affected by their value $w_{ij}=c_{ij}\ v_j$.

In the evaluation phase, the entry records are added to obtain the new state of the j cell:

$$v_j = \sum_{i=1}^{n} w_{ij}$$

The sought after result is obtained by cumulating the different states of each cell, as one goes through the iterations.

Which gives, at the end of the bow, the cumulation of the values of the $v^t_{\ j}$ state of the j cell at different moments.

$$s_j = \sum_{t} v_j^t$$

This cellular automaton routing scheme allows the calculation of the upstream surface by successive iterations. Furthermore, by the algorithm of a progressive upturn from a given point, we can mark off the watershed in a precise manner, the results being much better than those obtained with the functions of a classic GIS [CRA 01], [MIT 01].

9.4.2. Results from the simulation

The cellular automaton *RUICELLS* has the possibility of flowing from cell to cell, thanks to the outflow defined by the graph, whatever the kind of quantitative data. In this approach, the surface, which is the first information the cell contains, is principally used [DEL 02].

Following the same principle as for the outflow of water, the automaton induces the outflow from cell to cell, according to the flow direction and axes defined in the graph. At each iteration, all the cells evacuate themselves to the other cells that are connected downstream and receive the surfaces coming from those situated upstream. The iteration process completes itself once all the surfaces have transferred by the measuring point, concretely when the entire basin is emptied from its surface.

According to the option chosen at the beginning of the simulations, two processes are possible. In the first one, the circulation rhythm of the surfaces is controlled by the CRS. The speed, regulated by the computer's iterative process, is constant at all points in space. In this case, the cells' slope is not considered and only the incidence of the watershed's form in two dimensions is analyzed.

In the second case, the slope's effect is introduced in the simulation pattern by a slant in a formula that is affected to the cells and which modulates the surface transit speed. Thus, the simulated result translates the effect of the form at the basin water surface, but equally the impact of the relief.

This combined approach helps us to analyze the part of each variety of shape (two or three dimensions), and also to characterize the role of morphology in the hydrological response of a basin. At the end of the simulation, the model can display the surface outflow chart in the form of colored classes, according to the total surface flowing in each cell. This mapping presents the preferential outflow axes and the hierarchy of these axes. Two numerical files are equally conserved at the end of the stimulation. The first gathers the values used for this mapping, notably the surface amount passed in each cell. The second contains the surfaces passed to the outlet (measuring points) at each iteration. In some ways, this has to do with a hydrogram, but used at the surfaces.

Conveniently, the simulated curve developed solely from the distances at the measuring point is called a "surfaceogram", and the one which equally integrates the speed is called a "morphogram" to dissociate the simulations.

In order to further appreciate the surface behavior of the basin, some simple indices are taken from data charts (the flow of the surface, the flow of the point brought to the surface, etc.). Such indices are more synthetic and integrate the dynamic influence of the morphology better than the classical methods.

9.4.3. *Applications and contributions of the cellular automaton*

Some results are proposed here to illustrate our subject (Figure 9.5). They were taken from the study of three 3 to 4 order basins located in Seine-Maritime in Upper Normandy, northern France. The morphograms simulated at the final outlet of the three basins vary considerably. The form of the basin seems to play a minor role. The skewness of the network translates by a much bigger appearance than the morphogram of basin 1. The flow of the surface hardly exceeds 6 ha/mn but tangents this value during a long enough period. The morphogram of basin 2 is higher with a point that exceeds the 8 ha/mn.

These results bring to the fore the role of the reach organization. The network architecture of basin 2 seems far more efficient with a regular reach distribution which optimizes the distance between the executory and the points farther from the basin. On the contrary, when the network is very dissymmetric like the first basin, which presents a major axis, solely fed in the left bank by long thalwegs, the distance traveled by the water is much longer with a layout going from the north to the south at first, then from the west to the east. The surfaceogram takes a bigger form.

The shape of the basin, apart from the structure of its network, does not have any significance. The good distribution in the plan of the network and the distribution of the spatial connections are fundamental. This network's effectiveness can be concretized by different architectures. The well organized networks like the one of the Lézarde basin constitute the first type. The basin is drained by three main thalwegs, of equal lengths, which meet at a point close to the executory. The distribution of the distances at the executory, being the same from one sub-basin to the next, the morphogram takes a similar shape of the triangle with a very important point flow.

The behavior of basin 3 varies considerably from upstream to downstream. This upstream part determines itself the peak of the discharge simulated at the final outlet. The morphogram has a form close to that of the unitary hydrogram and the peak is nearer to 11 ha/mn, while this value decreases at the outlet of the basin due to reinfiltration of upstream parts.

Such behavior on basin 3 is explained by the network structure which is very efficient due to the convergence of all links in one point which favors a massive arrival of the surface at the same time. This effect is reinforced by the shape of the basin. Thanks to its triangular form, all the points situated along the waters' sharing line, are located at equal distance from the outlet.

This example proves that the basin's behavior is not stable through scales, except in the case of the networks which present a marked fractal character like the Lézarde basin, where there exists an obvious internal homothetic character [ROD 97].

Most of the basins present hydrological signatures that vary very rapidly within the basins. The outflow process is continuous in space and the methods of analysis based more often on a discretization of space (the counting of reaches) drive to biases in the interpretation. One of the first advantages of the cellular automaton, with the possibility that it can place an infinity of measuring points, resides in its capacity to reconstitute the global dynamic of the basin in its continuity and to detect, in a precise manner, the behavioral changing points and efficient parts hidden at the global scales.

In some cases, a small part of the basin explains the general signature of this. For Saint Martin, the downstream part serves as a transporting axis, but does not contribute to the development of the flow's peak.

These results lead up to the discussion of the notion itself of the study of a watershed. Choosing a reference basin materialized by an outlet is choosing to erase part of the particularities of the watershed and to study an average behavior upstream of this point. This average behavior is representative enough when the networks present an important internal homothety, but this choice can totally bias the interpretation when the basin becomes more composite and heterogeneous.

The necessary integrated approach of the network and of the previously described surface must be made through scales by favoring a continuous approach of space.

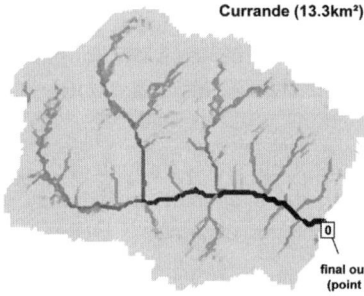

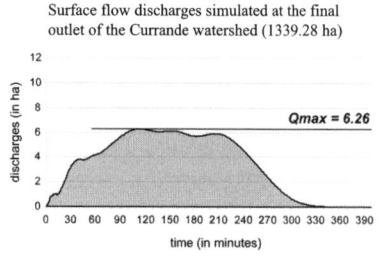

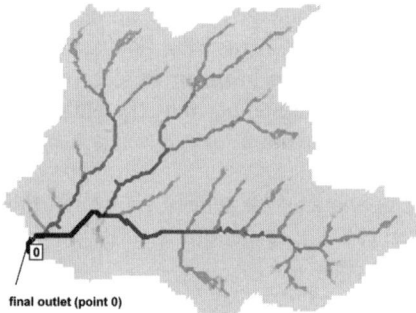

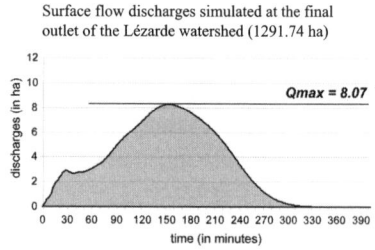

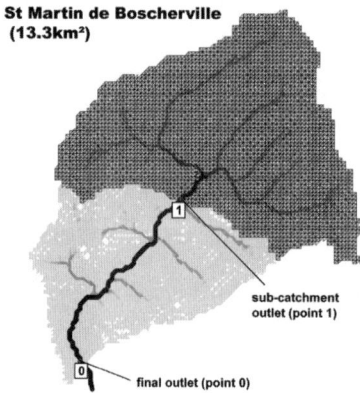

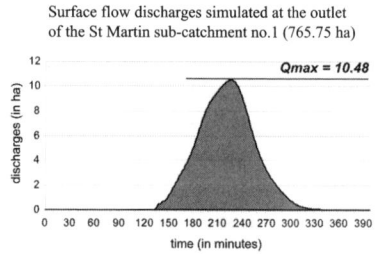

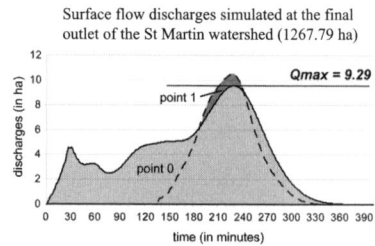

Figure 9.5. *Various morphograms simulated on three different basins*

9.5. Conclusion

Morphology has often been greatly forgotten in hydrologic studies. To be convinced, we must just refer to the very limited number of pages dedicated to it in all the works on hydrology. This disinterest is in part due to the incapability of the "static" indices of shapes and reach organization to translate the potential dynamics of a basin.

The methods emerging from the complex systems theory offer new perspectives. The previous sections have shown that a cellular automaton can be a good tool for geomorphologic investigation. Contrary to the classic indices that search for the "normality" of the organization of shapes (homothety), the automaton can "track down the morphologic abnormalities" which are truly explanatory of the hydrologic behavior. It consists of an analysis of the dynamics of the interactions between the morphological components (surfaces, slopes, networks, etc.) in a continuous spatial behavior. This dynamic character in the approach helps to further understand how the surfacic flow is developed through time and space. Until now, such a question could only be addressed indirectly by the study of flows which could only be done on the scale of large calibrated watersheds in order to dispose of a consequential network of measures. The approach proposed here is based on the simulation from the watershed's physical characteristics, and offers the possibility to extend research to all the waterways whether they are gauged or not.

9.6. Bibliography

[BEL 98] BEL HADJKACEM M.S., CHEVALIER J.J., ROBERT J.L., GOLD C., "Intégration de la méthode des éléments finis dans un système d'information hydrologique pour la gestion stratégique des eaux en surface", *Revue Internationale de Géomatique*, 1998, 8, (4), 301-318.

[BEE 93] BEER T., BORGAS M., "Horton's laws and the fractal structure of streams", *Water Resources Research*, 1993, 29, 1475-1487.

[BLÖ 95] BLÖSCHL G., SIVAPALAN M., "Scale issues in hydrological modelling: a review", *Hydrological Processes*, 1995, 9, 251-290.

[BRA 97] BRAVARD J.P., PETIT F., *Les cours d'eau; dynamique du système fluvial*, Armand Colin Ed., 1997, Paris, 222 p.

[CHO 72] CHORLEY R.J., *Spatial Analysis in Geomorphology*, Methuen and Co Ltd, London, 1972, 393 p.

[CRA 95] CRAVE A., Quantification de l'organisation des réseaux hydrographiques, PhD Thesis, University of Rennes I, 1995, 210 p.

[CRA 01] CRAVE, A., DAVY, P., "A stochastic 'precipiton' model for simulating erosion/sedimentation dynamics", in *Computers & Geosciences*, 27 (7), 2001, pp.815–827.

[CUD 00] CUDENEC C., Description mathématique de l'organisation du réseau hydrographique et modélisation hydrologique, Thesis, ENSAR, Rennes, 199 p. + appendices.

[DEL 99] DELAHAYE, D., GAILLARD, D, HAUCHARD, E. (1999), "Analyse des processus de ruissellement et d'inondation dans le Pays de Caux (France), intérêt d'une approche géomorphologique", in *Paysage agraires et environnement*, Stanislas WICHEREK (ed.). CNRS Editions, 209-219.

[DEL 01] DELAHAYE D., GUERMOND Y., LANGLOIS P., "Spatial interaction in the run-off process", *12th European Colloquium on Quantitative and Theoretical Geography*, St-Valery-en-Caux, France, September 7-11, 2001, www.cybergeo.eu/index3795.html.

[DEL 02] DELAHAYE D., Apport de l'analyse spatiale en Géomorphologie. Modélisation et approche multiscalaire des risques, research paper, University of Rouen, 2002, 471 p.

[HAU 01] AUCHARD E., De la dynamique non linéaire à la dynamique du relief en Géomrophologie. Application aux bassins-versants de la marge nord occidentale du Bassin de Paris, PhD Thesis, University of Rouen, 2001, 779 p.

[HEU 98] HEUDIN J.C., *L'évolution au bord du chaos*, Hermes, 1998, 185 p.

[HOR 45] HORTON R.E., "Erosional development of streams and their drainage basins; hydrophysical approach to quantitative morphology", *Bull. Geol. Soc. Am.*, 1945, 56, 275-370

[KIR 76] KIRKBY M.J., "Tests of the random network model and its application to basin hydrology", *Earth Surfaces Processes*, 1976, 1, 197-212.

[LAB 87] LA BARBERA P., ROSSO R., "The fractal geometry of river networks", *E.O.S Trans. A.G.U.*, 1987, 68, 1276-1283.

[LAB 89] LA BARBERA P., ROSSO R., "On the fractal dimension of stream networks", *Water Resources Research*, 1889, 25, 735-741.

[LAB 90] LA BARBERA P., ROSSO R., "On the fractal dimension of stream networks, reply to Taborton *et al.*", *Water Resources Research*, 1990, 26, (9), 2245-2248.

[LAM 96] LAMBERT R., *Géographie du cycle de l'eau*, Presse Universitaire du Mirail, Toulouse, 1996, 439 p.

[LAN 94] LANGLOIS P., "Formalisation des concepts topologiques en géomatique", *Revue internationale de géomatique*, 1994, 4, (2), 181-205

[LAN 97] LANGLOIS A., PHIPS M., *Automates cellulaires Application à la simulation urbaine*. Hermes, 208 p.

[LAN 02] LANGLOIS P., DELAHAYE D., "RuiCells, automate cellulaire pour la simulation du ruissellement de surface", *Revue Internationale de Géomatique*, 2002, vol. 12, no. 4, 461-487.

[LIU 92] LIU T., "Fractal structure and properties of stream networks", *Water Resources Research*, 1992, 28, 2981-2988.

[MAN 68] MANDELBROT B., WALLIS V.R., "Noah, Joseph and operational hydrology", *Water Resources Research*, 1968, 4, 909-918.

[MAN 95] MANDELBROT B., *Les objets fractals*, Flammarion, Paris, 1995, 212 p.

[MIT 01] MITA C., CATSAROS W., GOURANIS N., "Runoff cascades, channel network and computation hierarchy determination on a structured semi-irregular triangular grid" in *Journal of Hydrology*, 2001, 244, pp.105-118.

[MON 92] MONTGOMERY D.R., DIETRICH W.E., "Channel initiation and the problem of landscape scale", *Science*, 1992, 255, 826-830.

[NEW 90] NEWMAN W.I., TURCOTTE D.L., "Cascade model for fluvial geomorphology", *Geophys. J. Int.*, 1990, 100, 433-439

[PAL 98] PALACIOS-VELEZ O.L., GANDOY-BERNASCONI W., CUEVAS-RENAUD B., "Geometric analysis of runoff and the computation order of unit element in distributed hydrological models", *Journal of Hydrology*, 1998, 211, 266-274.

[PAP 86] PAPADAKIS C., KAZAN N., Time of concentration in small rural watersheds, Technical Report, University of Cincinnati, Ohio, 1986, 68 p.

[PER 87] BAK P., TANG C., WIESENFELD K., "Self-organized criticality: an explanation of 1/f noise", *Physical Review Letters*, 1987, 381-384.

[RIG 93] RIGON R., RINALDO A., RODRIGUEZ-ITURBE I., IJJASZ-VASQUEZ E., BRAS R.L. (1993), "Optimal channel networks: a frameworks for the study of river basin morphology", *Water Resources Research*, 1993, 29, 1635-1646.

[RIN 92] RINALDO A., RODRIGUEZ-ITURBE I., RIGON R., BRAS R.L., IJJASZ-VASQUEZ E., MARANI A., "Minimal energy and fractal structures of drainage networks", *Water Resources Research*, 1992, 28, 2183-2185

[ROD 97] RODIGUEZ-ITURBE I., RINALDO A., *Fractal River Basins; Chance and Self-Organization*, Cambridge University Press, Cambridge, 1997, 547 p.

[SHR 74] SHREVE R.L., "Variation of mainstream length with basin area in river networks", *Water Resources Research*, 1974, 10, (6), 1167-1177.

[SMA 68] SMART J.S., "Statistical properties of stream lengths", *Water Resources Research*, 1968, 4, (5), 1001-1014.

[STA 91] STARK C.P., "An invasion percolation model of drainage network evolution", *Nature*, 1991, 352, 423-425.

Chapter 10

Understanding to Measure…
or Measuring to Understand?

HBDS: Towards a Conceptual Approach for the
Geographic Modeling of the Real World

To what extent does modeling the "real world" make it possible to understand it? For us geographers, what do our models of the spatial complexity hide when they endeavor to make the world intelligible?

10.1. A forgotten face of the geographic approach

It is true that, like any other discipline of scientific nature, geography has continued to clarify its foundational concepts in order to make them more functional. Yet in this respect, we can only be struck by the highly unequal nature of the interest shown by the geographers in the different concepts that Pinchemel strove to list around 15 years ago. Evidently, several of these concepts such as "space", "distance", "diffusion", "interaction" in the domain of the general processes, or even "peri-urbanization", "mobility" in the domain of the themes, to name a few, have been subject to extensive questioning since their emergence. At the crossroads of the methodologies and the themes, certain avenues opened up such as that explored by Dumolard's *Geotaxinomy* [DUM 81], but some other equally important approaches have not been subject to comparable investigations. This is particularly so in the case of "geographic information" and "modeling".

Chapter written by Thierry SAINT-GERAND.

10.1.1. *The causality in question*

In fact, despite the profuse literature that refers to it, "geographic information" and "modeling" have often not been considered by the geographers for what they are in the first place: conceptual structures that geography implements to fill in and synthesize the dimensions pertaining to its way of thinking. Here we obviously do not mean to undermine the significance of the work that has been carried out for several years as much in France as in other countries in these two key domains of spatial analysis. It is simply an observation that in geographic literature, there is a big disparity between the considerable number of publications focusing on the *existence* of these two concepts, that is, the factual information that their implementation through various processes makes it possible to obtain, and the very small number focusing on their *essence*, that is, to the study of the internal logics that structure any spatial representation of the "observable world" conveyed by these concepts. In other words, much more work has been done in the domain of the spatial models of the working and *evolution* of the observable world than in the domain of the spatial models providing this *description*. Even today, for example, researchers are more at ease *measuring* the interaction between cities with the help of a small number of comprehensive "indicators", than establishing an *overall spatial description of all the elements* involved in the phenomenon of interaction itself. With Thom, let us recognize that "*predicting* is not *explaining*". Let us add further: "nor even making explicit". This nuance is not without importance. Being capable of predicting the behavior of a phenomenon would allow us to believe that we know its causes and processes intimately. Is this certain? A famous example throws light on the question in a somewhat disturbing manner.

Just as Newton admitted with good grace that he could not explain the gravitational force whose existence he had however demonstrated, the spatial interaction models derived from his work attest, without really explaining, that the exchanges between cities can be estimated by a function, integrating their size and respective distance in particular. Sure. Even so, is it not reckless to fully assimilate the inner structure of functioning of a phenomenon with its external structure of behavior? Isn't the meaning given to the words "information" and "explanation" as per the accepted definitions of these types of models (respectively numeric variability of a "comprehensive" indicator and percentage of this variability rendered by the model) simplistic? What about spatial complexity? These are difficult, disturbing questions (even sacrilegious for some) but pertinent if we claim to evaluate cause and effect relations through forms of facts supposed to be representative of their functioning.

Whatever the case, the preoccupation with the functional aspect prevailed over the preoccupation with the descriptive aspect such that the conceptual questions posed by the two methodological foundations of analysis, that is, "geographic information" and "spatial modeling" – the geographer's tools for apprehending "the spatial complexity" – were largely abandoned.

This observation is all the more paradoxical as the applications based on a modeling have been spreading in geographic literature at an increasing speed for the last few years. Progressively, geography has extended its palette of methods and techniques thus making it possible for several of its founding concepts (location, stranglehold, distance, scale, combination, interaction, neighborhood, etc.) to be implemented through the processing functions of what constitutes its own conceptual "atom": *the localized data.*

10.1.2. *The concept in the light of the technique: "collisions" and misadventures of a couple in disharmony*

The form and the effect of this evolution, accomplished in a few decades, can be understood only by putting the methodological and technical instrumentation levels related to each era into perspective. The history of information and modeling in geography, peppered some time ago by vigorous debates, to say the least, on the quantitative approach, and nowadays on geographic information systems (GIS), shows, whether we like it or not, the importance of the evolution of technical resources in the way researchers' thinking has developed. Everything happens as if certain concepts remained dormant, waiting for the discovery of a "technical feasibility" liable to unlock its implementation and subject them to thematic experimentation testing. Herein perhaps lies the explanation for a striking paradox of the 1970s. While systemic thinking and the conceptual methods that it requires were already being developed elsewhere, geographers, in the quest for an overall vision of space, remained attached to the modeling methods – quantitative and functional – that they had just appropriated, whose usage had to be mastered and whose contribution had to evaluated. This led many among them to continue to think about geographic space by assimilating it to a statistic space.

This historical overview throws light on the origin of the blazing controversies that emerged at the time regarding application of quantitative methods to geography. For some, the opposition came from a refusal to introduce a mathematical tool in a human science mainly based on the direct observation of elements that are sometimes hardly measurable. However, for others, it came from quite an understandable reluctance to admit three hardly insignificant presuppositions, which imply certain uses of statistics with respect to geographic phenomena. They are as follows:

1) a spatial relationship between objects always corresponds to a statistical relationship between their attributes;

2) the description of a spatial phenomenon can be summarized as statistical scores;

3) comprehensive and abstract indicators are enough to give information regarding the functioning of a space.

In addition, this methodological orientation, which, as we understand clearly, by nature, removes the form of the objects themselves and often even the configuration of their distribution, potentially involved three risks:

1) in the spatial reasoning; substituting *ipso facto* a table of measurements with a map, in other words, amputating the processing of geographic space from what constitutes its own specificity: its geometric and topological components;

2) empirically choosing the descriptors of the considered phenomenon, without backing up this choice with a more formalized method than the single "experience" that the author has on his subject;

3) deliberately removing certain domains of information on a phenomenon, insofar as the data pertaining to these domains is unavailable or not exploitable.

The concern for efficiency and the absence of methodological alternatives made some geographers gradually pick up the habit of only indirectly considering (this is not to say sidestep) the forms of the space in the description and the quantitative processing of the geographic phenomena. The thematic cartography, still largely manual, offered response times that were too long to systematically help in the spatial validation of digital results. An insidious slide thus occurred in the representation of the real world, from the spatial/cartographic mode to the digital/statistic mode.

This orientation had the merit of responding to a growing social demand, by unblocking the use of gigantic statistical surveys pertaining to the entire national territory (general population census, general agriculture census, etc.). However, it is nonetheless true that spatial grids defined as per the criteria pertaining to the administrative management were quite often used as a canvas meant for sampling, or even formatting the data derived from unrelated logics.

A posteriori, we can wonder if this mode of approach, which saw statistics take precedence over spatial analysis around the 1970s did not really owe its success to these intrinsic qualities alone, that is, its capacity to quickly provide comprehensive results and figures at social demand. In fact, compared to this age of information technology, computers were hardly efficient except in the domain of calculations, such that the statistical path was more or less the only way out available to researchers who desired to expand their methodological horizon beyond the ambitions of conventional geography. When, with Infographics and the Data

Management System, computers made it possible to carry out graphic representation and thematic representation of objects, and then associate the two in "automatic" cartography systems, a decisive process was accomplished in the direction comprehensively representing spatial complexity.

This progress in data *management* brought about the development and then the integration of relational, topological and geometric models. Thus was born the "geo-relational" model that GISs use even today to give the researcher the means to model spatial phenomena in localized databases.

While the use of computerized software models as they are is a technical matter and their application to some or other spatial question a thematic matter, transcending these models to understand their underlying logics and redeploying them in a systemic methodology of geographic modeling of the real world remains a conceptual matter, which we must now approach.

10.1.3. *The conceptual modeling of the geographic phenomena: a necessary prerequisite, why and how*

This issue leads us to scrutinize the very foundations of geographic perception and the properties specific to the information that it generates, which constitutes a sort of "internal composition law". It remains fundamental in my eyes, even though its declared objective, modeling space to describe the real world, could seem less ambitious than the objective of functional modeling, whose practical virtues as well as theoretical difficulties, when it resorts "*ex abrupto*" to quantitative methods, have already been mentioned.

Does not geography bring the researcher face to face with a spatial complexity for which simplistic explanations turns out to be caricatured, if not incomplete? Is not geographic space the archetypical case wherein, as indicated by Alain Boutot [BOU 93]: "the number of parameters governing the evolution of a system is such that it defies ordinary quantitative analysis...In a case of this type, quantitative analysis appears to be an impossible ideal. It is absolutely helpless and must make way to a structural analysis, which is perhaps more modest but is the only one capable of providing a beginning of intelligibility to the phenomenon, that is, of *reducing the arbitrariness of its description*". The question that emerges is not so much that of a methodological alternative but rather that which complements a spatial reasoning. In addition, in line with Mandelbrot, Thom and the morphologists, who do not claim any methodological exclusivity, I will simply replace "*make way for*" by "*be preceded by* a structural analysis". This simple substitution of terms considerably modifies the state of the things in the matter of research protocol. The objective of course remains identical: understanding the functioning of the real world through the image that a space gives it. However, the process differs in its progression. Functional analysis becomes subsequent to the explanation of the mental representation of the phenomenon, its components and their relationships. By doing so, it benefits from the act of intervening in a well "prepared" conceptual field, enabling a detailed identification and use of indicators that are to be taken into account. It is liable to reduce the role of "residues" and "noises", consequences of indicators, which are perhaps pertinent but neglected due to their marginal role *in the digital space*.

Carrying out the conceptual modeling of a phenomenon before processing the indicators, which this overall view will have made it possible to unearth, thus clearly seems to be the main and ultimately unavoidable task. It does not exclude, quite on the contrary, other more mathematical (statistical, geostatistical) or graphic (cartographic) forms of modeling but rather throws light on their implementation. A data structure that is well conceived from the beginning will consequently often avoid dealing excessively with calculation formulas, with the risk of turning into black boxes with opaque results. This reflection becomes crucial, in any case, in a spatial analysis process having to exploit the functions of the GISs.

In fact, contrary to a still extensively widespread idea, constructing a GIS, whatever the domain of application might be, is not limited to coupling computer aided cartography (CAC) software – however sophisticated it might be – with a localized database manager or a statistical tool.

This is because the information, before being produced is, above all, a process, and the concept of GIS requires first of all, the conceptual modeling of the considered spatial phenomenon. It means listing its components (objects) as per their level of definition, their nature, the semantic, spatial or temporal repositories that they are derived from as well as the relationships that turn these components into a system whose behavior and dynamics are to be apprehended. This modeling finds its formal expression in a schema commonly called the "conceptual data model" (CDM) and its "operational" expression in the structure that is deduced from it for organizing spatio-thematic databases.

The conception phase is fundamental in the constitution of a GIS insofar as developing a data structure requires the researcher to explain in detail his vision of the phenomenon studied. It takes precedence over the technical realization phase, which is certainly a significant consumer of human, material and software resources, but also one which represents only the development of the "extension" of the information that the CDM must already contain in "intention". It is thus not a computer process in the technical sense of the term but, above all, a thematic information process. By using it, the researcher is obliged to formulate his conceptual universe, that is, the concepts, the level at which the repositories are taken into account (reality, space, time). This is not an insignificant boon. In a nutshell, he specifies the perception of the observable world for which his subject presents an example.

This decision to describe a phenomenon comprehensively and in terms of concepts *before* processing them analytically makes one see the advisability of an intermediary stage: the choice of data organization. It is during this structuring phase that it is particularly important to take into consideration certain specificities of the geographic phenomena and the information associated with it. The GISs are, above all, data models.

10.1.4. *The GIS: a special spatial information system*

The protocols to be implemented when constructing an information system have abounded for a long time in specialized literature, notably as a result of work done by Tardieu [TAR 95], and Mélèse [MEL 72]. The "phasing" that makes it possible to go through progressive stages of the ideal project for operational realization is explained here in detail. Has everything been said, for all that? Are the norms of the protocols unchangeable? Is a GIS an information system just like any other? Far from it, as two remarks are called for:

1) GIS is a very particular form of information system due to the role that the spacialization of the phenomena to which it is applied plays in structuring the input data and the output information. In this respect, it moves away from the majority of other information systems that deal with the management of companies, banks, industrial processes and such like wherein, evidently, space does not have this structuring role.

2) Two main orientations are distinguished in the use and hence the configuration of the GISs. The great majority of the GIS applications couple, at a relatively elementary level, an interactive thematic cartography with localized database requests. They are, in fact, spatial documentation systems often related to jobs, making it possible to apprehend phenomena *in* the space. Their domain is situated especially at the territorial management level (DDE, urbanization services, etc.).

While it is true that nowadays we do find, isolated in the literature, a few avant-garde books like those by Léna Sanders [SAN 96], but on the whole, apart from a few exceptions in the domain of geomarketing, only a tiny minority of GIS applications are inscribed in a true spatial analysis project, that is, aiming to apprehend phenomena *through* the structures of their spatial inscription and the resulting relationships. These applications are derived from geographic formalization. In 1989, Sylvie Rimbert wrote [RIM 89]: "In the field of GIS...

research and applications have interests that are more complementary than common. The planning organizations have, above all, human and economic management concerns, which they must take stock of regularly; researchers must have concerns pertaining to, above all, structures and relationships of the terrestrial observations whether they are taken stock of or not."

Thus two uses of GIS do exist, even though an unfortunate amalgam perpetuates about them.

For each project, there is a different protocol. While in the two cases, the list of tasks remains unchanged, it varies considerably for the hierarchy between the same tasks. The applications of the management type calls for a very thorough organizational phase, relying on a team of complementary specialists, because they must make it possible to manage the functioning of a great number of services and activities involving multiple structures and personnel. The objects, functions, flows and concepts on the other hand result from well identified "job domains". They benefit from an easy tagging (often preestablished nomenclatures), coming straight from the daily pattern.

Things happen very differently in research. GIS projects are (alas!) often the act of a single individual (i.e. the PhD student), or, at the most, a few (in the case of statutory researchers). Their "universe of discourse" is obviously more open: the associated objects and protocols, even the concepts, are not necessarily all preestablished, even less "ready for use". By definition, research progresses – at least in part – in the unknown. A preliminary conceptualization thus cannot be avoided without the risk of a bias in the data structure: the consequences may be scale discordances, heterogenity in the precision levels, absence of pertinent semantic parts that will lead sooner or later to the blockages in the data processing and their interpretation. Furthermore, is not knowing that we do not know, not the main and the most punishing ignorance.

In the context of spatial analysis research, each GIS project finds itself confronted, for the domain that it is pertained to, with this elusive nature of the aforementioned "real world". Each project requires us to find its adequate form of practical restriction of the real, that is, its sub-set of "observable world". It means setting, every time and in the best way possible, the prehension of the data to the prehension of the space which we wish to associate it with. Several protocols and therefore, several visions have long been recognized: the isarithmic approach that considers the variations of states in a continuum, the meshed approach and the choropleth approach, which conceives spatial entities, wherein space is defined as continuous and homogenous: these entities are called "spatial objects", including geographic objects, which are a special case [CHE 95].

Applied to the modeling of a geographic phenomenon in a localized database, this approach integrates one of the main characteristics of the perceptive process in geography, that is, the granularity principle of observation, commanded by the scale of identification and description of objects. This principle is similar to that of a filter with variable focal lengths, grid widths and logics.

The informative material thus taken from the observable world in the form of spatial objects and localized data that is related to them possesses a level of "granulometry" particular to the overall setting. This is chosen deliberately by the observer from among all those possible: this setting defines its conception/perception scale and fixes the validity thresholds of the information that emanates from it.

Describing a phenomenon with the help of localized data thus presupposes, at the base, its positioning at a given level of the system of multiple correspondences between the different levels of definition of its observable reality and the different levels of spatial definition. It is the specificity of a conceptual data model in a GIS

10.1.5. *The geographic object: logic makes the entity*

Developing such models in the respect of the specific objective that has just been stated elicits a major difficulty that has already been raised by numerous authors (Joël Charre, Christiane Weber, Sylvie Lardon, etc.) since the geopoint symposium of 1994 [GEO 94]: how do we overcome the inherent dichotomy between two families of components: that of spatial entities and that of semantic entities?

In fact, the conceptual models retained nowadays for structuring the localized data rely almost exclusively on concepts taken from infography (raster, vector, "spaghetti", polygon, etc.), implemented either as they are, or through the intermediary of neighborhood mathematics (topology implemented via graph theory) or even of relational geography. This type of approach favored the development of powerful processing function of the formal envelope of the spatial objects (generalization, aggregation of zones, interlevel, intersections, distance calculations, etc.). However, it left the processing of its thematic content somewhat in the background, relegated to the role of an external, inert "tail attributes", implying a significant hypothesis in terms of representativity of the models induced, according to which the semantic formulation of a spatial object would be independent of its spatial formulation. We can consider that this aspect only presents the visible face of a fundamental problem, which is much more vast and which has, in fact, remained unresolved until now in the analysis of the observable world through the space: the explicit introduction in a spatial reasoning, of a level of conception that is not only spatial but also thematic and hence semantic and temporal. Not a very long time ago, the problem appeared such that some researchers considered [TOM 78] localized data as unfit for conforming to computerized models. Recent work has started to open up ways, such as those pioneered by the Temps-Espace group of the GDR CASSINI, but operational solutions are still some way off.

In fact, taking up this challenge leads us to look for methods of modeling that the computer specialists call a "semantic network", which is a special form of hierarchical, entwined and multidimensional structure that is in the end nothing other than the basic structure of the reductionist approach prevailing currently in the databases.

Endeavoring to surpass this present limit in order to adjust the processing functions of spatial granularity with the functions of conjunctive processing of its thematic and temporal granularity opens a path for theoretical researches and practical experimentations. It is time to embark on this path, and is even desirable to find geographers interested in it if we wish to develop the performances of the GISs in spatial analysis. It would seem that the lack of interest in descriptive formalization of the spatial phenomena explains the glaring absence in the current GISs of certain types of functions useful in spatial analysis. This includes those related to the redefinition of geographic objects, to the acknowledgment of immaterial spaces, or even to the regulation of spatial attraction models at the time of scale changes (Huff model, for example).

The problem that emerges does so at a general level: how can we introduce a formalized geographic reasoning in the organization of the databases?

10.2. Formalizing a spatial reasoning in databases

Every processing presupposes a phenomenon structure and a data structure. Ignoring their preliminary scrutiny almost always leads us to stumble upon incoherencies later, but in this case of course it is too late.

10.2.1. *Operational structures for the geographic modeling of the real world*

On the conceptual level, that is, beyond the constantly evolving technical aspect, the definition of the database that Gardarin [GAR 89] proposed 10 years ago remains in force, "a set of data structures according to their properties, assembled for modeling a given universe". Yet, it has moreover been demonstrated [SAI 02] that, explicitly or otherwise, all data derives its sense from the object to which it is linked and which it describes by the single act of bearing a modality of one of its attributes. This object does not have, by itself, a reason to be recognized except with reference to a phenomenon with which it is associated and for which its acknowledgement is justified by observed proofs and formulated hypotheses. By doing so, the object is an unavoidable referent and its identification, a preliminary to any collection and formatting of the data that fills it, is a methodological prerequisite. Its identification positions the level of semantic and spatio-temporal conception at which the analysis is placed. This is particularly important in the domain of localized data management whose pertinence, semantic and geographic precision, as well as validity, are closely related to the spatial object that they describe. It is then a question of adapting the geographic phenomena to categories of spatial objects (or even eventually non-spatial ones) established in accordance with the hierarchized levels of conception/representation, clearly defined with respect to their mutual relationships that are demarcated without ambiguity in the field of thematic and cartographic generalization.

There is no denying that until now, while researchers and engineers working in this field agree with each other without too much of a problem on "the objective to be attained", a certain ambiguity still reigns on the methods for ensuring such an objective is attained. Despite the progress accomplished during the last 20 years thanks to computers, numerous problems can still be found in this field.

10.2.2. *Preliminary research into the data structuring methods: a historical overview*

In fact, recent history of the database conception methods demonstrates the succession of several periods that are strongly conditioned by the level of software engineering on one hand and the evolving state of computer hardware on the other hand. Succinctly, it can be considered that until 1975, when the rarity of the screens and the slowness of the infographic management considerably restricted the endeavors of the human-machine interface (HMI), priority was given to the structural analysis of the "processings" and not to the structural analysis of the "data". The latter found itself relegated to the filing cabinet. Between 1970 and 1980, new ideas flowered in the dominion of the work carried out by J. de Rosnay [ROS 75]. In the field of formalism and structuring rules, these new ideas have a common point: conceiving data and processing on the same plane and structuring the application development process into 3 successive levels.

10.2.2.1. *The conceptual level*

This considers, above all, the nature and the structure of the phenomenon or phenomena studied, outside of all considerations related to the needs or to the technology. It means an overall, comprehensive approach, aiming to model, most often graphically, the conception of the sub-set of the "observable world" that the information system is responsible for translating. This phase of work leads to the constitution of a "conceptual data model" (CDM) sometimes also called "application schema", which remains, subject to updates that may turn out to be necessary in the future, the fundamental referent, with every database evolution having to be carried out in agreement with it.

10.2.2.2. *The external level*

This refers to the definition of logical entities on which the user reasons when operating the system. Also called external schema, or sometimes logical model, this structure must systematically be brought to the conceptual data model in order to verify that it is consistent with the fundamental structure of the application.

10.2.2.3. *The internal level*

Here the retained logical structure is concretized into technical and physical "settings" such as registration types, file types, field lengths, access modes (sequential, direct, indexed sequential, etc.). These elements are in keeping with the current capabilities of software engineering and hardware technology.

These distinctions are still in force today, as updated graphic representations (see Figure 10.1).

On the conceptual plane, several families of thought succeeded one another. Their differences and their similarities can succinctly be recalled here.

10.2.2.4. *The hierarchical model*

Here objects are inscribed as per an arborescence logic commanded by the "father-son" principle: every element is accessible, for example, at the time of a request, only when taking the unique path defined by the vertical sequential order of the stages of the filiation process that produced it, which excludes any collateral consideration at whatever level it might be. This logic only supports perfect interlocking relationships between objects.

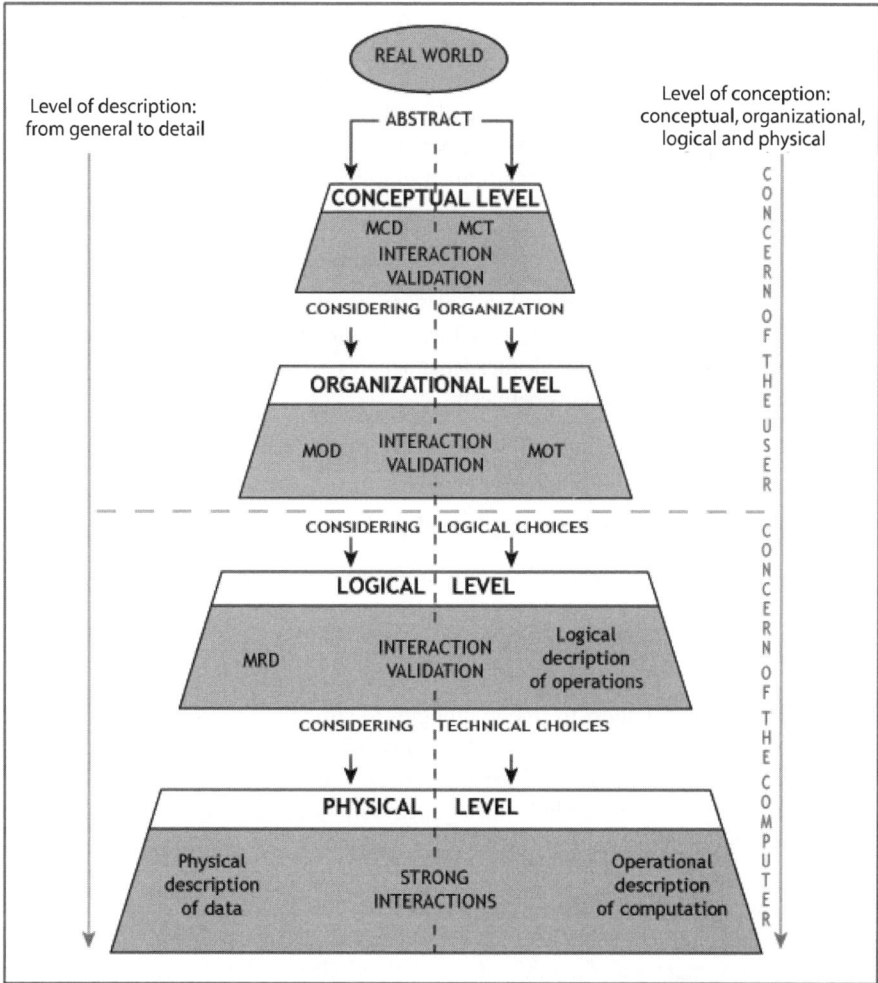

Figure 10.1. *Stages in the development of an information system (from [GAB 01])*

10.2.2.5. *The network model*

This model brings about a notable step forward in the efficiency of spatial complexity management. It is based on the concept of "linked lists" of pointers attached to the objects. These lists make it possible to reconstitute the composition links through which certain functional relationships can sometimes be addressed.

Following a very similar line of thought, Peter Chen developed the "entity/association" model that Codd would later finalize in the so-called "relational" model. This model, still in use, is based on the concepts of set theory, implemented with the help of algebraic algorithms developed from Wenn diagrams and different types of correspondences between elements that emanate from them (univocal, bi-univocal, simple to multiple, multiple to simple, etc.).

At the same time, Jacques Mélèse with the AMS model [MEL 72] makes an important contribution to the future of the systemic process in the database constitution protocols. Developed as part of (and for) company activity, this formalism treats the organization as a hierarchy of modules that, each at its own level, contains hidden functional interdependencies whose structure corresponds to the same schema of interactions between:

– "the inputs and the outputs respectively of the technological flow and the information flow that run through the module;

– the nature of the transformation carried out by module;

– the steering that is applied to this transformation for regulating it, and for satisfying it;

– the essential variables that mark the performances of the module."

In this context of a quest for methodologies to structure data wherein we find the first references to the concepts of "network", "allocated information" and "distributed databases", Hubert Tardieu with the "individual" formalism [TAR 79] lays the foundation of a so-called "third generation" method called MERISE, still in use today in the economic sector. MERISE is based on the general systems theory Von Bertalanffy presented it in the 1950s. The method does not try to structure only the data entering the information system but also the system itself, particularly the domain of the interdependencies between its functional constituents: organization, activity, environment and finalities. Its true specificity lies in the manner in which the *evolution* of the system is taken into account. Hubert Tardieu states that "the possible evolution classes will be identified and it will be guaranteed that the chosen representation remains partially or totally invariable when a certain evolution class is produced". This management of the upgradeability/evolution of the system, which leads to the integration of a maintenance stage, once the information system has become operational, clearly distinguishes MERISE from other methods, which on the other hand advocate managing the upgradeability/evolution within the framework of a constant redesigning of the system.

The success of MERISE in the entrepreneurial milieu curiously had no comparable echo in the research field, and we can speculate about the causes for this lack of interest among physicians, chemists or ecologists in experimental science, and geographers, sociologists or economists in human sciences. In fact, all are, *a priori*, liable to be "customers" of systemic methodologies for establishing their databases. This "rejection", it appears, is due to two types of reasons at different levels. The first can be qualified as circumstantial, the second as fundamental. As far as the diffusion of MERISE is concerned, a certain handicap resides in the slowness of the implementation, an offset inherent to the rigor and thoroughness of the protocols. This slowness combines with a basic difficulty already mentioned here, namely, the orientation of the method towards the resolution of problems of organization and management of the companies, domains that are much more stereotypical than those of the research. In addition, MERISE bases its structuring protocol on the "existing" while, on the other hand, the researcher needs to base his approach on a theoretical reflection broader than only the data available at the beginning.

With this rapid glance at the field of structuring methods of the information systems and the associated databases, it is observed that all the proposed methods are more or less inspired by a "systemic" vision. They consider the logical organization, the management and data processing as a "functional whole". The question is to know whether these can be retrievable, as they are for constructing the GISs.

In fact, probably because they were developed to serve mainly the corporate world [GAB 01], none of these methods take into account what remains specific to a geographic object, namely, the organic link between its semantic dimension and its spatial dimension. These modeling tools, however powerful they might be, only partially cover the need for integration in the management and processing of the geographic, geometrical, topological and thematic aspects of the objects that materialize or represent space. They do not rely on the fundamental and underlying logic of cartographic modeling of the real world wherein measurement levels (qualitative scales, quantitative scales), the conception level (variable between the elementary approach and the global approach), the scale of spatial perception (variable from the weakest one – 1/10 of certain DAO. charts – to the strongest – 1/33,000,000th of the planispheres) and the temporal scale (variable from the instant to the pluridecennial, even millenarian or more) are involved.

10.2.3. A methodology adapted to research: hypergraphic modeling by Bouillé

Taking advantage of his twin competencies as computer specialist and researcher in Earth sciences, Bouillé [BOU 77], looked into the problems caused by the digitization of geological maps. He then had the idea of directing towards the characteristics of the spatial data, the relational theory of data management which, till the time of his work, had hardly been applied to anything other than non spatial data. By using the integration of the set theory and the graph theory, Bouillé thus opened a new path for constructing data structures, taking into account at the same time, the global nature of the phenomenon or the phenomena processed, the essence and the existence of the objects describing this phenomenon or phenomena, attributes describing these objects as well as their relationships, the whole as per a graphic formalization explaining clearly the researcher's fundamental referents.

10.2.3.1. *A postulate and its active principle*

The most impressive innovation of the HBDS (hypergraph-based data structure) resides in an *a priori* surprising principle: deducing the data structure directly from the structure of the phenomena themselves, and not only from the problems that emerge. In fact, by combining systemic orientation, prevalence of the thematic aspect and topological planar transcription of the classes of objects, this method channels the path that the researcher takes in the semantic network of his topic. This path is hypothetico-deductive and certainly often starts at a high level of abstraction. However, he benefits on this account from a breadth of view beneficial for the pursued objective: constructing a synoptic model. By trying to model not only the stated research problems, but also the overall thematic from which these problems emerge, the approach is placed at the level of a holistic theoretical description of the real world. The problems are considered as subordinate sub-structures to the real phenomenon itself, and must be treated as such. The hypothetico-deductive perspective strongly distinguishes HBDS from MERISE on this point. MERISE depends more on the inductive logics for structuring more factual information systems that are strongly related to a known "existing job" such as those necessary for the companies.

With this objective in mind, the logic of the set theory uses four components: property, element, set and relationship; the term relationship being taken here in the sense of a link and not of a table as in the traditional relational terminology. These four terms are respectively termed the attribute, object, class and link. The systemic aspect of this modeling is rendered by a visual presentation with recourse to graphs and hypergraphs. Due to the fact that a graph essentially illustrates a system of relationships [BER 70], a hypergraph will constitute a higher level graph, produced by generalizing the underlying graphs: a new combination of points and arrows indicating a new combination of dimensions and links: a change in the conceptual scale.

Such a protocol makes it possible to benefit from the mathematical fundamentals of set theory and topology (neighborhood science), through graph theory, in order to conceive relationships between objects, classes and attributes as much in the vertical direction (decomposition of a complex object from row r into n simple objects from row *r-1*) as in the horizontal direction (transfer of attributes between distinct semantic objects by spatial operators).

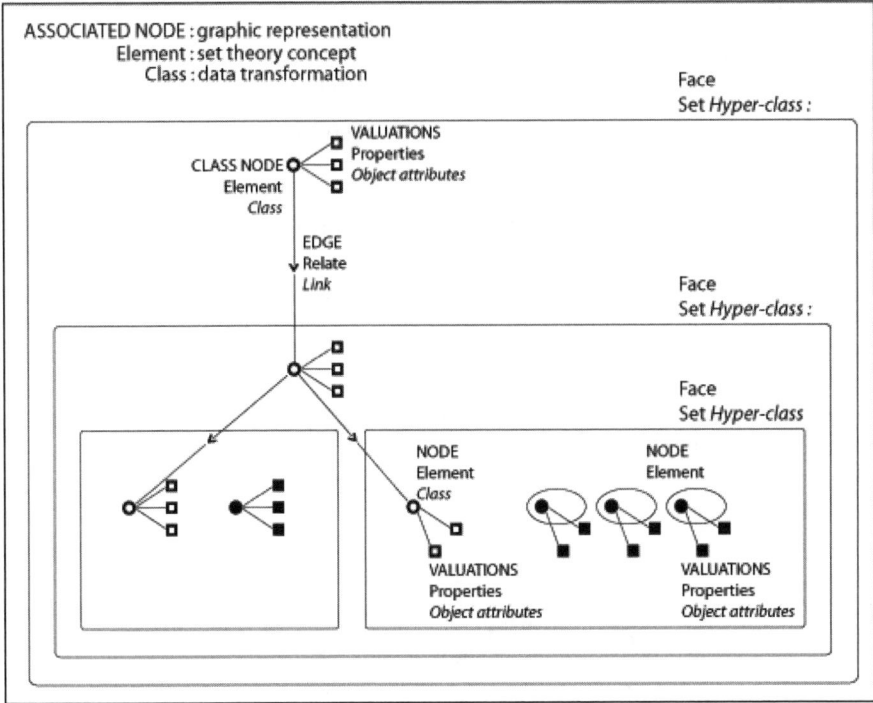

Figure 10.2. *Conventional HBDS graph*

The hypergraphic "vision" relies on four concepts [BOU 77] acting as logical agents of data structuring. These concepts can be divided into six distinct levels: six types of abstract entities, necessary and sufficient for describing all the types of conceivable combinations in the instances of these concepts in a spatial approach.

10.2.3.2. *Fundamental concepts*

The four fundamental concepts of HBDS (*hypergraph-based data structure*) are: *objects*, *classes*, *attributes* and *relationships*. Six types of abstract entities follow from it: the class of object, the attribute of class, the attribute of object, the class of link, the link between classes, the link between objects.

The concepts of the method, in the graphics of the conceptual data model (CDM), become peaks, classes, valuation and links. Their semiology is as follows:

– the vertices of the hypergraph represent elements (or objects);

– the edges of the hypergraph demarcate the sets (classes); they are associated with vertices representing them;

– the vertices are bearers of valuations representing the properties of the elements and the sets (or the attributes of objects and classes);

– the arcs between peaks represent the links between elements (or objects) and the links between sets (classes).

Graphically speaking, a hypergraph is presented in the form of a network of vertices (nodes), links and edges. It summarizes the description of the components of a phenomenon in a topological organization into *hyperclasses* and *classes* of objects connected by a network of *hyperlinks* and *links*. All these elements have values or, in other words, possess attributes that define their signification, determined by the author of the model, for example in the framework of "domains", that is, lists of modalities or limits of values pre-defined as description limits (see Figures 10.2 and 10.3).

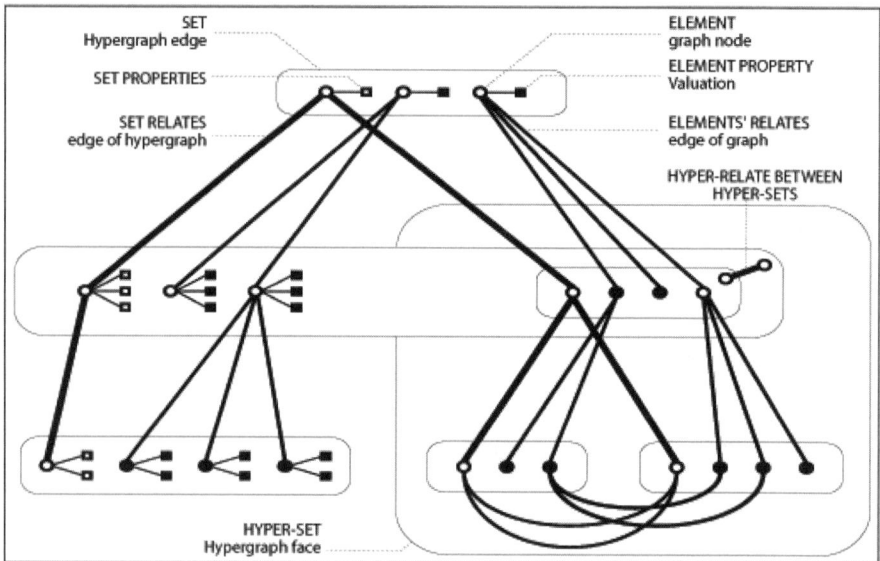

Figure 10.3. *Graphs, hypergraphs, links and hyperlinks according to Bouillé*

All these denominations find their correspondence in different available nomenclatures for describing the observable world in generic terms.

REAL WORLD	ENTITY	CHARACTERISTIC	RELATIONSHIP	CLASS
GEOGRAPHIC SPACE	GEOG. OBJECT	DESCRIPTOR	RELATIONSHIP	CLASS
DATA STRUCTURES	OBJECT	ATTRIBUTE	LINK	CLASS
HYPERGRAPH	VERTEX	VALUATION	ARC	EDGE

Table 10.1. *Correspondences between hypergraphic concepts and geographic, relational and concepts of the set theory [HAM 96]*

10.2.4. *Spatial concepts and planar law for a hyper(geo)graphic reasoning*

The method proposed by Bouillé can be studied from two angles. The first is derived from computer methods and their mathematical expectations. It concerns pure computer specialists. The second considers the conceptual logic inherent to the computer structures presented and makes it possible to conceive useful adaptations for organizing the localized data. For the geographer, it is natural when seen from this angle that this method is useful since it prepares the path forward in the general formalization of geographic phenomena directly applicable in the GISs.

In fact, the method offers the possibility of integrating into the recognition process of spatial entities, the notion of conception scale, an essential notion, which leads us to specify the level of generalization of the phenomena (and the objects that constitute them) by the number of dimensions taken into account. As a result, it processes a constraint identified since a long time by Dollfus [DOL 78] in his work on the geographic space: geographic objects have a sense only at the level where their identification made it possible to identify them as entities of a homogenous space. Consequently, either their limits are obvious, for example, visually discernible in the case of physiognomic entities of vegetation, or they must be conceived by defining the range of the object dimensions or, in other words, the setting for all the perceptive filters mentioned above.

The "observable world" is considered as an entwined hierarchy, that is, a structure imbricated with entities mixing spatio-semantic composition relationships (partial or total affiliations), representation relationships (attributory and/or cartographic) and functional relationships. This type of reasoning leads us to visualize the spatio-semantic network of a geographic phenomenon in the form of a structure skeleton (arborescences and boxes) convenient for interpreting.

The sequence of the operations is presented in Figure 10.4.

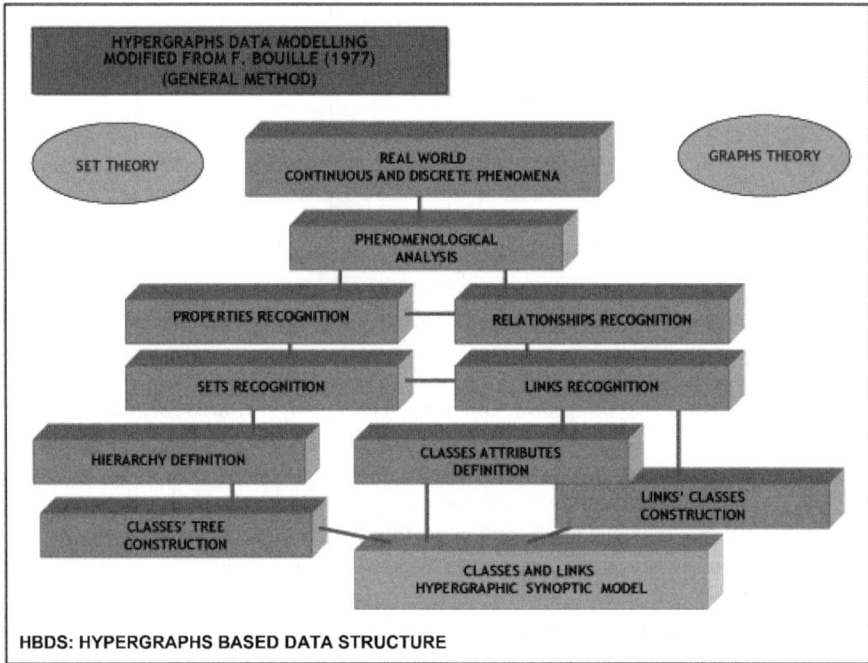

Figure 10.4. *Fundamental stages of hypergraphic structuring [BOU 77]*

Though barely mentioned by François Bouillé in his thesis, one of the major strengths of this design is because of its staged and progressive process making it possible to control at each moment, the coherence between two logics conducted simultaneously:

– a native *systematic* logic of description of a phenomenon from its most generic level to its most specific level;

– a *systematic* and geographic logic – to be introduced by the geographer – of description of the same phenomenon, from its most global and relational level to its most analytical level as much in the domain of its semantic as in that of its spatial morphology.

Under this geographical angle, the four general concepts, that is, the objects, the classes, the attributes and the links, can be applied to the spatial objects and to their modeled form, that is, the cartographic objects in their choropleth meaning (punctual, linear, polygonal), intermeshed or even isarithmic (continuum) to the dimensions, descriptive fields and relationships (composition, functioning or even neighborhood) that characterize them. A phenomenon derived from the "observable world" can be described in "intention" by a topological diagram type conceptual data model (CDM) with *n* dimensions called a *hypergraph* wherein all types of objects (material or otherwise) and types of possible links identified in the mental representation of the researcher are explicitly stated from the most global level to the finest level. The "extension" of the phenomenon as in the terminology of the relational databases is constituted by the data itself, associated with the objects (spatial and non-spatial) present in the structure.

This form of knowledge representation is called prototypical [CHE 99]. It calls upon the notion of "*frame*" or descriptive framework and the semantic heritage of the objects operates by subsumption (generalization/specialization). The essential mechanism is related to the classification.

This logic makes it possible to establish, for example, a primary generic model of the spatial object, applicable in the cartography (see Figure 10.5).

While it is true that this fundamental structure can appear to be basic, it visualizes how the signification – composition – construction are involved in a geographic object. With this as a base, all sorts of variants can be observed, from the most general case (topographical map for example) to the most focused cases from the thematic point of view, as will be seen later.

As examples, Figures 10.5 and 10.6 show their variations, which have guided the IGN in structuring the B.D. Carto to the 1/50,000th part.

In its principle, the graphic representation of the hypergraphic method is easy to apprehend. As the examples show, the synoptic schemas are clear, at least until the number of lines and links is limited. On the other hand, they become extremely compact when the problem requires the recognition of multiple hyperclasses, classes, sub-classes and associated links. The presentation of the schema on a single plane, the partial overlay of edges indicating composition relationships, the interlacing of links and the use of one single color end up making the clarification principle of the process inoperative. Changes become necessary in four directions:

– introducing the modeling approach of the model with a simplified schema presenting the guiding principle of the thematic reasoning implemented;

– using the principles of cartographic semiology to visualize the classes and the links: different colors and different semantic links; within each hyperclass, value variations of the basic color as per the internality level of the class, subclass or object;

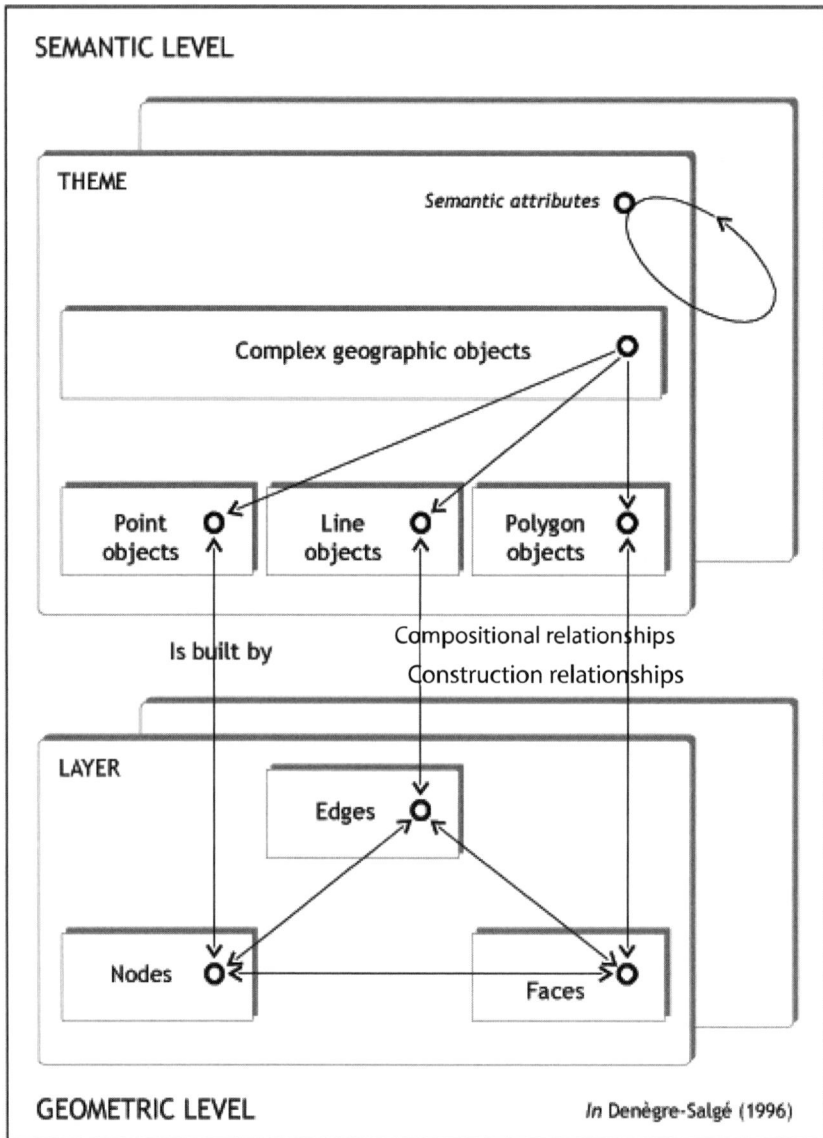

Figure 10.5. *Generic model of geographic object as per the hypergraphic acceptation*

– introducing effects of perspectives in the schema, in order to reinforce the perception of the hierarchical levels of organization;

– decomposing the model into two stages: first, the big hyperclasses that structure the mental representation of the studied phenomenon; then, in a second document, the details of the model;

– blurring, or even masking the evident (composition) or dispensable links, in particular certain recurrent links of functional dependencies.

The objective of all these provisions is to guide the perception of the model towards the logic of the principles on the basis of which it is established: a relational reasoning descending through stages of abstract concepts up to the spatial objects within an overall systemic vision.

The example given below to illustrate this method, developed with respect to the GIS related to the industrial risks at Notre-Dame-de-Gravenchon, refers to most of these changes. Other graphic improvements are being studied.

However, the main evolution, clearly geographic, which must be introduced does not pertain as much to the graphic presentation of the method as to its reasoning itself.

It develops a specification of the general form presented above, introduced in order to take into account the characteristics of the geographic phenomena mentioned above (see Figure 10.5). Its objective is to ensure an overall coherence between the layers of geographic objects retained in the GIS, the main difficulty for many applications.

This principle of HBDS use in the structuring of a GIS consists of firstly demonstrating the complexity of the phenomenon observed in as many data entities as the systemic analysis of the phenomenon retains to describe its composition and its functioning within the framework of the universe of discourse by its author. The progression takes place from the top to bottom: from the abstract to the concrete, from the generic to the specific, from the general to the particular, from the non-spatial to the spatial.

A principle commands that each layer can host only homogenous geographic objects, considering the 8 dimensions that are characteristic of them:

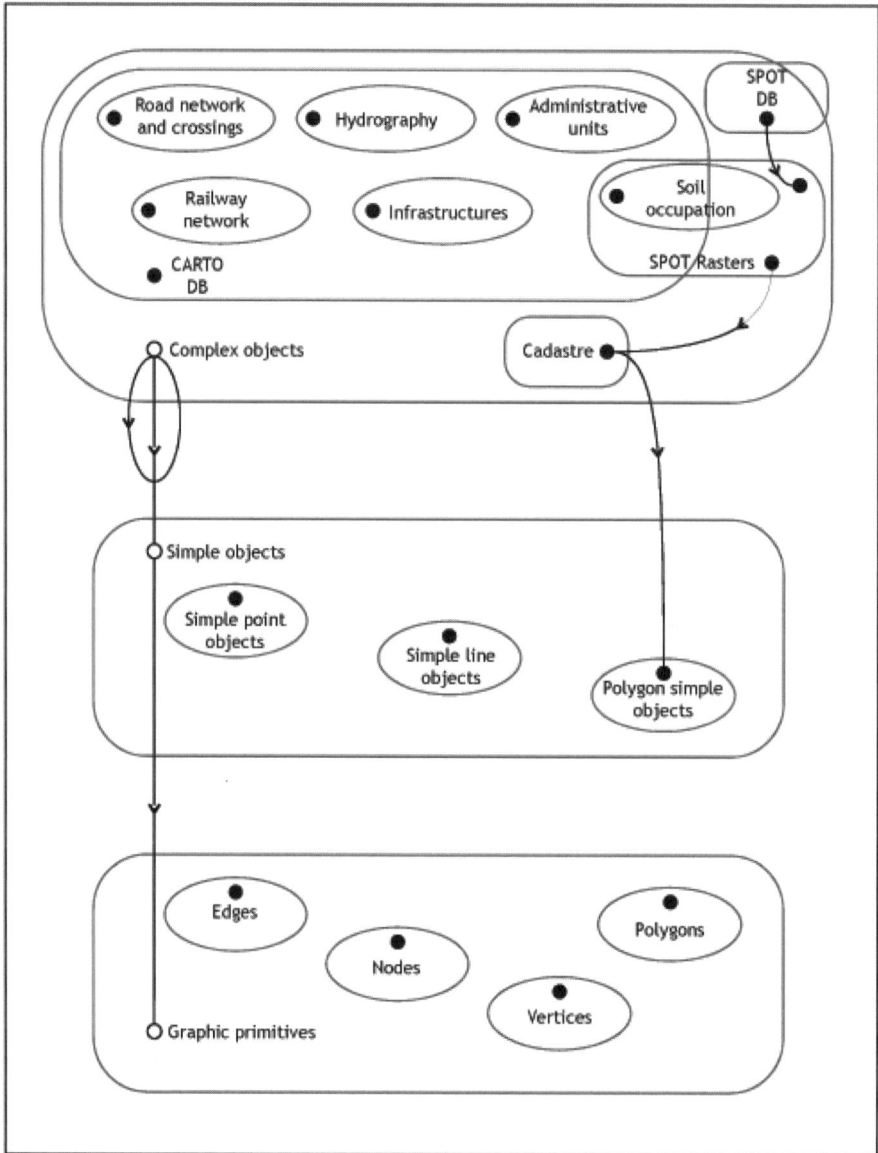

Figure 10.6. *The model retained by the I.G.N. for structuring the data of the BD Carto (1/50,000). Movement from the generic model to the general model*

1) Premise, through affiliation to one of the 3 types of forms: point, line, or polygon.

2) Semantic, through association of thematic attributes specific to its position in the conceptual, qualitative and quantitative scales.

3) Location, through situation in a geographic repository (x,y, (z)) common to all the layers (georeferencing).

4) Temporality, through positioning in the time: t0 ->Tn.

5) Spatial scale.

6) Geometry, through the tables and associated graphic primitives required for its infographic representation.

7) Topology, through conformity to the planar topological graph of the neighborhood relationships between objects of the same layer and by extension between objects of different layers (planar topology and multiplanar topologies).

8) Quality (metadata: sources, validity, precision, achievement mode, updates, etc.).

Gradually, these entities are distributed and detailed by conceptual scale levels on distinct graphic layers. The recognition of the layers and hence their constituent type objects is then accomplished as per a principle taking on the value of a law:

"To every level of semantic definition of a class of spatial object corresponds a single planar topological graph per type of premise considered (point, line, polygon)".

And its reciprocal (if necessary, because not all classes of semantic object take the form of a spatial object):

"To every level of spatial definition of a class of semantic object corresponds a single planar topological graph per type of premise considered (point, line, polygon)".

In other words, each theme, for a given level of conceptual scale (semantic, spatial and temporal), can only be represented on a single plane, through objects semantically (attributes) and spatially (premise) homogenous and without any intersections or overlap between them (Figure 10.7).

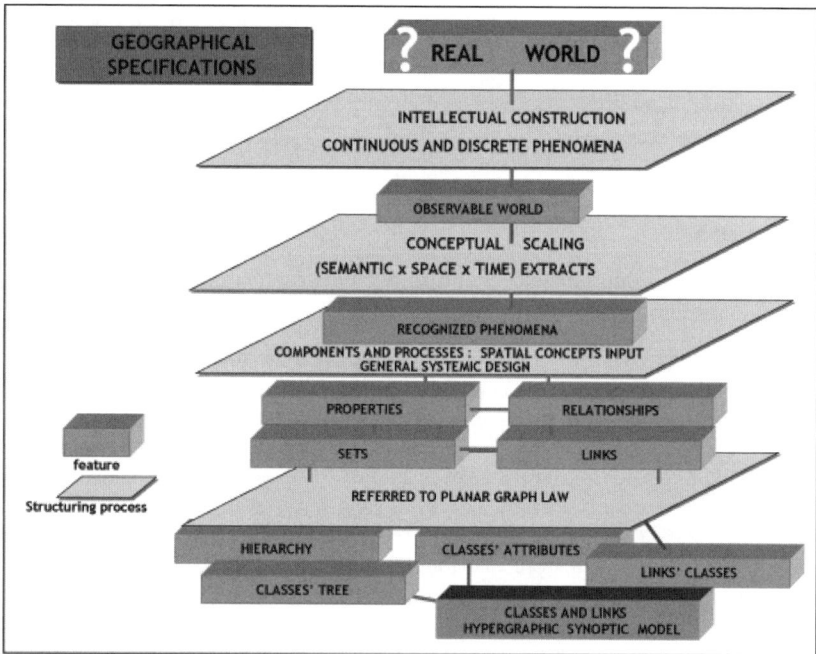

Figure 10.7. *Geographic adaptation of the data structuring as per the hypergraphs [SAI 02]*

This protocol respects the different specifications of the law of internal composition R x S x T (observable reality x space x time), on which the continuum "geographic perception-localized data-geographic information" relies.

The protocol lends itself to the union of two logics, which are *a priori* divergent and could have been believed to be incompatible; a logic of geographic structuring of the phenomena, above all based on their spatial form, and a systemic logic of conception of the themes from which these phenomena are derived. The first constructs, according to the set theory, a descriptive model discretizing the geographic phenomena through successive stages until their most analytical level. The second puts these systematic variations in perspective of the four fundamental domains characterizing a system: organization, complexity, globalization and interaction [DUR 02].

In doing so, while the structure of all the models preserving complexity require attentive reading, the detailed data structure, directly resulting from the application of the law of the topological planar graph, becomes very simple and very versatile; a pile of layers of independent objects, but virtually linked by their properties in the field of topology and georeferencing.

In the spatial analysis phase, the GIS operators then have coherent, simple, in a nutshell adequate data structures for exhibiting spatial relationships and through them, the part played by the functional relationships for which they are indicators.

In its development, the intellectual process operated in HBDS is "descending", that is, it starts with a representation of the real world through abstract concepts and leads to a concrete description, up to the level of the spatial objects themselves. This has a considerable advantage: it makes it possible to explain the conceptual networks underlying the modeling of spatial phenomena in an information system which makes it possible to preserve the organic link between the mental representation process and the computer modeling process. Thus, the principle that J. S. Bruner applies to the knowledge when he affirms "knowing is a process, not a product" [CLA 90] can be applied to the information by specifying which repositories and logic preside over the data structuring and consequently where the limits of the universe of discourse of the system can be found.

Through these characteristics, the hypergarphic modeling provides the opportunity of combining two classically distinct visions of the world: the "phenomenological" vision, which considers the global form of the systems, and the reductionist vision, which dissects its components. On this account, it opens up promising perspectives for connecting within the GIS certain geographers' practices, refuted not so long ago for being contradictory.

From an epistemological perspective, HBDS was one of the first signs announcing the arrival of the concept (in the computer sense) of an "object", which had to wait these last few years to be implemented. In fact, the hypergraphic reasoning defining the type of objects constituting the classes and the associated links is not without relationship with the one that presently forms the basis of the languages called "languages of classes". Let us recall that the latter are based on the following concepts:

– object: entity equipped with a structure, data and rules of behavior;

– class: description of a family of objects with the same structure and same behavior;

– field of a class: declared variable for describing the object that refers to the notion of attribute;

– method of a class: procedure establishing the behavior of objects, which refers to the rules of definition and manipulation of the objects of the class (defined by the level of spatial, temporal and semantic scale);

– sub-class and super class, which refers to the notion of interlocking into sets and sub-sets of the mathematical theory that bears this name.

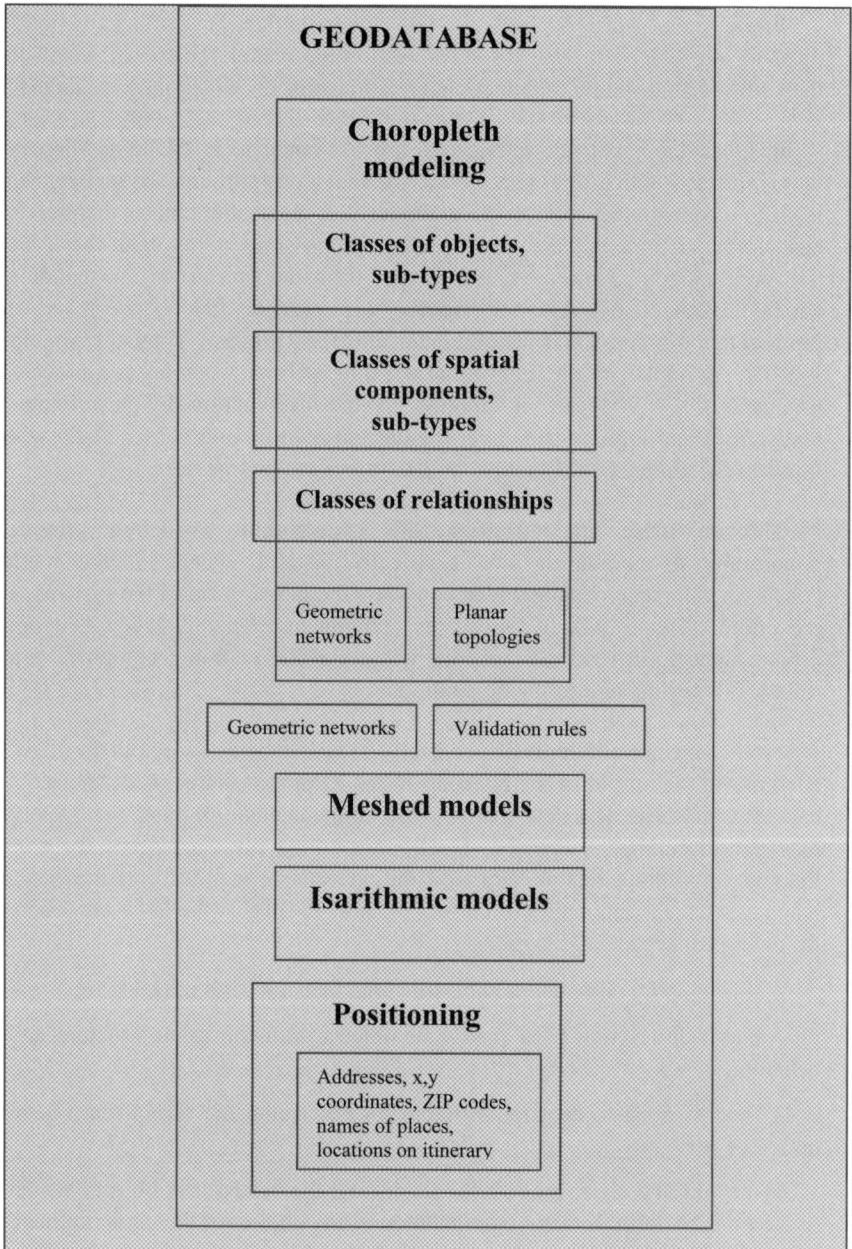

Figure 10.8. *Structure of the geodatabase (from [ZEI 01])*

Thus, it is not a matter of chance that today software sets appear with their conceptual and computer architecture closely integrating the hypergraphical principles and those of the "objects" programming. Morehouse and Scott [MOR 85] have in fact demonstrated that these principles were already hidden in the name of the "geo-rational" model in the first versions of Arc-Info. A very successfully completed example today is the concept of geodatabase that ESRI has just developed for ARC-GIS 8, which probably represents the most advanced level to date in the GIS in the field of integration of technical and conceptual structure.

Introducing HBDS in this manner in a choropleth (or even eventually isarithmic) modeling of the space in layers of distinct planar graphs, spatially structures classes wherein the objects are distributed by semantic category according to explicit scales. This protocol preserves the thematic readability of the database. In addition it radically lightens the next phase of defining the logical data model: the essence is directly provided by the layers of spatial objects themselves. As for the spatial links, which are mainly conveyors of functional links, they no longer have to be systematically mentioned: virtualized by this structuring, they will appear only at the right moment, when they exist, in accordance with the spatial treatments applied to the data.

By means of a few methodological precautions mentioned above, this coupling makes it possible to use the spatial essence of the objects (implantation, form, surface, location, neighborhood) as a tool for combining semantically different themes. It finally makes it possible to attain the real objective of the GISs, namely to carry out the synthesis of modeled spaces and through this synthesis represent the different processes of geographic combinations that preside over the constitution of spatial complexity.

10.3. Example of thematic application: the industrial risks at Notre-Dame-de-Gravenchon (lower Seine valley)[1]

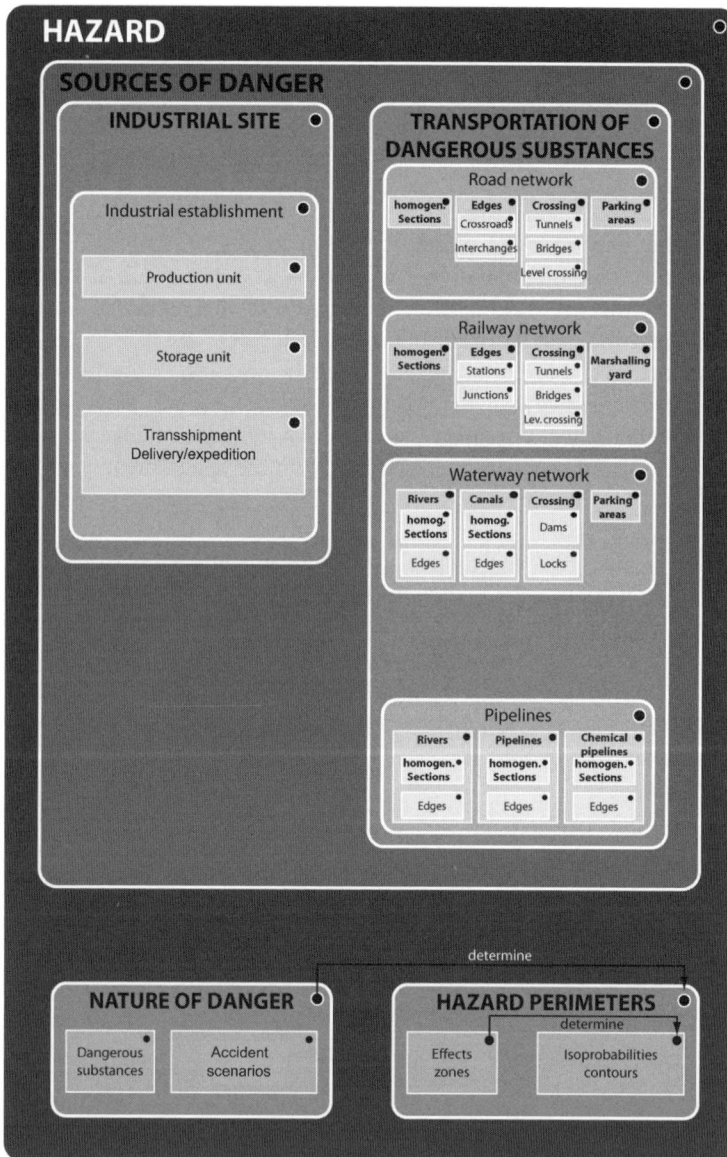

Figure 10.9a. *Hazard and vulnerability, the 2 fundamental hyperclasses: hazard*

---

1 This application has been realized with the collaboration of Eliane PROPECK [PRO 00] PRO 01].

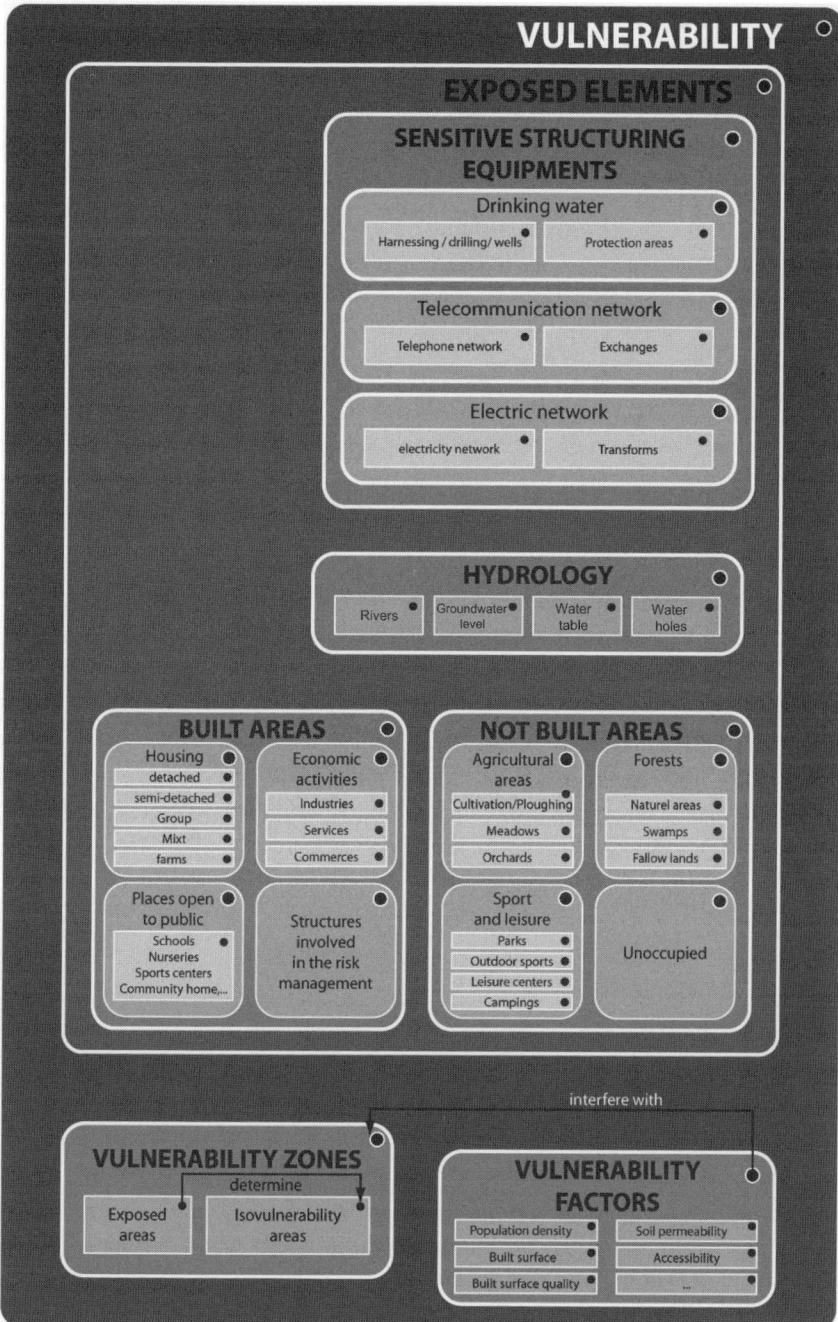

Figure 10.9.b. *Hazard and vulnerability, the two fundamental hyperclasses: vulnerability*

10.3.1. *Identifying the specific and central concepts*

Risk is the conjunction of a hazard and a vulnerability: these two concepts – general and abstract – are the first two "entries" in the reasoning. They provide the first hyperclasses that will be progressively divided into classes and sub-classes of more and more precise concepts and objects, up to the "ground" level fixed by the study. In Figure 10.8, the different classes of objects associated with the major technological risk are represented.

10.3.1.1. *Hazard*

A risk appears when there is a *source of danger*, an activity liable to have defaults, to cause an accident whose repercussions can be felt far beyond the place of the accident. The *nature of the danger* – fire, explosion, toxicity – is related to the nature of the products and the procedures implemented leading to the establishment of different accident scenarios for which occurrence probabilities can be evaluated. The set thus defined leads us to determine *the "hazard perimeters"* corresponding to the effect areas of the potential accidents (zones of lethality or irreversible wounds) and to the hazard iso-probability areas, with spatial hierarchies of hazards (areas where the probabilities of effect of an accident are the highest, by addition of the different probabilities of accident and damages in a place).

10.3.1.2. *Vulnerability*

This is defined by the *exposed elements* and the *vulnerability factors*, vulnerabilities corresponding to the entities mentioned (dwelling areas, economic activities, networks, etc.) and to iso-vulnerability areas (areas belonging to the same level of vulnerabilities as per predefined criteria).

The recognition of the risk perimeters is based on the guiding principle of the phenomenon's modeling: the risk is born from the conjunction of hazard and vulnerability. The risk perimeters are thus obtained by crossing the hazard perimeters and vulnerability areas, with these perimeters being distinguished from all the preceding segments by the fact that they result from all the factors constituting a risk.

10.3.2. *Identifying the peripheral concepts*

Identification is a matter of potential but non-specific actors in the hazard-vulnerability relationship, per category and level of scale. Almost all the geographic phenomena have links with the unavoidable "contingencies" (sometimes called "substrate") related to the structures of human origin (administrative and organizational) and biophysical structures (data of the surroundings). For this application, a hyperclass organizes them into two classes distinguishing the segments of territorial competency and the zonings of ecological factors. These are the "incidental structures".

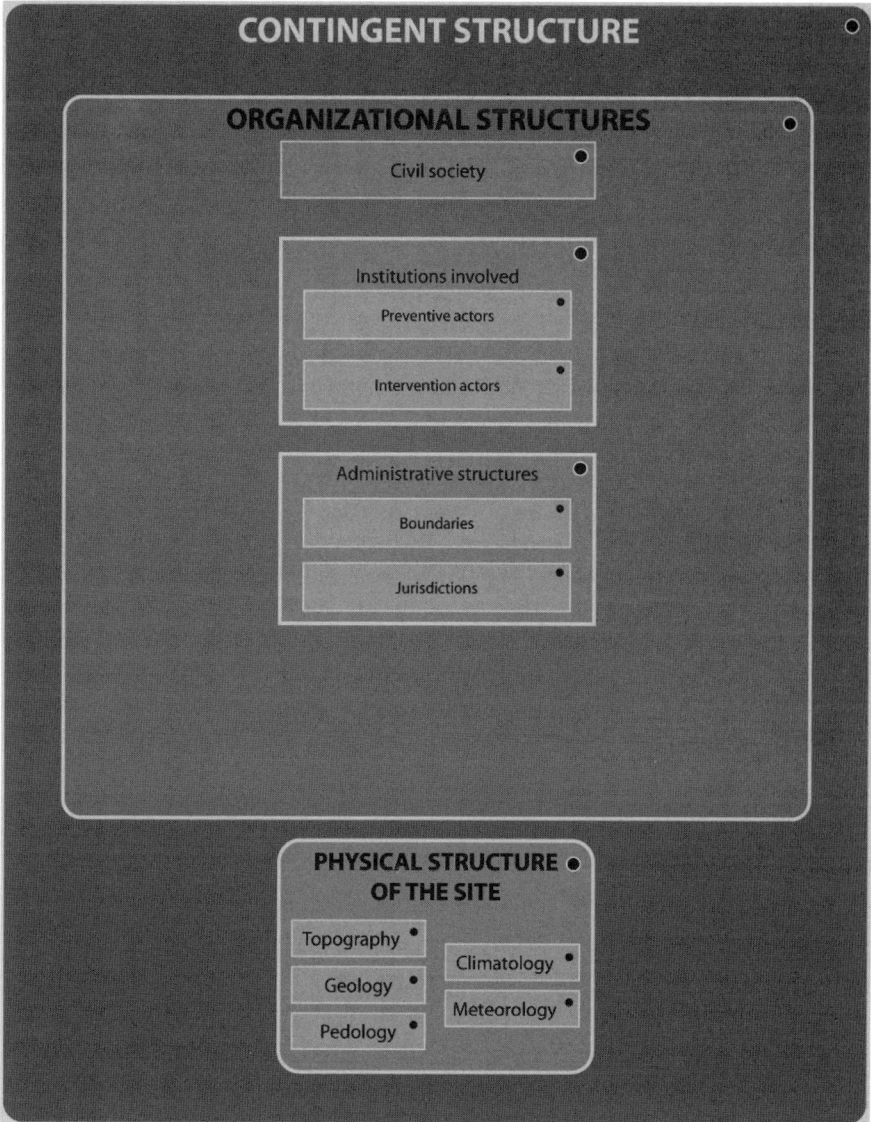

Figure 10.10. *Incidental structures of major technological risks*

10.3.3. *Formalizing the spatial synthesis of danger*

The essential purpose of the study is the identification of risk perimeters through spatial analysis, that is spatial morphologies resulting from the encounter between hazards and vulnerability in all their dimensions. These perimeters constitute a hyperclass. The hyperclass of risk perimeters (Figure 10.11) is subdivided into two classes corresponding to zonings specific to each of the domains: emission and propagation perimeters and areas exposed to damage.

With the first "conceptual clear out" having been fixed, what remains is to configure the disposition of the synoptic model such that the partial overlap of hyperclasses or classes of distinct nature visualize the composition relationships and the possibilities of combinations to be verified at the time of spatial processing.

This assembly makes it possible to draw the synoptic diagram in a simplified manner, making the visualization of composition links useless (Figure 10.11). Then comes the progressive details of the classes and sub-classes as well as links that are particularly representative of the conception of the entire phenomenon retained in the model.

As for every information system, a validation stage makes it possible to verify *a posteriori* (but before moving to the software!) that the data structure coming from the CMD is functioning well by checking in particular:

– in the CMD: respect of the logical and formal rules resulting from the geographic orientation given to the hypergraphic modeling, notably with respect to the law of the planar topological graph;

– in the LDM (logical data model, that will not be presented here), by checking the existence of juncture keys (valuations) between logically connectable objects as part of the "database" type conventional surveys;

– in the cartographic model, by verifying that the type of control (pinpoint, areal, linear, or even 3D) as well as the scale of representation retained for depicting the objects on different layers can figure. It means ensuring that the spatial relations liable to be given prominence by the spatial operators, assume an authentic spatial signification and are not the product of a computer operation disconnected with the sense of the data.

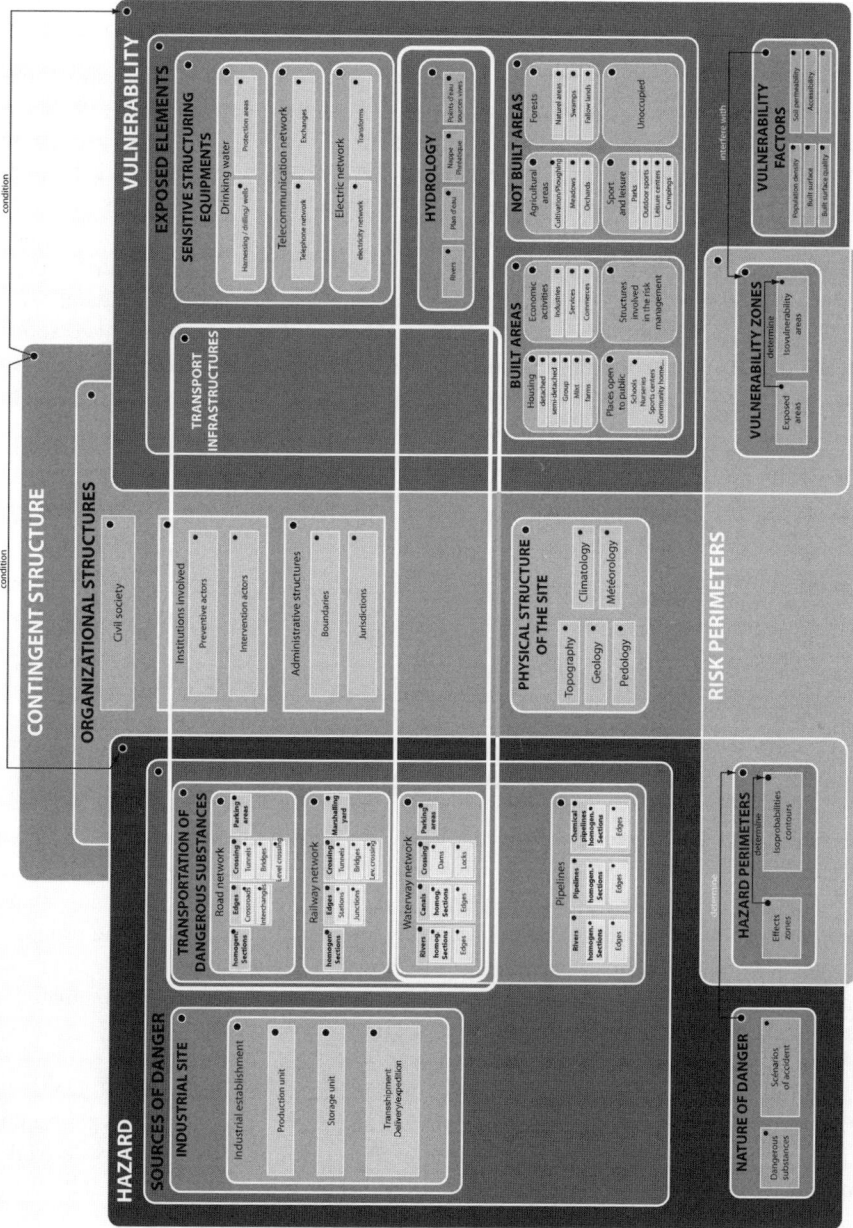

Figure 10.11. *Hypergraphic modeling*

10.4. Back to the sources...

While examining this synthesis of the different elements of data structuring method applied here, we realize that it means nothing else than going further in understanding this old but judicious definition of the map, given years ago by Professor Fernand Joly from the benches of the University of Paris VII: "a plain simplified and conventional representation, whether of the whole or part of the terrestrial surface, established as per a similarity relationship defined by the scale". We may find it curious, if not paradoxical, to see a definition resurfacing here, a definition that others will judge as being commonplace, banal if not outdated. And yet... Traditional, commonplace? Maybe! But it is not the less dynamic and robust for all that, to the point of remaining valid and surviving without meeting any opposition, the evolutions in cartography commonplace, invasions of infographics, audacities of anamorphoses, chorems and other mental maps, all of which were offered a solid base by this definition, even when it was as a counterpoint.

In the light of this definition, what does this effort at a theoretical formulation followed by a methodological formulation constitute? Above all, a transfer of concepts, recalibrated in order to become operational in a perspective of spatial analysis under GIS. The map reveals its depths: the "plain representation" becomes a "planar topological graph" and the "terrestrial surface" becomes the "real world". As for "similarity relationship defined by scale", this escapes from the prison of geometric properties of space and redeploys itself in a logic of conception/perception/representation scales interfacing the semantic, spatial and temporal dimensions of the "observable reality".

At the stage of progress reached today, the method is operational: theses, research and papers in geography as well as in history, archaeology, public health and many other specialties have recourse to it. Here we see a two-fold advantage: first concrete, with the creation of consciously structured databases for the proposed subject; second scientific, given the unsuspected dimensions of a thematic approach that is almost always revealed by the emerging dialogue between the thematician and the methodologist at the time of its application.

For all that, does it mean that the work is finished? Most certainly not. Under its graphic presentation angle, improvements are still to be made to the hypergraphic "design", notably to make the type of spatial object from which a class is derived – point, line, polygon – emerge without overburdening the reading. Above all, in the quest to get closer still to the systemic preoccupation that inspires it, the immaterial dimensions of spatial phenomena such as those related to the representations or the cultural factors, the "hidden face" of the Earth [SAI 02] remain to be introduced in this modeling method.

10.5. Bibliography

[BOU 77] BOUILLE F., Un modèle universel de banque de données simultanément partageable et répartie, thesis (major: mathematics, minor: IT), Paris, University of Paris VI, 1977.

[BOU 93] BOUTOT M., *L'invention des formes*, Editions Odile Jacob, Paris, 1993.

[CHE 95] CHESNAIS M., *Gérer l'information géographique*, Paradigme, 1995.

[CHE 99] CHEYLAN J. P., *et al.*, "Les mots du traitement de l'information spatiotemporelle", in *Représentation du temps et de l'espace dans les S.I.G.*, R.I.G. vol. 9, no. 1, 1999.

[CLA 90] CLARY M., *Modèles graphiques et représentations spatiales*, Anthropos, RECLUS, 1990.

[DE ROS 75] DE ROSNAY J., *Le Macroscope*, Paris, Editions du Seuil, 1975.

[DOL 78] DOLLFUS O., "L'espace géographique", *Que-sais-je?*, PUF, Paris, 1978.

[DUM 81] DUMOLARD P., "L'espace différencié: introduction à une géotaxinomie", Paris, *Economica*, 1981.

[DUR 02] DURAND D., "La systémique" coll. "*Que sais-je?*", PUF, Paris, 2002.

[GAB 01] GABAY J., *MERISE et UML, pour la modélisation des systèmes d'information*, Dunod, 2001.

[GAR 89] GARDARIN G., *Bases de données et bases de connaissances*, Addison-Wesley, 1989.

[GEO 94] GEOPOINT, *Proceedings of the Seminar on GIS*, University of Avignon (France), 1994.

[HAM 96] HAMEL P., *SIG du Grand Caen*, PhD thesis, GEOSYSCOM, University of Caen, 1996.

[MEL 72] MELESE J., *Analyse modulaire des systèmes*, *AMS*, Editions Hommes et Techniques, 1972.

[MOR 85] MOREHOUSE, S., "Arc-Info. A georelational model for spatial information", *Proceedings of the Auto Carto 7 Conference*, Washington DC, 1985.

[PRO 00] PROPECK E., *et al*, *La cartographie dynamique, méthode d'analyse des phénomènes spatio-temporels. Application à l'expertise des risques technologiques*, Research report "Programme risques collectifs et situation de crise" of CNRS, 2000.

[PRO 01] PROPECK E., SAINT-GERAND T., "Modélisation des RTM: de la connaissance du risque à sa gestion ou objectiver le risque dans un SIG pour l'objectiviser", *Proceedings of International Conference on Risk and Terror*, Lyon, 2001.

[RIM 89] RIMBERT S., "Cartographie et SIG", *Mappemonde*, no. 2, 1989.

[ROS 75] de ROSNAY J., *Macroscope*, ed. du Seuil, 1975.

[SAI 94] SAINT-GERAND T., BERGER M., "Adopter ou adapter les S.I.G. pour la recherche en S.H.S.?", *Actes du colloque du groupe Dupont*, 1994.

[SAI 02] SAINT-GERAND T., SIG: Structures conceptuelles pour l'Analyse Spatiale, PhD thesis, University of Rouen, 2002.

[SAI 08] SAINT-GERAND T., "Réunir le matériel et l'immatériel pour découvrir la face (cachée) de la Terre", *Actes du colloque Geopoint*, University of Avignon, 2008.

[SAN 96] SANDERS L., "Dynamic modelling of urban systems", in FISHER, SCHOLTEN, UNWIN (eds.), *Spatial Analytical Perspectives on GIS, GISDATA*, Taylor & Francis, 1996.

[TAR 79] TARDIEU H., *et al.*, *Conception d'un système d'information; construction de la base de données*, Editions d'Organisation, Paris, 1979.

[TAR 95] TARDIEU H., *La méthode MERISE, Principes et outils*, Editions d'Organisation, 1995.

[TOM 78] TOMLINSON H., in Morehouse and Scott, "Arc-Info, a georelational model for spatial information", *Proceedings of the Auto carto 7 Conference*, Washington DC, 1978.

[ZEI 01] ZEILER M., *Modeling our World*, Redlands USA, ESRI.

Chapter 11

Complexity and Spatial Systems

11.1. The paradigm of complexity

Theories of *complexity* are currently in vogue and have tended to override non-determinist theories such as that of chaos or of instability which have themselves moved into the realm of determinist theories. The paradigm of complexity does not rest on well-established mathematical theories. Evidently this is not completely new ground as it incorporates previous advances, but to this day a theory of complexity does not technically exist, when compared to dynamic systems for example, which are founded on the theories of differential equations and partial derivatives. These theories are regularly employed in certain domains, most notably that of liquid, continuum and celestial mechanics and have been endorsed by a variety of prominent figures, including Poincaré, Kolmogorov, Forrester, Prigogine, Lorenz and Thom.

Numerous papers discuss the modeling of complex systems, such as [WEI 89], [LEM 91], [PEG 01], [DAU 03]; however, there remains a characteristic absence of any fixed concept of complexity. A complex system is defined in [WEI 89] as "a system formed by numerous different elements in interplay". In response to this, G. Weisbuch gives a number of different examples of non-complex systems – such as perfect gas: elements of this system (in this case molecules), even though numerous, are identical and rarely mix – or complex systems such as the human brain or computer systems where numerous elements from various different categories interact solely amongst themselves.

Chapter written by Patrice LANGLOIS.

These prominent figures however do not go any further than simply defining its characterization, its measure and its diversity. For Lemoigne [LEM 91], "a complex system is one which by definition is irreducible to a finished model". He also states that the notion of complexity similarly "implies the eventuality of the unpredictable, the plausible emergence of the new" and distinguishes itself from complication, for "that which is complicated can be summed up by a simple principle" [MOR 77]. It is this difficulty of defining this concept that has persuaded J.L. Lemoigne to declare that "complexity is perhaps not a natural property of phenomena" and is "applied by the model builder to representations which he constructs about phenomena that he perceives to be complex". Complexity would not therefore be ontological, however this does not bring us any closer to being able to define it. In his work on theories of complexity in geography [DAU 03] Dauphine catalogs certain characteristics of complexity, such as *variety* (number of components), *interactions* (or connections) and *level of spatial and temporal resolution*. He similarly addresses different theories that put into practice a certain type of complexity, where different forms of complexity are defined, without really broadening any further our knowledge of this notion.

However, the concept of complexity has been formally defined within the realm of computer science, and what is more in great detail (see for example [BEA 92], [AU 92], [FRO 95]) but this form of complexity actually encompasses a basic assessment (also referred to as *cost*) of the time necessary to calculate, or the amount of memory necessary to complete an algorithm in order to solve a problem. These problems can also be classified in terms of their increasing complexity; certain problems can be resolved in linear time, meaning that the time needed to resolve them is a linear (or affined) function $f(n)$ of the amount of memory n of the entry elements of the algorithm. Furthermore there are problems known as polynomials (P), where the time taken to calculate them is a polynomial function. There are also the non-polynomial problems (NP) such as the traveling salesman problem. The mathematician Alan Turing goes even further and shows that there are also problems and numbers that cannot be calculated (for example the general problem of determining whether or not an algorithm that ends in a finite number of stages is incalculable).

Another interesting approach developed in particular by Kolmogorov (1960) and expounded upon more recently by Chaitin [CHAI 87] reverts back to the concept of an *algorithmic complexity* of a sequence of numbers as if it were the smallest program capable of completely describing itself. If the program is separated into two parts, those with instructions and those with data, the size of the first will determine its *sophistication*. An infinite and random sequence could only be described as an extended enunciation, therefore possessing a void sophistication and an infinite complexity. By extension this definition can also be applied to a system that has already been formalized. This approach likewise allows us to take into account the diversity of the system in terms of both its objects (structural complexity) and in terms of its performance (functional complexity). The concept of complexity is therefore linked to the description (in this instance an algorithm) of the system in its entirety. It is evident that which is complex is difficult to describe and even more difficult to explain.

This definition is linked to Boltzman's theory of thermodynamic entropy, which measures the rate of disorder of a physical system. As a matter of fact, the more disorganized the system, the more its minimal algorithmic description equates itself to the extended enumeration of its elements. In addition, certain authors (such as Atlan) employ the notion of *informational complexity,* calculated using Shannon's entropy formula. This signifies the number of questions with binary responses that it is necessary to put into the system, in order to describe it perfectly.

From the theoretical algorithmic complexity (generally incalculable), we can take a provisional complexity, equal to the smallest algorithm that we can write at any given moment in order to solve a problem (in finite time). It is in this manner that problems often judged complex at a given moment might thereafter be simplified.

Complexity is the opposite of simplicity, and something that is complicated is something that is unnecessarily complex, which we must therefore be able to simplify. A complex mathematical demonstration is quite often a long one, at the very least difficult and in need of a great deal of mental vigor. Even if it is sometimes transcribed in as short a form as possible, it musters a great deal of fundamental knowledge that would bring about a very long and complicated result if it was necessary to clarify it in a series of logical confirmations. A complex algorithm can also be measured by its length, a method even more efficient that an actual demonstration, for an algorithm must be totally explicit.

As a result of this, the notion of complexity according to Kolmogorov or Chaitin [CHA 03] is linked to the notion of minimal clarification. It is tempting to move from clarification to explanation, whilst stating that explanation is a minimal clarification in proportion to its size. To explain is to compress the description so as to render it as simple as possible without losing any information. On the contrary, if an algorithm associated with a system is equal to or even longer than its sequential description, then this algorithm is not an explanation. It is thus necessary to simplify this description by a theoretical, logical reasoning to highlight repetitions, symmetries, etc. Unfortunately it is evident, as proved by Chaitin, that the general problem of knowing whether an algorithm is of minimal simplicity or not in order to solve a given problem, is an answerless question. Consequently a system judged complex could one day find a simple explanation, unproven up until this point.

Hence, there is no evidence that everything could be explained, nor that there exists a compact description for every system, nor for everything. Take the case of real numbers for example. Only a certain few have a concise description, for example solutions in the form of algebraic equations ($\sqrt{2}$ is the solution for the equation $x^2 - 2 = 0$) but the majority of real numbers can only be described by their decimal expansion, an expansion which is also unfortunately infinite. As a result the majority of numbers are forever inaccessible; they are however the fruit of the human brain.

By analogy, there is a strong possibility that the explanation of everything in this world (and all the more in this universe as a whole) could not reduce itself to a few simple laws or to a certain amount of regulations that allow for the simulation of the reality of the entire universe, contrary to the beliefs of Wolfram [WOL 82] and certain other researchers of the Santa Fe school (the theory of the whole). It is possible that if many universal phenomena could be explained in a simple manner, the majority of facts and phenomena would remain unspecific and therefore impossible to condense, and so without any explanation. Why does the location and shape of a pebble found at the bottom of a garden path have a simple, compact and scientific explanation? Some of its properties can be explained (it weighs 124 g, it is a piece of flint, etc.), however only the list of the millions of specific interactions it carried out with its surrounding environment in the process of its history have shaped its present state. Any method of clarifying the list of these interactions would be the only way of explaining its current state. This list could be even more complex than the pebble itself, each pebble having its own unique description. We have thus left the arena of scientific explanation connected to the universal, and crossed the threshold into the world of the specific, the unique, the individual, the random. We have not however entered into the world of the incomprehensible because observation and description still remain possible due to our senses; we can see the pebble, feel it, weigh it, categorize it and can therefore give it a partial yet significant presentation; the brain does not need a complex reconstruction of the object to "understand" it. This object can also be described with every method of language, literature, painting and so on; these domains appear thus as a supplement to a scientific description.

In order to fully understand complexity, it is necessary to consider it as a model, a representation, or a description of an object more than the actual reality of the object itself. Indeed, if it is supposed that complexity is measurable it is more than likely infinite, as where exactly does the enhancement of nature end?

The scale of complexity according to "Density of Description" ranges from the simple to the complicated and the complicated to the incalculable. Few properties are simple, but they are frequently universal, if not at least shared by a great number of objects. On the other hand, there is the incalculable, the infinitely complex. This is in all probability the case with the majority of things that are specific and can only be described by themselves, by their extended description, element by element; it is however necessary to describe all of these elements in the same manner, a job that is likely to be endless.

The objective of science is therefore to abbreviate complexity to an even lesser extent by simplifying to the utmost, the explanation of phenomena and of things. This poses the problem of the actual meaning of the scientific approach concerning the study of complex systems. The risk is perhaps researching in order to simplify that which is complex in the reality, trying to reproduce it as close as possible to the original form, in an approach in which explanation would be less important than the reproduction of reality. When would the quality of mimicry become a kind of scientific proof? The science of complexity would therefore be destined to wallow in its own black box. The increasing strength of computers and the evolution of software equipment furthermore render the complexity of simulation models possible. This could lead us to the paradox of "an inversion of complexity", which would consist of developing models more complex than the observed reality, in order to accomplish a better copy of it.

Whilst not forgetting that our approach is clearly not the research of complexity itself, we are however still destined, when faced with the fascinating and infinite complexity of reality, to try to reduce it through a simple formalism in order to understand it better. The systemic approach often uses a process of reductionism, which consists of splitting the system gradually starting with its global aspect, into the simple parts (or subsystems), whilst explaining at each level the interactions between these parts. However, certain complex systems, biological, social and so on are characterized by the emergence of global behaviors that are of a completely different order from those of lower levels. Comprehension of these properties, self-constructed from an individual to an aggregated level, cannot be easily applied to a reductionist approach. Simulation therefore consists of starting with simplified rules governing individual behaviors, to restructuring the behavior of the system at an aggregated level. As a result, interest consists more of attributing a meaning to these rules than reproducing the reality, as close as possible to the original. A pedagogic example of this can be seen in Shelling's model, which shall be discussed in the following chapter.

Another ascending approach still used today is the constructal theory developed by Adrian Bejan [BEJ 00] and promoted by [POI 03] which attempts to counter fractal theory. By advancing through the progressive aggregation of components, certain global properties are optimized instead of continuously splitting them according to a rule of self-similar disintegration as with that of fractal theory. We are thus currently witnessing an altercation between the reductionist approach and the holistic, synthetic and ascendant approach. Both these theories are not in conflict with each other; they may in fact mutually benefit each other.

The notion of complexity has been analyzed to retain only a very general definition connected to the notion of information. The scientific determination is to always simplify in order to explain, starting with the blurred explanation of the reality; it therefore appears that the notion of a system demands a more precise approach from the perspective of its use as a simulation "machine".

11.2. The systemic paradigm: from the combinatorial to emergence

The establishment of a model in the majority of cases underscores the notion of a system. Is the modeling process not just the construction of a system? The term system is used in many different contexts, which must be distinguished. We perceive the real and the actual of this universe first and foremost by our senses and then by our scientific observations; we never will know it in full. In attempting to understand or to explain even a portion of this universe, we can initially identify it as a real system, which is at this stage not yet formalized or explained, but only demarcated, more or less clarified by an observation, associated with one problematic, one scientific project. The real system exists outside of us, independent of the observations we have made about it or the awareness we have of it. In order to become scientific, it has to go beyond the level of individual observation, and find a social existence that originates with multiplicity and independent of external yet confirmed observations. This existence would materialize with the formalization of a theoretical framework, expressed in any kind of symbolic language, natural, algorithmic, graphic or mathematical. Claude Bernard stated as early as 1865 "systems exist not in nature, but in the minds of men".

11.2.1. *The systemic triangle*

The system gradually begins to define and refine itself, through the scientific approach of establishing a model, moving between observation and experimentation to become an object pertaining to the world of knowledge. In order to accomplish this, the inextricable complexity of reality has to be simplified in different ways. To begin with, we can simplify by making a problematic and a hypothesis, where we are bound to make thematic choices, and the study thus becomes limited, not only in length but in depth. As a result we can only retain a limited portion of reality by observation. Consequently we can only retain a scattered image of this harsh reality. Another factor also comes into play here, a form of simplification discussed previously in relation to complexity, that of scientific formulation. Attempting to re-enter observation by an economy of thoughts, this consists of a familiar theoretic framework (eventually leaving to make it evolve), through a concise and simplistic formalism. This allows in this way a connection of this reality with a scheme of familiar semantic relationships, which subsequently will give them meaning.

In addition, the evolution of technology, in particular computing, affords today's researcher the possibility of prolonging the purely intellectual construction of the system through a hardware construction (as with a laboratory experiment, a model or a machine) or even by a virtual reconstruction, purely informational and software orientated. These constructions allow for a more precise representation.

This possibility of simulating the performance of a system on a computer offers great flexibility, as it allows the testing of a great deal of calibrations, and the diversification of the initial conditions, even those of an unrealistic nature, and thus to observe the consequences, an experimentation that cannot be carried out in reality (meteorological simulations, social simulations, etc.). This thus enables the model to evolve in an incremental manner through confrontation and successive validations whilst being observed [GDDL 04].

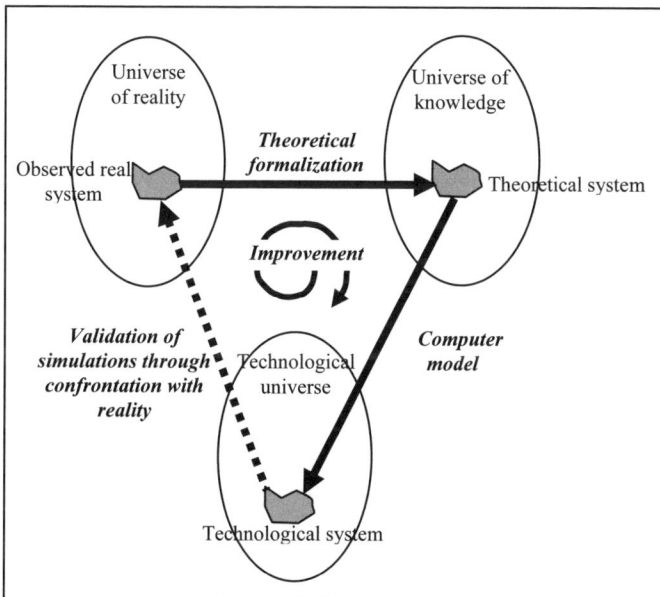

Figure 11.1. *The three systems*

These two sides of the same observed system, theoretical and technological, must not be confused. They generate two modeling methods, theoretical and computational. If the principal aim of the theoretical model is to understand the real system, the computational model may have several aims, either attempting to better imitate the real system (possibly by methods other than those of the theoretical model) in a practical approach without necessarily wanting to understand it better (meteorological, climatic, hydrologic previsions, nuclear experiments, traffic regulation, etc.) or continuing and extending the construction of the theoretical system, respecting its simplifications, in order to sustain the theoretical approach of understanding. This allows for a validation of the behavior – or of certain properties – of the theoretical system, without necessarily applying it to reality. The establishment of a model often finds an intermediary between these two approaches, unfortunately recognizing that the more a model borders on reality, the less explanatory and thus the more complicated it becomes; however, the more explanatory it is, the more it distances itself from this reality, due to its conciseness.

11.2.2. *The whole is greater than the sum of its parts*

The notion of the systemic paradigm rests in part on a frequently expressed affirmation that in any system, the whole is greater than the sum of its parts. This proposal asserts the fact that the set theory is not sufficient to formalize the idea of system, as it does not suffice to realize the gathering or the sum of the individual behaviors to obtain the behavior of the set. This therefore demonstrates that the basic notion of interaction is lacking.

This proposal also implies that, in a reductionism approach whereby the problem is subdivided, the overall wealth of relationships and interdependence that exists between the system components, is lost, which would thus justify a holistic approach. It should be noted that this cutting out is not necessarily destructive. On the contrary, by individualizing consecutive levels of subdivision of the system, it enables a clear formalization of the interactions between lower level components, hence enabling the establishment of conduct for the higher levels of the system. In a back and forth methodological movement, from one level to another, top to bottom, both the simplistic decoupage and the overall cohesion that emanates from the reconstruction of interactions can be noted. An emergence can thus appear to come from individual behaviors. Breaking down in order to simplify it is not therefore sufficient; however, it offers no reason not to continue to do so.

11.2.3. *The whole is less than the sum of its parts*

If the set theory is not equivalent to that of systems, it nevertheless makes it possible to formalize it for the most part. It is already noticeable that the set of components of a set E is richer than set E itself and this consequently introduces the combinatorial notion, the main origin of the systemic complexity. The number of components of a set of n elements being 2^n, for 10 elements there are $2^{10} = 1,024$ components, but this quickly becomes astronomical, for 100 elements there are around 1.27×10^{30}. Thus, the important thing to note here is not the set, but its organizational structure, hence its components. Indeed if we were to consider the system to be constructed from "basic building blocks", its formalization in terms of the set theory is natural, the system being a set E, constituted from elements e_1, $e_2...e_n$. These elements may themselves be considered as elements of lower levels, and so on. This brings us to the notion of a set hierarchy formed by a sequence of partitions, each one being more refined than the last, within the same set. This is a standard organization of a system's elements, but it already encompasses a great deal of complexity, as the combinatorial of these hierarchies is much greater than the components of a set (for a set of 10 elements, there are 2.5 billion different complete hierarchies!). These objects can be formalized by introducing the notion of a trellis, which is a generalization of the notion of hierarchy.

The hierarchical or "vertical" organizational structure is not the only arrangement present in a system. The "horizontal" structure between the levels of the same hierarchy also plays a fundamental role. Once again, the set theory, through the notion of relationships, provides the essential tools for the formalization of the links between the components of a system, and allows for among many things an outline of a neighborhood graph.

11.2.4. *The whole as a structure of its components*

The system changes its status here. From the set evoked earlier, it now becomes a *structure*, which is, in an outline approach, a set endowed with relationships. The vertical and horizontal links which define the structure, allow for a hierarchy of levels between the elements, and the definition of the links (neighborhood, communication, dependence, etc.) between elements of the same level. This structure generally serves to direct the interactions between the objects of the system.

It is imperative to pause for a moment here and consider the notion of structure, because according to the terms of Raymond Boudon, "the structure seems both indispensable to all the human sciences, as witnessed by the mounting frequency of its use, but despite this it is hard to define". We only have to refer to the *Encyclopedia Universalis* to be confronted with four definitions concerning this topic:

1) complex organization (administrative structure);

2) the manner in which things (whether abstract or concrete) are organized into sets;

3) in philosophy, a stable set of interdependent elements whereby each one is only what he is in and by his relationship with the others;

4) in mathematics, a set supplied by certain relationships or laws of composition.

We note how these definitions are in relation to our subject. The first definition reintroduces complexity. The second brings us back to the notion of an organized set, discussed at length in sections 2.2 and 2.3. The third, in a simplified version, reverts back to the theories of structuralism (Saussure, Merleau-Ponty, Piaget, Levi-Strauss, etc.) but does not contradict the mathematical formalization suggested by the fourth definition. What is more this last definition returns to the contemporary stance of defining an object not by its intrinsic properties, but by its exterior connections. Its function is defined by that which it consumes and produces externally, and not by its content nor by its inner workings. This is more particularly the systemic paradigm of the black box.

Since the end of the 1950s, the reference to the concept of *structure* is quite general in the field of human sciences. As a result, structuralism developed in the 1960s is not a school of thought we can easily identify. It is both multi-disciplinary and trans-disciplinary, indeed interdisciplinary. Human sciences seek to clarify a concept of structure through structuralism, but we can actually observe a thematic approach of this concept, which is a product of heterogenous rationalities, through the diverse disciplines of human and social sciences. For that reason, it is often considered a polymorphous concept. Using this, a reformation of knowledge is sought, bridging gaps between the sciences. We can see from earlier works a willingness to renew forms and representations, and to clarify the links and relationships between different structures. Structuralism attempts to legitimize human sciences in bringing it closer to the so-called "hard" sciences, in order to bypass the clash that currently exists between the scientific world and the academic world. As a result of this the "concept" of *structure* brings forth the illusion that there exists a unity between the various paradigms of human sciences.

In addition, if we consider the etymology of the word "structure", we observe that it is composed of "structura", to construct. It is really a matter of studying a construction of knowledge, and in this manner to reconcile the sciences between themselves using the same concept. Yet, as we previously observed, each discipline attaches a different form to the concept of structure.

Structuralism has certainly modified each discipline strongly influencing their evolution, by renewing the representations and by reducing the partition into disciplines. This does not however intend to create homogenity of methodological and epistemological principles that apply themselves evenly to all human sciences. There is therefore no common definition of *structure* that could be applied to human and social sciences. It is interesting to note how the field of mathematics approaches this notion. A great diversity of meanings also exists in this domain, but each one is very accurately defined, as it must be with mathematics. Each mathematical structure nevertheless defines functional links between the elements of a set. In the case of an algebraic structure, the operations (addition, multiplication, etc.) define the links between elements. However, it is mostly the properties of the operations that are important (commutative and associative properties, etc.). The example of the group structure[1] is symbolic as it is both simple and still plays a fundamental role in mathematics and physics by translating certain properties of invariance and symmetry in natural phenomena. This refers to Euclidean invariance by the displacement group (translations and rotations), to Poincaré's group which defines the rules of invariance in the theory of relativity, or to the renormalization group for the invariance of the physical observables in the quantum theory. The group structure is enriched if other operations such as multiplication are added. We therefore witness the prosper of a multitude of algebraic structures with falsely embellished names such as modulus, ring, body, algebra, vector space, topological space, Hilbert space and so on. All of these structures play an absolutely essential role in intellectual wealth for both mathematics and physics. If we can establish a permutation between two sets of objects (often in very different domains), in respecting furthermore their respective algebraic structures (isomorphism), we may apply all the acquired results from one domain to the next. Moreover, each of these domains shed light on the other under a new representation whereby the mutual understanding of each other is improved.

To return to complexity, we notice that mathematical structures concern mainly "simple systems" in the sense that the elements, even if they are often in infinite numbers, are both completely different, yet identical in their properties relative to the structure. They all have the same "behavior" with regard to the rules of operation. Reality is evidently more complex, being formed from elements as diverse in content as in behavior. It must then be simplified to the extreme to make use of these structures. Their use nevertheless allows for a correct explanation of very general complex phenomena that does not rely too much on these variations. For example in the study of gravitation, it is of no use to know the composition or the color of the objects in action; all of these objects are solely characterized by their position, their speed, and their mass. We can use a very general mathematical theory

---

1 A group is a set G on which is defined by an internal operation called addition, marked +, that possesses the relevant properties:

1) associative property: for all the elements a, b, c, of G, a+(b+c)=(a+b)+c;

2) existence of a neutral element: marked e, which verifies, for all the elements a of G, e+a=a+e=a.

Every element x of G possesses a symmetric x' (also known as opposite and thus marked –x) such that x+x'=x'+x=e. For example the set Z of the relative integers supplied by the addition is a group (which is moreover commutative, because for all the relative integers a and b, the addition verifies: a + b = b + a).

to formalize and explain this phenomenon, without keeping all of the complexity of reality, useless for explaining the gravitation.

11.2.5. *The whole as an emergence of its parts*

The notion of *emergence,* linked to the theories of auto-organization, corresponds to the idea that in any given system, there may appear an unexpected but significant configuration, not explicitly inscribed in the operative rules of the system. If the notion of emergence is frequently used in a sociological (methodological individualism) or in a historic context, it is delicate to handle in formalized systems, as it implies a judgment by the observer, who interprets a configuration with respect to an external representation of the system. As a result, this notion can be classed more in the range of interpretation than that of simulation itself, which for the moment brings this notion outside the scope of this study.

11.3. Moving towards a more formalized definition of the notion of a spatial system

11.3.1. *First definition of a system*

The notion of structure, as employed in the construction of the notion of a system, is obviously mathematical in essence. At first glance, a system is a set of objects S supplied by two structures: an organizational structure that defines the spatial links between objects, and an evolutionary structure that defines temporal links, that is the dependence between the system in time t and the system at times t', preceding t.

For example, a simple yet proven organization refers to objects regularly disposed in a discrete space of dimension 2, where the nodes (i, j) are defined by integer coordinates, that is a squared mesh, and taking as an organizational structure the links of contiguity between objects (4 or 8 neighbors). As a simple evolutionary structure, we will take, as with space, a discrete time (t is an integer), and the state of the object at time $t+1$, calculated only by the state of its neighbors at the preceding time t. This is the concept of a cellular automaton. This structure could be generalized taking into account more distant neighbors, not only in space but also in time.

Many geographic systems, due to their complexity, cannot be modeled mathematically in such a simple way. We must therefore attempt to define the notion of a spatial system or a geosystem with a more effective aim, that allows for a computer construction in the use of simulations that could take account of the diversity of objects, of their organization and of the interactions composing them and allowing them to evolve, but without forgetting most importantly the general position in which the modeling process is inserted. Simulation must achieve results as close as possible to the reality that we are trying to understand.

To move past the notion of elements of a set, too abstract to manipulate a geographic system, we must try to generalize through the notion of a geographic object. The relationships between these objects will be discussed later through the concept of spatial interaction. After having described how a system operates globally, the notion of a spatial system or geosystem may be defined more precisely.

11.3.2. *Geographic objects*

An *object* is said to be *geographic* if it is localized, fixed, preferably unchangeable, demarcated and identified as being different to others. An object is relative to a scale, a defined temporality and materiality, three properties that are encompassed in the notion of spatial and temporal granularity. A geographic object must also be preferably significant, that is to say connected to a well-established set of specifics, particularly with a reference to a specific way to formulate the question, or to an established practice. It is therefore practically the equivalent to the concept of a place. In that sense, a cloud, a car, a pedestrian or a drop of water are not therefore geographic objects, either because they fluctuate too much in the scale of an allocated time period, or because they are too small for the spatial scale that is used. They are often referred to as *individuals* or *particles*. This does not however stop them from being essential for the comprehension of geographic phenomena as they are mediators of spatial interaction. The study of the behavior of a pedestrian or a motorist concerns psychology or sociology, but the mass effect of these behaviors, in relation to the streets, houses, and places of work and so on directly concerns geography. These concepts of geographic objects and particles will be specified even further.

The establishment of a geographic model can avoid going down to the actual particle itself. Instead of considering every tree individually in a parcel of forest, we often retain a single variable storing the number of trees, or if we need to be more precise, we store a vector of values giving the number of trees, or the percentage of surface used for each type of species, or for each age range, and so on. However, progress in storage capacity and information processing, if useful for the actual problem, allow us to zoom in on the individual tree and establish a model for its behavior. Even if every tree is summarily described, we can easily give each individual a personality of its own, as well as a rhythm of differentiated growth, which will influence – and also be influenced by its environment – and provoke, at the parcel's level, the emergence of a characteristic that would have been impossible to identify in a global description of the parcel.

11.3.2.1. *The choice, form and organization of objects*

A difficult question, for the formulation of any geographic problem, is how to know which relevant objects are most likely to express the problems, to formalize and then treat them. It poses on the one hand the question of the scale and on the other hand the question of the possible partition on a given scale. In principle the geographic space may be envisaged as a spatial continuum possessing rather homogeneous areas in terms of description, with ruptures, discontinuities, and borders, between these homogeneous areas. On the other hand, there are several layouts for describing and observing this space, the purely radiometric layout (remote sensing view) can be considered, or the land use (*Corine land-cover,* for the European space), geomorphologic, politico-administrative, sociological layouts, etc. Each of these layouts may have their own relevant partition. The implementation of a Geographic Information System (GIS) makes it possible to store, superimpose and combine these different geographic partitions, geometrically incompatible in the same space.

However, behind the term of partition there lie in fact two or even three different notions. There is firstly that of a significant "spatial unity" which will create the geographic object, and that of "granularity", which constitutes the most refined partition, beneath which the contents are no longer differentiated. These are typically the pixels of an image, but they can quite possibly be urban blocks, agricultural parcels in a rural landscape, or even a geometric mesh to describe a digital elevation model (DEM). This granularity breaks the space down into "spatial grains" that are precisely located in space, stable in time, with a topology of contiguity, forming a connected and compact partition of the space under observation.

In certain cases, however, it is necessary to consider "grains" of another nature, because space is not only a spatial and temporal continuum. Space may also contain non-connected grains, topologically separate, of undefined forms and variable locations, even if transient in time. They can be molecules of water circulating from the atmosphere into the sea, having passed through rivers, ground water, glaciers and ice floes. They can be the grains of sand that mould and re-mould the relief of the desert, the cars circulating on the road network, the inhabitants of a town, etc. These fluctuating grains, which will be referred to here as "particles" or "individuals", do not contain the same properties as the geographic objects. Geographic objects are therefore made up of grains. We will see later on how these objects are linked to particles.

Let us see firstly the necessary information for the structuring of geographic objects as much in their individuality as in their spatial organization.

Geometry demarcates the limits of an object by its different contours. For an area each contour is often represented by a closed polygon, which can be represented by the sequence of the coordinates of its vertices. This object can be constituted of *n* disjoint parts that will be defined by *n* exterior contours. In each connected part, there can be found holes that are defined by internal contours. It is necessary to distinguish between internal and external contours, in particular for calculating surfaces and for the location of algorithms (does point P belong to object Q?), which requires structuring the different contours in a tree structure that is more or less complex. This choice of description known as a *vector description* of geometry tends to define the object by its shape and limits. Its content constructs itself afterwards by identifying those grains that are within these limits. Nonetheless the object could also be defined firstly by its content and then by deducting further its limits by the aggregation of grains of the required content. Such a description (called a *raster* or *image*) is characteristic of image processing whereby an object is defined by a set of pixels i.e. by its interior, and not by its limits. A third method of defining geometry is the description using a *mesh* of space, which intermediates between both preceding methods. This description has the advantage of being able to define geometric space in a continuous way, without necessarily differentiating between objects. This becomes useful when we have quantitative spatial information (altitude, temperature, etc.) at certain points in space (the nodes of the mesh), information which is continuous in reality, i.e. existing at every point in the real space. The mesh therefore allows for the definition of a spatial interpolation, meaning a continuous function piece by piece (each piece being an element of a surface or a mesh) with continuous conditions on the edge of each element as well as on the edge of the domain. In such a representation, the partition into spatial objects can therefore be carried out, either by a demarcation of each contour line or by a morphologic study of the surface by underlining singularities: local minima, summits, crests, thalwegs and saddle points. The morphologic objects are constructed by structuring them around these singularities, outflow networks, plateaus, slopes, mountains, basins, etc.

Topology is the structure that allows for an organization of proximity or neighborhood links between geographic objects. For example, in a zonal partition (such as an administrative partition) it is interesting, in order to manage the interactions, to automatically recognize the direct neighbors of each zone. To achieve this we no longer structure the zones by their polygonal contours, but by border lines (arcs) between zones (from which the name ARC/INFO for a well known GIS derives). Each edge is linked to two vertices V1 and V2 and separates two contiguous sides S1 and S2. The set of edges is therefore structured according to a model of a plain topologic graph. This graph of edges can also be used to structure a transport network. It is interesting in that case to know, for each vertex, in which order the edges attached to it exit. This information can also be organized starting with the edges, retaining for each one which is the following arc, turning to the left around its final vertex, and which is the previous edge on its initial vertex, turning to the right. We therefore obtain a structure referred to as a "combinatorial map" [DUF 88], [GRO 89]. These two topological structures may coexist in order to build a level for describing geographic objects, either a linear network, a zonal partition of space [LAN 94] or a 3D mesh to organize a DEM.

A *hierarchy* of the interlocking objects may also be necessary when successive levels of scale are present in the system. A perfect example of this is the hierarchy of an administrative system: districts, counties, regions, state, etc. Each layer contains the objects of a given scale and "vertical associations" of inclusion enable the interlocking organization to be stored or calculated.

The *content* of the object is as essential as its shape; that is the description of its internal characteristics, material or informative. Two levels of structure can be identified. Firstly, the part of the description which is common to all objects in the same class and which may be omitted in individual descriptions, and secondly the part that differentiates an object from the others (its individuality). The description of objects according to a hierarchy of "object-orientated" classes provides the best method for recognizing the progressive differentiation of the object characteristics, from the most generic classes to the most individual objects.

This hierarchy of description must obviously not be confused with the hierarchy of organization by the interlocking of elementary objects into objects of a more complex nature.

11.3.2.2. *The behavior of objects: agent, actor*

The behavior of an entity (whether object or particle) is defined by the way which it evolves in time. This depends on its structure and its internal functions (think of the anatomy and physiology of any living being) that define its intrinsic capacities. The state of the entity evolves according to its interactions with its environment. To go further in formalizing the behavior of an entity, we must refer to the multi-agent paradigm (see Chapter 13) as it offers the opportunity to describe the entities of a system as *agents*, which may be active, intentional, cognitive, capable of learning, etc. They can even become *actors* if they can influence the behavior of other agents; modify the aims, the rules, and direct orientations, etc.

11.3.3. *Interactions*

Francois Durand-Dastes [DUE 84] defines *spatial interaction* as the condition whereby "the contents of places react to one another through a series of reciprocal relationships". In order to specify this concept we will say that a spatial interaction, defined on a set of geographic objects (those of the system), is the macroscopic result of the action of "microscopic particles of interaction" that progressively transform the objects of the system. These particles may be individuals, material objects, ideas or information, which transform themselves, increase, move through space, and the global effect they produce between objects is called interaction. The concept of interaction affords us the opportunity to link two conceptual levels not previously linked in our study, those of the geographic objects and those of the particles (or individuals).

The growth of a town over one or even several centuries results in the continual movement of a swarm of city dwellers-particles, who are born, grow up and then die. During their short lives they move between their house, their work and their leisure each and every day. However, that which interests geographers most is that they build and then demolish roads, houses, factories, and thus modify urban land use and spatial development.

When there is a flow of people going from their homes to their place of work, this provokes (among other things) a simultaneous reduction of the population in residential areas, and a consequential increase of the same number in employment districts. This transition is characterized by an increase in traffic flow on the transportation networks. This translates at a geographic level of the town, as an interaction between the residential areas and the areas of employment, an interaction that can express itself in different ways according to what is actually being studied. If the research project is the number of people actually present in different zones at any given time, we can witness a sort of phenomenon of a daily wave flowing back and forth between these two zones. In the long run, people attempt to cut down on traveling, but also seek the most favorable place for each activity. The dynamics of the zones themselves appear, where interaction acts as an energy field crossing over between residence, work and leisure zones (and the agricultural zones around the town). The different types of zone seek to both align themselves with one another in order to economize on transport, and to distance themselves from each other in order to find better conditions. We therefore witness the phenomenon of urban diffusion, irregular in its amplitude and its location, whether pulsating or not. Interaction thus appears to be like a game of mutual influence between places of different usages and more or fewer neighbors, influences that progressively encourage these places to modify their usage.

These interactions may be formalized by a combination of processes of demographic or economic growth, usage conversion, centrifugal and centripetal migration, differentiation and even segregation, (and so on). The combination is sometimes more favorable to either one or the other of these processes, according to internal conditions to which the process may suddenly become very sensitive, or to external events which are usually impossible to predict or to take into account. As a result, the evolution of the system presents itself more like a multitude of possible trajectories that can become apparent at any moment, than that of a unique and determinist trajectory readjusted from time to time.

Let us take for example the modeling process of epidemics. It can be studied at the level of a geographic system based on the main cities of the world. The "epidemic" phenomenon reverberates from town to town. One town is considered to be affected if it contains at least one infected person. Comprehension of the evolution of an epidemic rests on the size of the population of each town and on the interaction between them, specified by the flow of individuals from one town to the other (gravitation model). Knowing that any individual has a certain probability of being infected, the process may be simulated. However, a more advanced comprehension can be obtained if we take into account a second system, at a more refined level, where objects (hosts) are now human beings. The particles of interaction are pathogenic germs that reproduce themselves within an individual and travel between individuals to infect them. This allows for the establishment of a model of interaction, materialized here by the process of contamination between individuals, adapted from the process of propagation of the infected agent (incubation period, contagion period, method of transmission, etc.). With this example, there are three possible levels of hierarchy for objects upon which a system can be structured, that of the towns, that of the individuals, and that of the pathogenic germs, and there are two processes of diffusion that bring each of these two levels into play, town-individual and individual-germ.

Interactions can therefore be structured here by three rules that connect two consecutive hierarchical levels of a system known as *macro* and *micro*, transcribed as A and B and formed respectively by the objects a_i and particles b_j:

– a first rule T (for *transition* or *acquisition*) that treats the evolution of the state s_i, of each object a_i, starting with its previous state and the entering b_j;

– a second rule S (for *exit* or *propagation*) that deals with the evolution of the number of b_j present in a_i, and destined to leave, starting from the state of a_i and of the entering b_j at the previous stage;

– finally a third rule M (for *circulation*) that manages the transport of particles. It distributes the b_j that have just left the objects a_i towards the entryways of the connected objects.

11.3.4. *The functioning of a system*

The functioning of a system must be envisaged at different levels. There is a type of global functioning, but also the elementary structures that allow for the production of behaviors and diversified interactions of objects, according to the situations to be modeled.

Concerning the global functioning, theoretical formalization of reality lends itself more to total parallelism, that is to say that objects are both "living" and interacting at the same time, in a continuous manner, each at their own rhythm, but they are also capable of synchronizing their actions upon interaction. This necessitates that they are all aware of their spatio-temporal location, referred to in a common benchmark (same origin and same unity of time and space).

However, computer simulation leads to another formalization, conforming more to the limits of present-day technology. We should take note here, in terms of formalization, that current computers do not operate in a parallel way, and that their time is necessarily discrete and their memory limited. As a result, if there are n objects to deal with, the period of time being inevitably discrete is broken down into n iterations, and at each iteration an object is calculated.

On a PC a sequential operation, called synchronous functioning, will simulate the parallel functioning. It is therefore necessary to store the whole configuration of the system at time t and to construct the new configuration at time $t+1$, from the memory at time t. As a result each object evolves as if time had been stopped between the times t and $t+1$, and so as if iterations were all calculated at the same time.

Functioning can also be *asynchronous*, in this case it is not necessary to store the state of the system at the preceding moment because the n iterations are considered to correspond to different moments, the object i being treated at time $t+i/n$. In this case, some objects j of the environment of i are already calculated (if $j<i$) and the others have not yet been (if $j>i$). A stocktake of all objects is constantly being redefined in an unpredictable manner at every point in the process. To avoid a long-term bias linked to the order of storing of the objects, a running order for all the objects is defined again at random, with each step of time.

Finally the general functioning can be completely *at random*, that is to say that the objects to be calculated are chosen by successive, random, independent draws.

Furthermore, in a complex system, interactions are numerous and fundamental, it is therefore necessary to decide at what time they should be taken into account. Generally speaking, in synchronous mode, at each step of time, after a phase of calculation of the state of the objects, there intervenes a phase of updating the exchanges between the objects of the system.

11.3.5. *A formal definition of a spatial system*

We can now propose in a more precise manner than before a definition of the notion of a geo-system as a set of localized objects in the same spatio-temporal referential. The *spatio-organizational structure* defines the content, the geometry, and the topology of the set of objects. The *evolutional structure* defines the behavior of each class of objects and the process whereby they interact. Each defined process between objects, produces a type of interaction, formalized in the form of flows (of particles or individuals) between objects (watersheds: water flow, migration: flow of population, economic process: flow of goods and money, telecommunication: flow of information, etc.) and as a result transforms the global configuration of the system. The system is open; it interacts with outside objects, which can be regrouped under the term of environment that constitutes an even more vast system that is non-descript. The environment is taken into account by entering movements (the intrants) and exit movements (the extrants) and possibly by mechanisms of exterior control also.

Furthermore, a system appears to be a recursive, self-referential concept, meaning that each object of the system is either itself a system (and described as such in the system) or a terminal object. This terminal object cannot be broken down, and is not described in a systemic mode, but described by a motor, which is a black-box algorithm, simulating its behavior, meaning that we are interested in what it does more than how it actually does it. What's more, the system as a whole has the characteristics of an object pertaining to a system that contains it (its environment). The fundamental systemic circularity comes into view: system-object, object-system, which confers to it a hierarchical structure. In conclusion, the following definition may be proposed:

Definition of a geo-system

It is an intelligent or technological construction supposed to describe a portion of reality explicitly limited between two levels of scale and knowledge. A system is formed by two structures. The spatio-organizational structure is composed of hierarchical objects, which are themselves systems, or terminal objects, (possibly composed themselves of grains). The evolutionary structure is composed of interaction processes (determinist, stochastic or mixed) acting between objects (possibly through the intermediary of particles or individuals), which transform their content and their organization. The objects and the processes evolve in the same spatio-temporal referential. A system is limited to the outside by the environment that encircles it and limited to the interior by its terminal objects, of which we do not seek to understand the functioning, but which are each functioning as a system in any case.

After such a definition that provides an extended yet very precise theoretical framework concerning the notion of system, we will now see how computing enables us to decode the notion of system, through functional technological models, capable of performing real simulations. We will introduce in the following chapters, firstly the notion of cellular automaton, that are historically older, and subsequently the notion of a multi-agent system.

11.4. Bibliography

[AU 92] A. AHO, J. ULLMAN, *Concepts fondamentaux de l'informatique*, Dunod Paris, 1992.

[BEA 92] D. BEAUQUIER, J. BERSTEL, P. CHRETIENNE, *Eléments d'algorithmique*, Masson, Paris, 1992.

[BEJ 00] A. BEJAN, *Shape and Structure from Engineering to Nature*, Cambridge University Press, 2000.

[CHA 87] G. CHAITIN, *Algorithmic Information Theory* (1st edition), Cambridge University Press, 1987 (3rd edition, 2003).

[CHA 03] G. CHAITIN, "L'univers est-il intelligible?" *La recherche*, no. 370, pp. 34-41, December 2003.

[DAU 03] A. DAUPHINE, *Les théories de la complexité chez les géographes*, Anthropos, Paris, 2003.

[DUF 88] J.F. DUFOUR, "Spécification progressive d'une algèbre pour manipuler les cartes topologiques orientées", in *Actes de PIXIM 88*, pp. 61-80, 1988.

[DUR 84] F. DURAND-DASTES, "Systèmes et localisations: problèmes théoriques et formels", *Colloque Géopoint* 1984, Avignon, Groupe Dupont.

[FRO 95] C. FRODEVEAUX, M.C. GAUDEL, M. SORIA, *Types de données et algorithmes*, Ediscience international, Paris 1995.

[GRO 89] C. GROSS, Opérations topologiques et géométriques sur les multicartes combinatoires; application à la cartographie thématique, thesis, University of Strasbourg 1, April 1989.

[GDDL 04] Y. GUERMOND, D. DELAHAYE, E. DUBOS-PAILLARD, P. LANGLOIS, "From modelling to experiment", *GeoJournal*, 2004, vol. 59, iss. 3, pp. 171-176(6), Kluwer Academic Publishers.

[LAN 94] P. LANGLOIS, "Formalisation des concepts topologiques en géomatique", *Revue internationale de géomatique*, vol. 4, no. 2, pp. 181-205, 1994.

[LD 02] P. LANGLOIS, D. DELAHAYE, "RuiCells, automate cellulaire pour la simulation du ruissellement de surface", *Revue Internationale de Géomatique*, pp. 461-487, vol. 12, no. 4, 2002.

[LEM 91] J.L. LEMOIGNE, *La modélisation des systèmes complexes*, Paris, 1991.

[MOR 77] E. MORIN, *La méthode – Tome 1, La Nature de la Nature*, Paris, 1977.

[PEG 01] C.P. PÉGUY, *Espace, temps, complexité, vers une metagéographie*, Paris, 2001.

[POI 03] H. POIRIER, "La théorie constructale, clé des formes parfaits", *Science et Vie*, pp. 44-65, no.1,034, November 2003.

[WEI 89] G. WEISBUCH, *Dynamique des systèmes complexes, Une introduction aux réseaux d'automates,* Ed. InterEditions/Ed. du CNRS, Paris, 1989.

[WOL 02] S. WOLFRAM, *A New Kind of Science*, Wolfram Media, 2002.

Chapter 12

Cellular Automata for Modeling Spatial Systems

12.1. The concept of the automaton and its modeling

The evolution of computer power in the past few years has facilitated the emergence of simulation methods at the expense of the analytical resolution of mathematical models. Indeed, cellular automaton simulation allows us to free ourselves from the resolution of partial differential equations, by explaining these equations in discrete terms of time, space and condition. Thus, the performance of office computers, and the development of theories and techniques of simulation like cellular automata or multi-agent systems allow us to attack these increasingly complex systems, with a quite fine discretization of space and time.

Moreover, the difficulty, or the impossibility even, of performing experiments in the social or environmental field to test hypothetical theories, adds to the interest of simulation, which allows the rapid realization of numerous tests, supported by graphical results, often connected to a geographic information system (GIS) which allows the easy comparison of the result with the observed terrain.

Nonetheless two paths seem to diverge quite substantially in this domain. The quest to complexify the models risks losing the essential objective of the research, that is, to explain.

Chapter written by Patrice LANGLOIS.

To explain is to articulate, in the easiest possible way, a phenomenon in the framework of a rational theory. However, the eagerness to continually simulate reality more precisely through models is also the wish of practitioners who can also make weaker predictions such as in city planning, meteorology, weather forecasts, etc. Unfortunately these two approaches are not always compatible, as often the closer a model resembles reality, the less it can actually explain.

12.2. A little bit of history

World War II acted as a stimulant for many new scientific developments. It led to the birth of the computer, made necessary due to the huge number of calculations needed for the creation of the atomic bomb. It is also the period of the development for cryptographic methods used to decode the German's secret messages. In this environment, mathematicians like the American John Von Neumann (1903-1957) and the Englishman Alan Turing (1912-1954) became pioneers. Von Neumann worked on the design of the first computers at the Los Alamos National Laboratory. There he invented cellular automata in the late 1940s. It was also Von Neumann who developed game theory in 1944 [VNM 44]. Turing, for his part, invented automatic decoding systems to decode German encrypted messages. He also participated in the development of ideas for what became computing, and most importantly, through his theoretical work, he invented the concept of the virtual machine, the Turing machine. He participated in the mathematical revolution of the 20th century following Gödel's results concerning incompleteness where he showed in particular that numbers and functions exist that are incalculable and possess unsolvable problems. All of these developments question the grand theoretical program imagined by Hilbert during the previous century, who ambitiously set out to codify mathematical reason in a general system of axioms and rules of inference.

It is in this context that the concept of the cellular automaton emerged [FAT 01]. Von Neumann tried to invent an electromechanical machine with this capacity but the level of technology at the time was insufficient. One of his colleagues, Stanislaw Ulam (1909-1984) who worked on recursive geometrical objects, gave him the idea of a formal construction, using the computers in the Los Alamos laboratory to operate a cellular system subject to simple rules. After this the cellular automaton was born.

Von Neumann developed a virtual auto-reproductive machine which had the properties of a universal calculator, although he did not publish it in his lifetime, perhaps thinking that it was too complex (29 states) and that it did not follow the "natural" rules of physics concerning invariance by rotation and by symmetry.

It was not until 1970 that a much simpler cellular automaton was made public, John Conway's game of life. It was publicized by Martin Gardner in the American Scientist. In 1982, Conway and other researchers proved that the game of life also possessed the properties of a universal calculator. It is thus far the simplest cellular automaton constructed with this property, since there are only two states and moreover it verifies the properties of invariance by isometric transformation. We will not here develop these mathematical properties, as they are quite difficult. For more details see [BER 82] or [POU 85].

In order to present the concept of the cellular automaton in a didactic fashion we will not be following the historical development of this concept. We will begin with the most elementary concept of the automaton in its finished state before formally constructing that of the cellular automaton and seeing examples applied in geography.

We will address here only the automata in discrete time and state, even if the continuous automata associated with mathematical techniques such as the Laplace transformation can play an important role in certain areas of geography such as hydrology.

12.3. The concept of the finite state automaton

A finite state automaton is a mathematical object, and we will first present it intuitively to better understand its formalization after that. We must imagine a device which has at least an input channel, an output channel, connected to a box containing a self-powered mechanism. An input channel receives one by one (sequentially), coded information with symbols that constitute the input alphabet. Likewise the output channel produces symbols written in the output alphabet. In short, the box contains the means of internal representation, a memory capable of containing a symbol called the *state* of the automaton traced in the alphabet of states. The three alphabets, input, output and state contain only a finite number of symbols. The value of an input or an output can be logical (binary), quantitative (integer, real), qualitative or purely symbolic (encryption according to a discrete alphabet like before). It can also constitute a vector of elementary inputs (or outputs) when there are several input or output channels. Once again we give these more or less complex values entering or exiting the automaton the name symbols, without specifying their nature.

The internal mechanism breaks down into two functions. The first function is the capacity to read the symbol at each input (which we call the vector or more simply the input) and to modify the state of the automaton contingent on the input and the previous state. This is the *transition function*. The second function enables a symbol to be present outside in the output channel, calculated in contingence with the input and the state of the automaton. This is the *output function*.

The input, output and state symbols can, in the most general cases belong to different alphabets. However, in a simplified version, used in particular with cellular automata, the same alphabet is used for the three and the output mechanism is reduced to its simplest form and consists only of producing the state of the automaton on output. Therefore the mechanism is reduced to a single transition function.

Consequently it is apparent that a finite state automaton is an elementary system which can serve the construction of a complex system by a series of connections or in parallel with several automata. The outputs of some are connected to the inputs of others. To be able to connect several automata, it is necessary to synchronize them through the definition of a common time and a control mechanism synchronized between the automata and their connections. We then obtain a network of automata [WEI 89].

12.3.1. *Mealy and Moore automata*

We are now able to formally define the idea of an finite state automaton, or Mealy's automaton, as a structure $M = (S, A, B, T, H)$ where S is the state alphabet, A the input alphabet, B that of the output, T the transition function which is the application of $S \times A$ to S, and finally H the output function, which is the application of $S \times A$ to B.

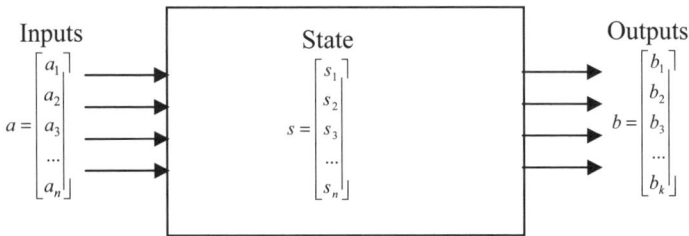

Figure 12.1. *General outline of an automaton*

Mechanism: at the discrete point t, for the automaton in the state $s(t)$, the arrival of an input value $a(t)$ makes the automaton pass into an other state $s(t+1)$ by the application of the transition function T, and calculates the output $b(t)$ by applying the function H. The automaton will be outlined by these two equations:

$$s(t+1) = T(s(t), a(t))$$
$$b(t+1) = H(s(t), a(t))$$

which are those of a *dynamic deterministic system in discrete time*. If the output b does not depend on the input a in H, then it is Moore's automaton.

12.3.2. *An example of Moore's automaton*

Now a simple example of an automaton is presented to show how such an object can be manipulated. The following Moore's automaton is called an adder. It is an automaton that takes two input numbers in binary code and sends back their total in output, also in binary code. Each number n is composed of k bits and is noted $n = n_{k-1} \ldots n_i \ldots n_3 n_2 n_1 n_0$ which symbolizes its binary spelling, the succession of 0 and 1. In total n is considered as a word of k letters written with the alphabet $\{0, 1\}$. m and n represent the two inputs and r the output of the automaton containing the total of m and n. The automaton reads the two numbers sequentially, starting from the right, in other words at the beginning with the least heavy bits. At each stage i it processes the bits m_i and n_i and calculates their total, r_i. The automaton also needs a state, s to memorize the carry digit (0 or 1). The alphabet of input, output and state is therefore the same, $A = B = S = \{0, 1\}$. The two functions of transition T and of output H are defined according to the binary addition table: $0+0 = 0$; $0+1 = 1$; $1+0 = 1$, these three additions are made without a carry digit (in other words a carry digit of 0) and $1+1 = 0$ with a carry digit of 1. The transition function T therefore combines a carry digit s_i and an input $m_i + n_i$, a new state s_{i+1} which is the new carry digit after the addition of m_i and n_i. This is written $s_{i+1} = T(s_i \, ; \, m_i n_i)$. They are presented as follows: $T(0; 00)=0$; $T(0; 01)=0$; $T(0; 10)=0$; $T(0; 11)=1$; $T(1; 00)=0$; $T(1; 01)=1$; $T(1; 10)=1$; $T(1; 11)=1$. Also the output function H is defined by $r_i = H(s_i; m_i n_i)$ with: $H(0; 00)=0$; $H(0; 01)=1$; $H(0; 10)=1$; $H(0; 11)=0$; $H(1; 00)=1$; $H(1; 01)=0$; $H(1; 10)=0$; $H(1; 11)=1$. This is summarized in the two following tables.

T					H				
$m_i n_i$ / s_i	00	01	10	11	$m_i n_i$ / s_i	00	01	10	11
0	0	0	0	1	0	0	1	1	0
1	0	1	1	1	1	1	0	0	1

Table 12.1. *Transition function and output function*

The machine operates in the following fashion: in the first instant, the state (carry digit) s_0 is at 0 so the automaton reads the first two bits $m_0+n_0 = 1+1$. The table of function T gives the following state $s_1 = 1$, and table H gives the output $r_0 = 0$ which corresponds to "1+1=0 keep 1". Then we move on to the second bit $m_1+n_1 = 0+1$; with a carry digit of 1 which again gives $r_1=0$ and we keep $s_2=1$ and so on. The final result is $r = 01101000$ and the carry digit is zero. If the carry digit is not zero at the end of the k bits calculation, there is an overflow in capacity, and the result cannot be carried in k bits.

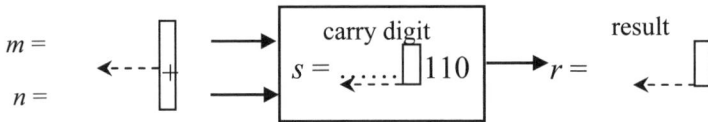

Figure 12.2. *The adder*

12.3.3. *Moore's automaton simplified*

Very often, and this will be the case for cellular automaton, the input function is reduced to the simple communication of the state towards the exterior (this is the identity function). The function H in this model has been omitted, it becomes apparent. In this case there is also $B = S$ as the output symbols are the states. Moreover, as the outputs of an automaton are often the input of another automaton, $A = S$ is also used, resulting in there being only one set of symbols for the inputs, states and outputs at the same time. A simplified automaton M is therefore limited to the data set S of states and of the transition mechanism T, therefore $M = (S,T)$.

12.3.4. *Logic gate AND: an example*

A logic gate can be considered like a Moore's automaton simplified to two inputs and one binary output (or Boolean). Here the transition function does not depend on the inputs or the state. For example, the Boolean operator AND takes state 1 (and sends it on output) if its tow inputs are worth 1; if not it takes the value 0.

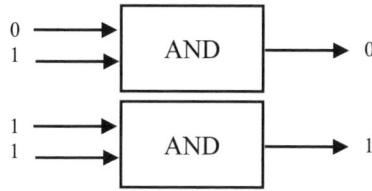

To define the transition function *AND* it suffices to give for each pair of inputs possible, the value of the associated output: therefore $AND(0,0)=0$, $AND(0,1)=0$, $AND(1,0)=0$ and $AND(1,1)=1$. These values can be summarized in a matrix of 4 columns and 2 lines. In the first line of T all of the possible inputs are placed and in the second line the associated outputs. This is therefore equivalent to writing the truth table of the logic operation *AND*.

$$T = \begin{pmatrix} 00 & 01 & 10 & 11 \\ 0 & 0 & 0 & 1 \end{pmatrix}$$

By the interconnection of automata which carry out the logic operation of basic Boolean algebra (AND operator, OR operator, NO operator) complex logical functions can be constructed and the arithmetical calculation of the binary numbers are deducted, which are at the root of the function of microprocessors. Therefore the adder can construct itself like a combination of logic gates.

12.3.5. *Threshold automata, window automata*

Threshold automata are very widely used, especially in neuron networks. This represents a simplified automaton with binary values. It has n inputs where $a_i(t)$ is the value of the input I associated with a weight p (a real number, called synaptic weight in the neuron network), p_0 being the weight of the state of the automaton and Θ a threshold of excitability. This excitation level is defined by a linear combination of inputs and state. If the excitation level is lower than the threshold Θ then the state remains equal to 0 (not excited). If not it passes to 1 (excited). Therefore, the new state (which is also the output) is calculated by:

$$a_0(t+1) = \begin{cases} 1 & \text{if} \quad \Theta \le \sum_{i=0}^{n} p_i a_i(t) \\ 0 & \text{else} \end{cases}$$

The window automata are also often widely used. The state becomes excited when the value of a linear combination of states belongs to an interval between a minimal Θ_{min} threshold and a maximum Θ_{max} threshold:

$$a_0(t+1) = \begin{cases} 1 & \text{if} \quad \Theta_{min} \leq \sum_{i=0}^{n} p_i a_i(t) \leq \Theta_{max} \\ 0 & \text{else} \end{cases}$$

The simplest example of a window automaton, used for example in the game of life, is where the weight p is worth 1, which means that the excitation level is simply the number of its excited neighbors.

12.3.6. *The automaton and the stochastic process*

In cases when the transition mechanism is no longer functional but random (which is frequent in social sciences) a generalization of the function $y = f(x)$ is used, which is called *transition probability:* instead of associating a single value y with each value of x, as the function does, a transition probability combines the total worth of several x values of y but these values only appear, given that x, in accordance with a certain probability $\pi(x, y)$. Thus, if for a given value of x the probability $\pi(x, y)$ is zero for all the value of y except one (which is therefore of certain probability) and we find the usual function concept.

For example, take a network of n automata that model the flow of transport (counting, for example, the number of vehicles) where the nodes each contain a stock, the overall stock of the system staying unchanged. Each automaton is connected to the others. Each possesses a state $s_i(t)$ which represents its stock at the time t. The probability $\pi(i, j)$ is the probability that an element of the site i passes into j. We can then proceed to progress this process using the Monte Carlo method. If the sizes are very big it is also possible to treat the model in a deterministic manner, the new stock is equal to the earlier stock with less outputs and more inputs, which is written simply by:

$$s_j(t+1) = \sum_{i=1}^{n} \pi(i, j) s_i(t)$$

if we call it $s(t)$, the vector line of n states at the moment t and T the matrix of the transition containing the $\pi(i, j)$, the calculation of $s(t+1)$ is made with the following matrix product:

$$s(t+1) = s(t).T$$

We will examine in a little more detail a probability diffusion model, Hägerstrand's model.

12.4. The concept of the cellular automaton

After having examined the concept of the automaton we can now move on to examine the concept of the cellular automaton as a network of automata in a finished state, all identical and dispersed regularly in space. The automata here are called cells, and the input-output connections between cells are the links between the automata in this space. Immediately it is apparent that this concept can be used in geography to model a spatial dynamic. The cells are like the pixels of an image but which also possess an evolution mechanism of their value.

12.4.1. *Level of formalization*

The concept of cellular automaton can be defined on at least two levels, which we will identify as "concrete" and "abstract".

The "concrete" level is the computing model (graphic, conceptual or algorithmic) which will be programmed. This model therefore has the objective of making a program work in a computer and producing results on a screen or in a file using the information we give it.

The "abstract" level is a purely mathematical definition, very simple in its structure; its properties are simplified in comparison to the "concrete" level. This allows the fundamental properties to be studied more easily. Nonetheless, in this simplification, certain characteristics are generalized distancing themselves from their concrete form. For example, in order to avoid the effects of borders which modify configurations during functioning, we consider, in its most abstract form, that the cellular space is an infinite network of cells. This makes it impossible to concretize in a computer.

These two definitions of levels are obviously useful but can be misinterpreted if the reader does not find their context in the description. Our objective here is not to advance the mathematical theory of cellular automata but to show applications that can be used in the particular field of geography. Nevertheless, this does not prevent us from profiting from the theory to properly define the "concrete" automata in concern with the rationality of the model.

We will use a formalized definition in the presentation of the concept of the CA to remain general and didactic. In the applications section the models used are much too complex to be able to formalize completely. They will therefore be described in a more intuitive manner so as not to forget the objective, which is here the application and not the theory. For a mathematical approach to the theory of CAs Nicholas Ollinger's thesis [OLL 02] "Cellular automata: structures" can be consulted.

12.4.2. *Presentation of the concept*

A *cellular automaton (CA)* is a network of Moore's automata simplified, interconnected and (in general) of identical types. Each automaton is called a *cell*[1]. These cells are organized within a *network* (of one, two or three dimensions, rarely more) where they occupy the nodes. They are connected to each other by a *neighborhood graph*, which makes up the network links. Each cell, at each moment, is in a certain state (a whole, a color, etc.), which belongs to a *set of finished states* common to all cells. The connections between the cell and its neighborhood allow the cell to "know" the state of its neighbors. Thus, the motif constituted by its own state surrounded by the states of its neighboring cells allows each cell, with the help of its *transition mechanism* to evolve its state.

The cellular network possesses a *structure* which simultaneously defines its global and local characteristics: global form and area size, network geometry, the topology of the linking edges: an infinite area, or a limited area without joining, or a finite area but unlimited due to a total or partial linking which can be looped in one dimension or for two dimensional in cylinder, sphere, torus, etc.)

Moreover, the cells are located and "drawn" in a geometric *space*; they have a form (2D: squared, rhombus, triangle, etc.). The joining of this group of forms makes up the *spatial domain* of the cellular automaton that must be connected (most often a rectangle for the squared cells). The functioning of the cells is linked to a common *time* for all cells. This time is discrete, it is represented by a variable integer t that is worth 0 at the start of the simulation and rises by 1 at each stage of the transition of the automaton.

Finally, it is necessary to define the *cellular model*, which understands the definition of the states and the transition mechanism.

We can now give the formal definition of a CA.

---

1 Von Neumann worked on the modeling of the auto-reproduction of life, using biological analogy.

12.4.3. *The formal definition of a cellular automaton*

A cellular automaton is a quadruplet $(\mathbf{Z}^d, S, V, T)$ where the integer d is the *dimension* of the CA, the finished group S is the *set of states*, V a series of n elements of $\mathbf{Z}^d$ is the *neighborhood operator* (or more simply the neighborhood) and the function T of S^{n+1} in S is the *local transition rule* (or more simply the transition).

Given certain cellular automaton A, we denote by S_A, V_A and T_A respectively the group of states, the neighborhood and the transition of the cellular automaton A.

This definition, a little abstract, needs a few details.

12.4.4. *The cellular network*

In the definition, the cellular network is identified as a direct product (Cartesian) $\mathbf{Z}^d$. It represents the indexation of cells forming a regular network immersed in the geometrical space at d dimensions $\mathbf{R}^d$. Thus, in one dimension, a line of cells forms the network, each indexed by an integer i. In two dimensions (d=2) the cells are organized in the nodes of a gridline, and $\mathbf{Z}^d$ is the group of indexes (i_1, i_2) of integers, representing the number of line and column of each node of the network.) We will frequently call the index an element $i = (i_1, i_2, ...i_d)$ of $\mathbf{Z}^d$.

In practice, the number of cells remains complete, it is limited to a connection area $D = [1, n_1] \times [1, n_2] \times ... \times [1, n_d]$. For example, for $d = 1$, $D = \{1, 2, ..., n_1\}$, for $d = 2$, D is formed by couples of integers (i_1, i_2) with $i_1 \in [1, n_1]$ and $i_2 \in [1, n_2]$.

12.4.5. *The neighborhood operator and cell neighborhoods*

The neighborhood operator is an application V which allows the construction of all the cell neighborhoods by the same method. It is formalized by a series of n translation vectors (the relative offsets of indexes) allowing it, as long as it is applied to a cell I, to make all the cells of its neighborhood. For example in one dimension, the neighborhood operator $V = (-1, 0, 1)$ allows us to obtain the neighbors of cell 72 by 3 offsets, towards the left, the centre and right by: $i \rightarrow i$-1, $i \rightarrow i$ and $i \rightarrow i$+1 therefore the neighborhood: $V(72) = (71, 72, 73)$.

In two dimensions, the neighborhood type $V_4 = ((0, 1), (0, -1), (-1, 0), (1, 0))$ called the Von Neumann neighborhood (see Figure 7.3) makes it possible to access the four cells situated above, below, left and right of the cell (i, j) in question:

$$V_4(i, j) = ((i, j+1), (i, j-1), (i-1, j), (i+1, j)).$$

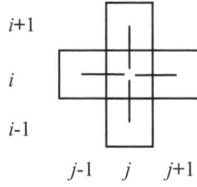

The neighborhood type $V_8 = ((-1, -1), (0, -1), (1, -1), (-1, 0), (1, 0), (-1, 1), (0, 1), (1, 1))$, known as Moore's neighborhood (see Figure 12.3) enables access to the eight cells situated around the cell (i, j) of reference:

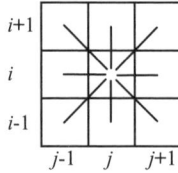

$$V_8(i, j) = ((i\text{-}1, j\text{-}1), (i, j\text{-}1), (i+1, j\text{-}1), (i\text{-}1, j), (i+1, j), (i\text{-}1, j+1), (i, j+1), (i+1, j+1)).$$

12.4.6. *Input pattern*

The input pattern a_i of a cell i is the vector of n states of cells of its neighborhood $V(i)$, therefore $a_i = (s_j)_{j \in V(i)}$. Thus, for the neighborhood $V(i) = (i\text{-}1, i, i+1)$, the pattern of its states is therefore the sequence $a_i = (s_{i-1}, s_i, s_{i+1})$.

The neighborhood of a cell may or may not include the cell itself. In the case where it is contained, a more concise notation of the transition mechanism T is allowed which only takes on input a_i instead of s_i and a_i. However, it can happen that the treatment of the cell state can be different from that of the neighborhood, and it is in this instance that they are differentiated.

12.4.7. *The local rule of the transition of the cell*

The automaton of each cell i is of the form $M_i = (S, T)$ where S is the group of states and T the transition mechanism which is written $s_i(t+1) = T(s_i(t), a_i(t))$ if the neighborhood does not contain the cell, or more simply $s_i(t+1) = T(a_i(t))$, if the neighborhood contains the central cell.

12.4.8. *Configuration and global transition mechanism*

The *configuration* of the CA occurs at the moment *t,* the application associates a state $s_i(t)$ with each cell *i* of the network. When there are *n* cells in one-dimension it is a vector of states $s(t) = (s_1(t), s_2(t), ..., s_n(t))$. The *global transition mechanism G* occurs when the application is dealing with an ordinary configuration *C* the configuration $C' = G(C)$ obtained by applying the local rule of transition to each network automaton.

12.4.9. *Configuration space: attractor, attraction basin, Garden of Eden*

With deterministic automata, if at time t_1 we again come across a configuration *C* already found at the time t_0 the series of configurations will repeat itself after t_1 in the same way it did after t_0 until it returns to *C*. The system between is therefore in a loop called an *attractor*. If the period of the loop (its length) is equal to 1, it is a *fixed point*, if not it is a *limited cycle*.

The group of configurations, which reach a given attractor, after an unknown number of iterations, is called the *attraction basin*.

In addition, it is interesting to know which configurations reach the same attractor or which bond more or less with each other. The configuration space is equipped with a distance that counts the number of states which differ in the two configurations (Hamming distance). If the network contains a finite number of automata, the number of configurations is also finite and in this way, the number of attractors and of basins is also finite and every configuration reaches an attractor after a finite number of iterations. The group of attractor basins is therefore a partition of the configuration space. However, if the network is infinite (general definition) a series of never converging configurations can exist.

We can also investigate configurations which can never be reached, that's to say those which can only be taken as initial configurations. These are called *Gardens of Eden*. Moore posed the question of the existence of the Gardens of Eden in 1962 in relation to auto-reproducing automata. Alvy Smith demonstrated the existence of these Gardens of Eden in the game of life in 1970.

It is obvious that the combinations of configurations are enormous, and this is why the theoretical study of the behavior of the CA is so complex, behavior which depends on both the initial configuration and the transition mechanism. It is quite an active field of research and the theory of cellular automata is beginning to take shape. Moreover, CAs are very handy tools for the study of discrete dynamic systems. One of the important theoretical questions being posed is the classification of CAs. After an exhaustive study of the 256 binary CAs in one-dimensional Wolfram proposed a classification of the CAs in four categories, inspired by the theory of dynamic systems [WOL 83] [WOL 86]. This classification was criticized, but injected enthusiasm into the area of study and has even been the subject of a recent PhD thesis [OLL 02] which address other problematic relative to the calculability of indecisiveness and in particular universal calculability (which confers with a CA the attribute of power to simulate any cellular automaton). Problems linked to chaos, instability, sensibility of initial conditions are also important questions concerning dynamic systems and information theory [MAR 01].

12.4.10. *2D cellular automata*

1D cellular automata will not be developed here (see [WOL 02], [WEI 89]). 2D, surface automata, principally used in geographical simulation, will be examined here. The cell space is most often a rectangular area, associated with a network of squared mesh. However, automata with triangular or hexagonal meshes can be found, to see more complexity (Delaunay's triangulation, Voronoï diagrams) in irregular networks.

The network geometry infers a type of neighborhood between the cells.

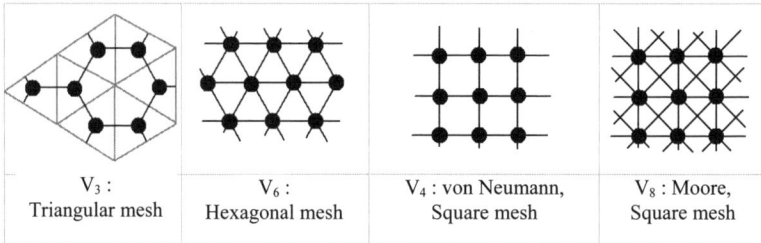

| V_3 : Triangular mesh | V_6 : Hexagonal mesh | V_4 : von Neumann, Square mesh | V_8 : Moore, Square mesh |

Figure 12.3. *Common types of neighborhood*

Figure 12.3 shows the most common types of neighborhood. For squared cells, two types of neighborhood are commonly used, V_4 and V_8. Types V_3 and V_6 are a little more complex concerning the level of indexation of points and of the definition of neighborhoods. The neighborhoods can also be defined more generally, from a particular metric space. A neighborhood is therefore formed from cells present in a disc of a certain radius centered on the cell.

For example, the *Manhattan metric* defined by $d(P_{ij}, P_{i'j'}) = |i\text{-}i'| + |j\text{-}j'|$ defines the disk in a diamond form, which is equal to V_4 for radius 1.

The *maximum metric* $d(P_{ij}, P_{i'j'}) = Max(|i-i'|,|j-j'|)$ also gives squared disks. It coincides with V_8 for a radius of 1.

For irregular meshing a neighborhood operator that would apply to all cells cannot be defined. The links of each cell with its neighborhood are specific; therefore each cell contains a list of these links in its structure.

To respect invariance by rotation and by symmetry, properties of great use in the world of physics, many cellular automata, instead of calculating their transition from the state of each neighboring cell individually (as generally happens) only the number of neighbors in a certain state (excited, for example, for binary states). Consequently threshold or window automata are often used. This is the case for the game of life.

12.4.11. *The game of life: an example*

The game of life [CON 70] is an emblematic automaton, it is very simple as regards its rules and yet at the same time complex regarding its dynamic. For this reason it has been widely studied (Conway, Gosper, Ray-Smith, etc.). In particular, the game of life possesses the universal calculator function and it is also the most simple, 2D CA with the auto-reproduction property (knowing that there are only three known CAs of this type, Von Neumann's 19 states and Codd's 8 states).

Description

Each cell is a binary, window automaton. Moore's neighborhood (V_8) is used. State 1 represents a living cell, state 0 a dead cell.

The transition $T(s, n)$ is the function of the state s of the cell i and of the number n of surrounding living cells, with $n = \sum_{j \in V(i)} s_j$. The following windows define the local mechanism:

$$T(0,n) = \begin{cases} 1 & \textit{if} \quad n = 3 \\ 0 & \textit{otherwise} \end{cases}$$

$$T(1,n) = \begin{cases} 1 & \textit{if} \quad 2 \leq n \leq 3 \\ 0 & \textit{otherwise} \end{cases}$$

[12.1]

which is set out by: if a cell is inactive and three of its neighbors are active it will become active, and it will only stay active if 2 or 3 of its neighbors are active.

All automata have this type of transition, that is to say whatever the function of the previous state *s* and the number *n* of living neighboring cells (to 1) can be defined using a table like the one below.

		n								
T		0	1	2	3	4	5	6	7	8
s	0	**0**	**0**	**0**	**1**	**0**	**0**	**0**	**0**	**0**
	1	**0**	**0**	**1**	**1**	**0**	**0**	**0**	**0**	**0**

The transitions can also be represented by a *transition graph* like the one below.

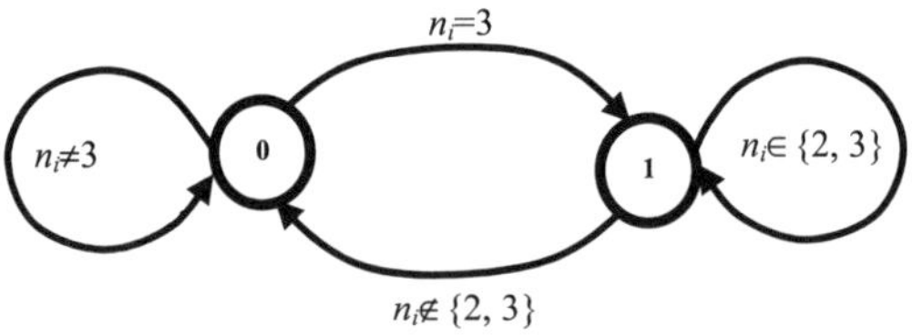

Figure 12.4. *Transition graph*

If we examine the number of automata of this type, each transition function possesses 2 x 9 = 18 possible couple values (*s, n*) to which 2 results can be attributed (0 or 1) which represents $2^{18} = 262{,}144$ possible transition functions. This is considerably more important than the 256 1D binary automata. Moreover, the behavior also depends on the initial configuration. If a quite small area is used, for example 10 x 10 = 100 cells, there are 2^{100} possible initial configurations, which represents a number of the command one thousand billion of billions of billions (10^{30}). By multiplying the number of possible transition functions the number of possible games to the command of 10^{65} are obtained.

The behavior of the game of life

Although totally deterministic, the long term configurations of the game of life, obtained according to the rules (1) are practically impossible to predict. Nonetheless some simple forms can be noticed, those that are stable when they appear isolated, like a 2 x 2 square or others that reappear according to a very short cycle, like a line of 3 cells which oscillate between a vertical and horizontal position. However, these forms are not absolutely stable. For example, they can collide with other mobile forms (named gliders) that have a quite short transformation cycle that is always moving, like that in Figure 12.5. The collisions produce other forms, making overall behavior more complex.

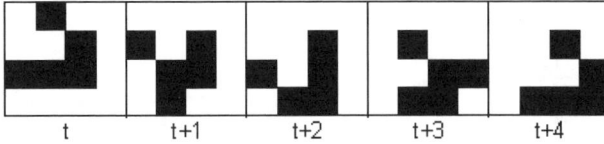

Figure 12.5. *Glider: here a cyclic figure moves towards the south-east. Using symmetry 3 others can be constructed, moving in the 3 other directions*

12.5. CAs used for geographical modeling

Now a few examples of the use of cellular automata in geography will be shown. Our objective is not to make a new discovery in this area as there are, at the moment, numerous teams working on this subject. There is an abundance of bibliographies, for example the CASA website (see websites in the bibliography) or in the French speaking community the work on cellular automata applied to urban simulation [LP 97]. For educational purposes we prefer to describe more precisely a few realizations rather than citing a myriad of works without really unveiling their content. We have chosen 3 examples from varied fields, 2 of which were carried out in our laboratory. The first is inspired by the Hägerstrand model on diffusion, the major geographical process that will provide the opportunity to present a probability model and its deterministic correspondent. The second (SpaCelle) is a mini "platform", that is, software which contains no programmed model but in which the user interface allows the user to describe the behavior of the automaton through a series of simple rules, explained in a spatial representation language. This model was initially developed to be applied to urban geography but its field of use is much more general. The last example is applied to physical geography (RuiCells). It is a much more complex model than the previous one, applied to hydrological risk management. It presses layers of geographical information (DTM, digital terrain model). The automaton is constructed on a mesh of the surface in heterogenous cells (punctual, linear and surface). It models the surface runoff according to precipitation, terrain morphology and land use.

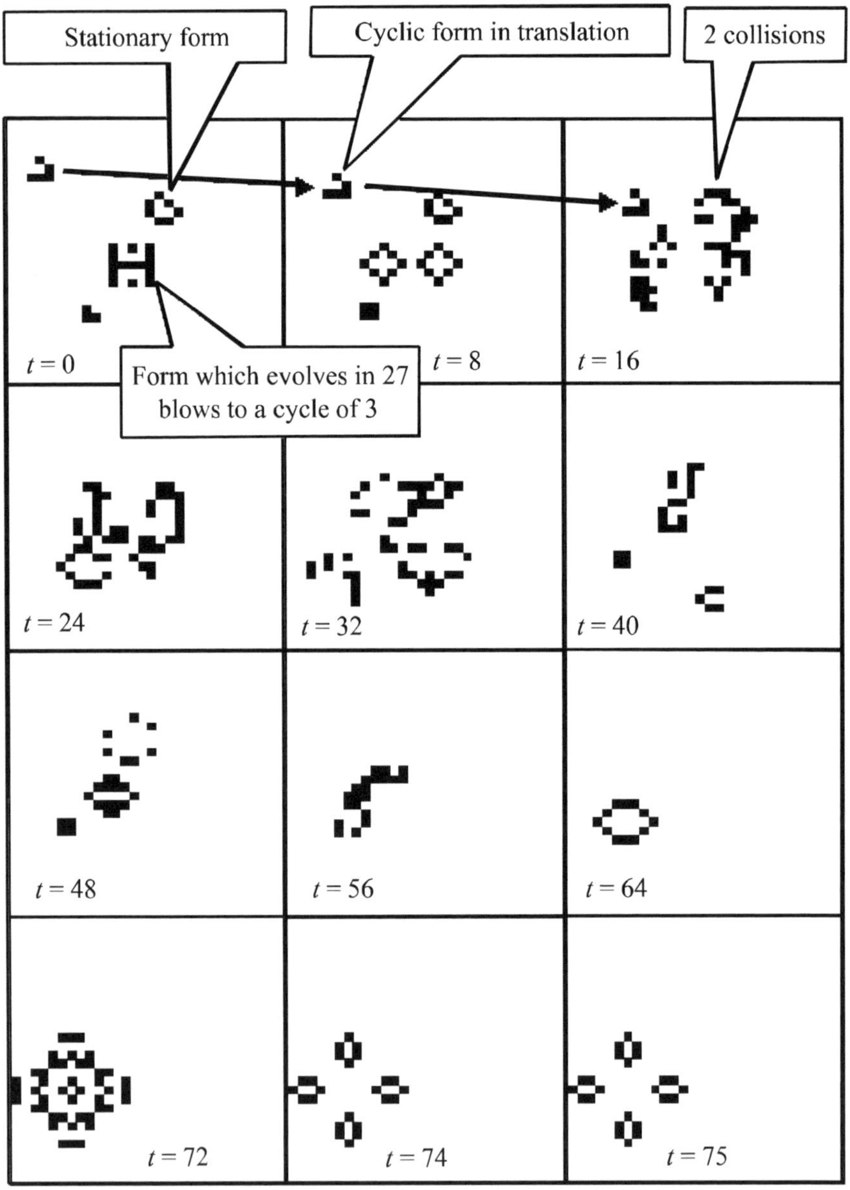

Figure 12.6. *Evolution (every 8 steps) according to the rules defined in (1) of an initial configuration of stationary forms and moving forms which collide (t = 16) to create a chaotic situation (t= 16 to 56) and then restructured (t = 64) into a simple form which converges towards a fixed form in t = 74*

12.5.1. *Diffusion simulation*

12.5.1.1. *The Hägerstrand probability model*

It can be said that the interest of geographers in cellular automata begins with Hägerstrand [HAG 67] as he models the diffusion of an innovative process (agricultural grants for the transformation of wooded areas in pastures in the Asby area, south central Sweden 1929-1932). Although we can improve the Hägerstrand model by using a multi-agent system, as Eric Daudé did [DAU 04], this model equally conforms to the paradigm of the cellular automaton which behaves according to the regular cutting of time and space. The process is managed by a local transition rule applied to a neighborhood.

The big difference with the formal definition of a cellular automaton lies in the fact that this is non-deterministic. Its special area is a rectangle 70 km x 60 km dissected according to a 5 km sided grid which gives 168 cells. Each cell i contains a certain number of individuals e_i (these are the agricultural operators liable to be funded; this number remains the same during the simulation). An individual can be in one of the four following states: *innovator* (having adopted the innovation at the time t = 0), *having adopted* (in the past), *adapters* (at the present moment), and *potential adopters* (that will perhaps be adopted in the future). We seek to model the number of operators of each cell that has adopted the innovation over a course of time. The state of one cell I is therefore represented by the number $x_i(t)$ of those having adopted the innovation. The cell also contains the number of operators, e_i, which remains constant. From these two values, we deduct $y_i(t) = e_i - x_i(t)$, the number of potential adopters. The diffusion process depends on the frequency of communication between the operator who has adopted and the potential adopters; it is a typical logistic model. Instead of using this model directly at a cellular level, we can simulate it at an individual level, that is to say, at the level of the operator itself by the random generation of an "adoption message" according to a certain probability model. This probability contact declines rapidly with distance and therefore proceeds essentially by neighborhood. For this we define a *contact field* from the neighborhood operator given in the definition of the CA. At each of these n shifts $V = (v_1,\ldots, v_k,\ldots, v_n)$ of the neighborhood operator we associate a probability of achieving a contact $P = (p_1, p_2,\ldots, p_k,\ldots, p_n)$. The vector P is the contact field whose sum of its elements is equal to 1 (like all laws of probability). The field of contact allows the random selection of a message which will be sent from the active cell i if it possesses at least a *having adopted*, to the cell j chosen at random. With a computer, random selection is carried out by a standard pseudo-random function giving a real number p in the interval [0, 1]. Such a selection allows us to choose the affected cell using the Monte Carlo method.

The local transition process takes place in the following manner: every time there is a *having adopted* (that is to say $x_i(t)$ times) in the cell i we proceed to send an adoption message. If the kth cell is chosen, the message is "sent" to the cell $j = i + v_k$. (2). This is followed by a new random selection q uniform between 1 and e_j (number of operators in j) if $q \leq x_j$ the message "falls" on an operator that has already adopted it and is therefore lost, if not it arrives on a potential adopter and the number of adopters is raised to 1. The same process occurs for all the cells. At the end of the iteration we update the state x_i, of all the cells i in adding to x_i the number of adopters calculated during the iteration.

The contact field can be defined homogenously, and is therefore constant throughout the area. Certain natural barriers (lakes and forests), which limit contact between the individual cells, are also taken into account. In this case the contact field is variable and must be defined for each cell or a balance of contiguity lines between cells must be taken into account.

12.5.1.2. Deterministic diffusion model

This diffusion model can also be treated on a cellular level if the number of individuals of each case is big enough that the large number law applies. The model therefore becomes deterministic.

The messages received in the cell i from its neighboring cells have been sent by each x_k having adopted the kth cell towards the $e_i - x_i$ potential adopters of the cell i. Their total is therefore $x_k(e_i - x_i)$. These messages come from boxes more or less unconnected to I, of which we know only a proportion r_k (lessening with distance) are fulfilled. We arrive therefore at a logistic formula of the number of adoptions in i achieved by messages coming from the kth cell of the neighborhood:

$$a_{ik} = r_k x_k (e_i - x_i)$$

The number of adopters $a_i(t)$ in i at the time t is given by the summation of the neighborhood terms:

$$a_i(t) = (e_i - x_i(t)) \sum_{k \in V(i)} r_k x_k(t)$$

and the transition function of the model is therefore:

$$x_i(t+1) = x_i(t) + a_i(t)$$

---

2 The sign "+" represents the translation applied to i of a vector v_k, which is a sum of two vectors. For example, if $i = (2, 5)$ and $v_k = (-1, 1)$, we will have $j = (2-1, 5+1) = (1, 4)$.

The probability contact field P is here replaced by the deterministic contact field $R = (r_1, ... r_k, ... r_n)$ of the realization rate of a message between a having adopted and an adopter at each step in time.

We present in Figure 12.7 a few results of simple simulations according to the principles defined above, initialized with one focus of innovators in the middle of the area. The first two images concern the diffusion probability: in (A) with the homogenous distribution of individuals (operators), in (B) with the random distribution of individuals. The two following images concern the deterministic diffusion: (C) with the homogenous distribution of individuals and (D) with the random distribution of individuals.

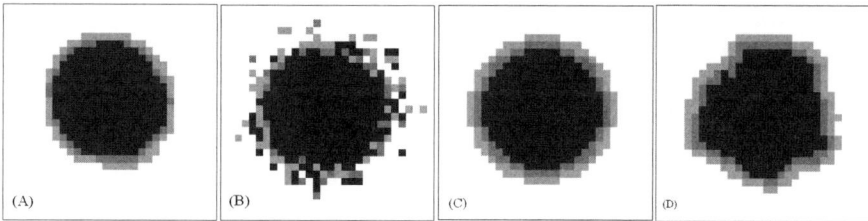

Figure 12.7. *The level of gray shows the rate of adopters at the end of several iterations (in four threshold classes 0%, 25%, 50%, 75%, 100%)*

The diffusion models are not exhausted with these few elementary examples.

12.5.2. *The SpaCelle model*

The SpaCelle model (the Libergeo base of models can be consulted) is a small platform of cellular automaton modeling. Thus the model is not defined in the software itself, nonetheless it functions on a certain number of general principles that we will call a meta-model; it is these principles that we will explain. The user must define the initial configuration, by input or importation of the states of cells along with the transition rules of the model it takes from a knowledge base. It can also define the form of an area, the cell geometry (squared or hexagonal) and the synchronization mode (synchronous or asynchronous or completely random). The state of each cell is qualitative (like a type of land usage) and it is defined by a key word and a representative color.

The performance is based on the *principle of competition*. It manifests itself between the "life force" of a cell and the "environmental forces" emanating from the other cells. When a cell is affected by a new state, we witness the birth of an individual (cell). It is therefore affected by a maximal life duration (in this state) depending on its class (ID: infinite duration, FD: fixed duration, RD: random duration according to a life expectancy and a standard deviation). Upon its natural death, an individual changes state and takes the definite dead state according to the rule of life of its class. The individual also possesses a life force worth 1 at its birth and decreases linearly to 0 at its natural death. However, an individual can die prematurely if one of the environmental forces affecting it is stronger than its own life force.

For example the rule of life "Pav > Fri = DA(100; 25)" signifies that the class "Pav" (house type) becomes "Fri" (fallow land) after its death and possesses a random life span (RD) according to a life expectancy of 100 and a standard deviation of 25 years.

The environmental forces are defined by transition rules built on the following syntactic model: "$State_1$ > $State_2$ = $Expression$". The term "expression" represents a *spatial interaction function* or a combination of these functions. A spatial interaction function is most commonly written in the form F(X; R) and makes it possible to evaluate, for each cell, the "environmental force" owed to the individuals type X in a radius R around the cell. For example, if X is "Ind+Com" this represents the under population of the cell type "industry" or "commerce". R is the radius of the disk defining the neighboring action of X on the cell. The function F represents the type of interaction calculated. There are 20 predefined functions. For example, the function "EV(Ind+Com; 5)" signifies "there exists, at least one individual type 'industry' or 'commerce' in the neighborhood of radius 5" and the function "ZN(Ind; 5)" ZN for zero in the neighborhood (N)) signifies that there is no industry in the neighborhood radius 5.

The phrase: "wild land can become a housing estate if there is already an estate or business area within a radius of 3 and if there is no industry within a radius of 5" translates itself by the following transition rule:

Fri > Pav = EV(Pav + Com; 3) * ZV(Ind; 5)

The conjunction "and" in the phrase is represented by a multiplication sign "*" while an "or" is translated by an addition sign "+".

The knowledge base consists of 3 parts: the definition of the states, the definition of the rules of life (if necessary) and the definition of the transition rules.

Whatever the basis of the defined rule, the mechanism of the model is as follows. For each cell on state s the system executes all the transition rules $R_1, ... R_k$ of which the first member $State_1$ equals s. It therefore evaluates the life force f_0 of the cell and the resulting forces $f_1,...,f_k$ associated respectively with the rules $R_1,...R_k$. It is always the maximum force that takes it. If several rules give the same maximum force, a random uniform selection is performed to choose the transition that will be kept among the cells of maximum force. If it is the life force f_0 that is retained, the cell remains in the state if it is one of the fixes f_i, (for $i>0$), the cell dies prematurely and its state becomes $state_2$, which is recorded in the right member of the rule R_i.

It can therefore be seen that the transition mechanism is deterministic. Nonetheless a random quantity exists in the case of an equality of maximum force between several rules or in the cells in which the life span is random. There also exists special interaction functions which trigger events, be it a moment selected at random or on a precise date. These chronological functions, by combining with the other functions, allow us to modify the system behavior randomly or from a precise date.

12.5.2.1. *Examples of modeling with SpaCelle*

12.5.2.1.1. The game of life

In the case of a very simple model like the game of life, it is sufficient to define two states (L= life, D= death) no life rule is necessary here and two transition rules are defined according to the formula (1):

$$D > L = NV(V;1;3)$$
$$L > D = SV(V;1;2;3)$$

The first rule signifies that each dead cell (D) takes life (L) when there are 3 active cells in its neighborhood radius 1 (the function (NV) stands for the number of neighbors). The second rule states that cells stay alive (L) only if the number of its living (L) neighbors is in the interval [2:3]. It is also necessary to specify certain choices not in the rule base like the neighborhood type (here Moore's type is used, 8 neighbors, induced by the max distance), the synchronized performance mode and the form of the squared mesh.

12.5.2.1.2. The Schelling segregation model

This model, typical of the sociology method of *methodological individualism* which uses three fundamental concepts: the concept of emergence, associated most often with aggregation, the concept of modeling, allowing the simplification of individual behavior, which in reality are all different, and finally the rationality concept of the actors who translate a hypothesis of behavioral intelligibility.

In a town made up of several social groups or communities (ethnic, religious, economic, etc.) the Schelling model shows how spatial segregation can appear without segregational behaviors at the individual level. Indeed, it shows that even if each individual has an elevated level of tolerance concerning the presence of "foreigners to their group" in their neighborhood we nonetheless see a separation or socio-spatial segregation emerging over time which transforms by the appearance of considerably more homogenous areas than individual tolerance may have lead us to believe. It is therefore a simple yet significant example of the emergence concept in a complex system. Even if the reality is very different, this model still shows that the whole, that is to say the collective behavior must not be directly interpreted as if the rules of individual behavior applied directly to collective behavior: the individual is not segregationist therefore the group is not either. It can be seen, through this example that an individual rule can produce, if certain conditions are brought together (in this case for example a sufficiently high population density and rules given for the highest level of toleration) an overall organized behavior of which the occurrence is certain at the end of the given time but of which the form is totally random.

We have developed this model in SpaCelle as follows. The cellular area represents a town where the cells (10,000 in number) represent the habitations. The town is composed here of three communities noted A (in black), B (in dark gray) and C (in light gray). When a house is not inhabited, the cell is in the state F (free) and left in white. There are consequently four possible states for a cell: A, B, C or F.

The initial configuration results in a random selection of a state for each cell among the four possible states. This gives a quasi-equal weight between the three communities.

The rules of transition are very simple, there are no life span rules and there are two transition rules, identical for each community:

– The moving-in rule; if a cell is free, a family from any one of the three communities A, B or C has an equal chance of moving in. The installation of a family is not linked to the freeing of another cell in such a way that it lowers the overall population density (function DE) as the model possesses a maximum overall population density (99% for example) over which moving in is no longer possible. There are therefore three identical moving in rules, one for each population. As a result, the probability of the installation of an individual is the same whatever the group:

L>A=DE(A+B+C; 0; 0.99)
L>B=DE(A+B+C; 0; 0.99)
L>C=DE(A+B+C; 0; 0.99)

– The moving-out rule; if a family living in a given cell is surrounded by too many "foreigners to its group" it moves, freeing the cell (the PN function is used for this which calculates the proportion of foreigners in the neighborhood radius 5. It is worth 1 if it is in the interval (70%–100%) and if not sends back 0). The move out is not directly linked to a move in but decreases the overall density and eventually allows a move in elsewhere. Therefore, move out is explained by a rule which has the same form for each population:

A>L=PN(B+C; 3; 0.7; 1)
B>L=PN(A+C; 3; 0.7; 1)
C>L=PN(A+B; 3; 0.7; 1)

This model is slightly different to the original Schelling model [SHE 80]. In fact that model dealt with only two different types of individuals (white and black), moreover it processed in a synchronous manner (all cells changing at the same time) and finally each move-out was immediately followed by a relocation elsewhere. Here we have taken three different populations (but this changes nothing in principle), the order of simulation is random, that is to say at each moment in time a cell is chosen at random to be independently treated from the cells already done. Finally, in our simulation the two types of action (moving-in and moving-out) are independent and are chosen uniquely in relation to the state of the cell which is chosen. According to the evaluation of the rules for this cell, if it is a free cell moving-in can occur. The behavior of the model is therefore subject to a minimum number of rules to create the dynamic of the system.

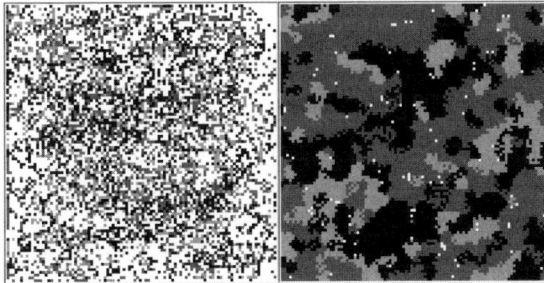

Figure 12.8. *Simulation of the Schelling model: (1) initial configuration, (2) after 50 steps in time*

12.5.2.1.3. Interpretation

We soon notice the emergence of an organization stabilizing after 50 time steps (each step corresponds to the processing of 10,000 cells chosen at random). The tolerance percentage 70% corresponds to a tolerance percentage more important than the average as if the 3 populations were equal in proportion ($^1/_3$ each) there are $^2/_3$ of 99% in the mean, 66% of cells which are made up of "foreigners". The random situation of departing positions means that locally (within a neighborhood radius of 5) the probability of reaching or surpassing 70% of foreigners is quite high. The moving-out occurs quite often. The box left free will only be able to stabilize itself with one of the other two populations which will progressively reinforce the homogeneity. We notice then a quite complex final result, where homogenous "areas" appear, composed of one single group whereas others are composed of a mix of two groups, but none appear perfectly mixed, a situation which does not however seem forbidden by the rules.

12.5.2.1.4. Urban development in Rouen over 50 years

In [DGL 03] we used this system to simulate the urban development of the agglomeration of Rouen over a 50-year period. This work is more realistic, it begins with a real observation situation in 1954 and with a set of 15 rules it reaches a simulated situation in 1994, which is compared to the present day. Overall the configuration is very close to the observed situation. The analysis of local differences underlines the behavior which is spatially coherent, others underlining the logical evidence outside the space.

12.5.2.2. *The limits and originality of the SpaCelle model*

This model only allows dynamic modeling whose rules of evolution are spatial (interactions with the neighborhood). It is therefore simplistic but it does allow a complex modeling from a not necessarily mathematical knowledge. For example, the rules set can be constructed from the analysis of a text. However, quantitative data cannot be introduced except in declaring these values in qualitative classes (this is a finite state automaton). Moreover an economic variable like the price of land cannot be taken into consideration at the same time as land use became the model takes only one type of information. For example, if the price of land is separated into different classes the land use types can no longer be used, unless the two are mixed in a quite complex way. Nonetheless the model allows us to realize experiments by simulation, which gives quite good results that will not be discussed here. The Rouen model tends to show for example that the cost of land results from localization and spatial interaction between these localizations since it is not necessary for modeling urban evolution. However, other exogenous factors, economic or social, cannot be taken into account. We can therefore more or less analyze the influence by the analysis of the differences between the model and reality.

This system, which allows the formalization of a dynamic by *sentences* (the rules), explained in a *knowledge-representation language,* quite close to natural language, is an original alternative to classic modeling which explains a system dynamic by equations.

12.5.2.3. *Recent evolutions of the SpaCelle model*

We have developed more general versions of this model where the cells can be of any polygonal form and thus can adapt directly to the cuttings originating from a geographical database. We have also generalized the types of cell state. Instead of defining the state of one cell with a unique exclusive quality, it is more realistic to define a state of multiple behavior. For example, land use: housing but with some business and a small amount of industry and roads.

In this extension, the cell state is represented by a series of n real numbers $s = (s_1, s_2, ..., s_i, ... s_n)$, which can have very different meanings:

1) The state s can be a vector of dimension n where each dimension i of the state associated with a modality (for example 1: housing, 2: industry, 3: business, 4: road, etc.) of the same qualitative variable (here the land use) and s_i represents the proportion or probability of presence (depending on whether it is in a deterministic or probability situation) of the modality i in the cell.

2) The state s can be composed of n different quantitative variables (for example 1: population, 2: GDP, 3: surface, etc.) and s_i is the value of the ith variable. We have used as part of the modeling of the development of the standard of regional life in the European Union (represented in a simplified manner by the GDP per inhabitant in purchasing power parity.) In using the rules of intrinsic rural growth but also of diffusion by neighborhood (known as horizontal interaction), we are thus underlining a competition between the two processes, the first pushing for divergence the second for convergence, of the standards of living between the regions. We search then to understand the influence of national and European aid, as well as the role of taxation (vertical interactions, rising for taxes and falling for help). This introduces a supplementary level of complexity to the studied system, both in the structure, through the necessity to use a multi-layered hierarchical automaton (regions, states, Europe) and through the dynamics, combining the horizontal and the vertical interactions.

The state s can also be a real matrix $p \times p$. We have modeled within the framework of diffusion in the behavior of French voters, using cantonal cutting as cellular meshing [BL 04] that state of the call is therefore defined by a behavior matrix $S = [s_{ij}]$, of $p \times p$ dimension where p is the number of candidates. The behavior matrix of a canton-cell evolves over time (between two elections) by diffusion of voting behavior from the polls and ultimately makes it possible to calculate a new vote vector V' (percentage of votes for different candidates) from the initial vector (observed) V by matrix multiplication $V' = SV$. A rule bearing, for example, a positive relative influence to a candidate i modifies the behavior matrix, in raising the term s_{ii} of the matrix and lowering the other terms of the column i so that the total remains constant (equal to 1).

12.5.3. *Simulation of surface runoff: RuiCells model*

This model, much more complex than the previous, was developed as a response to a concern over the understandings of intense phenomena of surface runoff which regularly provoke catastrophic damage in the form of mudslides in normally dry drainage basins. A more precise description can be found in [LAN 02] (the base of Libergo models can also be consulted).

12.5.3.1. *Inputs*

A certain amount of data coming from a geographical information system (GIS) is taken on input to the software to construct a part of the automaton structure, essentially the digital elevation model (DEM). The others are used for performance: precipitation table, vector card of land use, images, etc. These different inputs can be geometrically harmonized.

12.5.3.2. *Structure*

The model of this cellular automaton differs from the strict definition given earlier in the first part. In fact, the cells here are "drawn" on the DEM which represents an elevated surface. This surface is first meshed according to triangulation. If the DEM is composed of irregular random points (originating from, for example, a digitization of level curves) we construct a Delaunay[3] triangulation associated with these points (Figure 12.9a). If the DEM is a regular grid of points, we cut each square into two triangles, by choosing the lowest diagonal (in altitude) so as not to introduce artificial barriers to the flow (Figure 12.9b).

The cells are constructed from the surface triangle mesh. The triangles constitute the first type of cell, the surface cell. However, this is not sufficient to be able to suitably model the flow that naturally concentrates along the lines of the bottom of the valley (the thalweg). It is also therefore necessary to introduce linear cells which are the sides of triangles as they actively participate in the flow process. Finally, in diverse areas, a summit, a side, a triangle, even a group of these objects constitutes a minimal local altitude, a basin. We define therefore a punctual cell, which represents this local minimum. A specific punctual cell recovers also all that runs outside of the domain, in order to conserve the total volume of water of the simulation. These different cells are structured geometrically and topologically in such a way that each "knows" its surface neighbors.

---

3 The Delaunay triangulation is that of the interior of the circle circumscribed to each of its triangles contains no summit of triangulation.

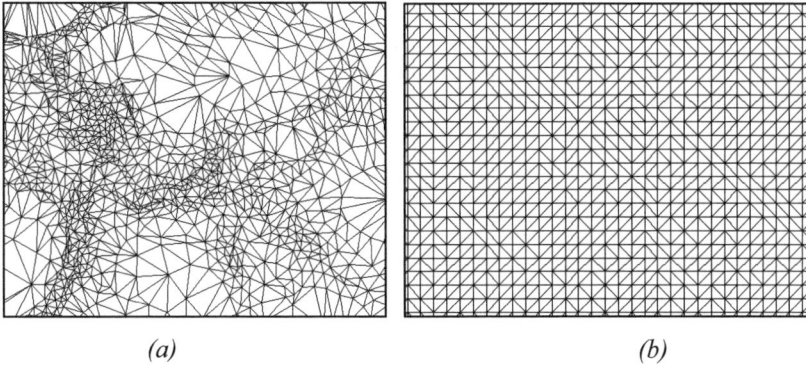

(a)   *(b)*

Figure 12.9. *Two modes of triangulation*

So that the surface runoff process can be modeled it is necessary to define within the cells a directed *flow graph* which indicates for each, in which other cell(s) it flows into and if the cell flows into several others, a sharing co-efficient of flow between the downstream cells needs to be defined, which is calculated through the cell geometry (form and incline).

When the flow arrives in the local minimum it is necessary to calculate the replenishment of the basin and define the spillway point and the receptor cell so that the flow graph is not interrupted. The graph must also be correctly directed in the horizontal zones towards the output zones without making loops. The algorithmic definition of this graph is a delicate part of the model. The cellular structure is routinely composed of many thousands or hundreds of thousand cells.

12.5.3.3. *Functioning*

The local functioning of each cell is operated by a "cellular motor" which is a hydrological model of surface runoff based on the discretization of differential equations in finite-difference equations (the user interface allows us to choose between several motors). To calculate the runoff it takes into account, for each period of time Δt, the volume of rainfall, the volume of water already present and the volume arriving from the upstream water cells, as well as possibly the water loss and infiltration. These variables makes it possible to calculate the volume leaving during functioning compared to the flow speed which depends itself on the slope and height of the water present.

The overall functioning of the automaton manages synchronously the circulation flow of water between the cells; the structure of the neighborhood used here being defined by the flow graph. At each iteration, which corresponds to a moment in time Δt, the automaton proceeds in two phases:

– the communication phases where the outputs (calculated before) are communicated to the inputs of the cells downstream,

– the transition phase where each cell calculates its new state x(t+Δt) which is the new volume of water in stock and its new output b(t+Δt) which is the volume running off downstream in function to its previous state x(t) and its input a(t) which is the volume coming from upstream and precipitation.

12.5.3.4. *Outputs*

Software can chart the evolution of variables in time, to draw out diverse charts and graphs like hydrographs of measured points defined by the user and precipitation curves. It also allows the calculation of the shape and surface of a basin, helps us to draw level curves and bigger trends, create charts with shadowing, or represent a basin in 3D.

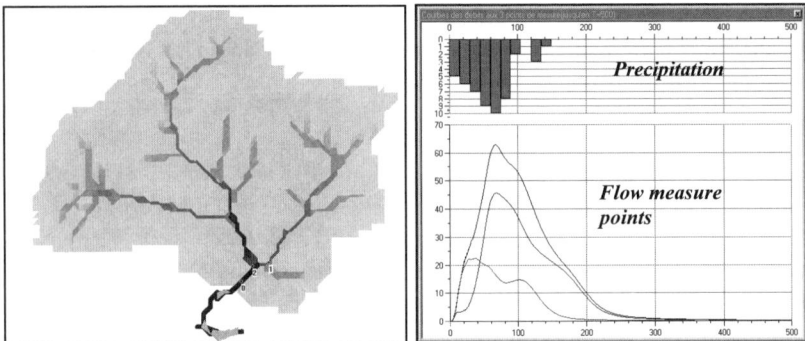

Figure 12.10. *Chart and hydrographs in different measuring points of a drainage basin*

12.6. Bibliography

[BER 82] E. BERLEKAMP, J. CONWAY, G. RICHARD, *Winning Ways*, vol. 2, New York: Academic Press, chap. 25, 1982.

[BUS 04] M. BUSSI, P. LANGLOIS, E. DAUDÉ, "Modéliser la diffusion spatiale de l'extrème droite: une experimentation sur le front national en France", *Colloque de l'AFSP*, Paris, 19 p., 2004, (see http://www.afsp.msh-paris.fr/activite/diversafsp/collassp04/collafspassp0904.html).

[CON 70] M. GARDNER, "Mathematical Games: The fantastic combinations of John Conway's new solitaire game 'life'", *Scientific American*, pp. 120-123, October 1970.

[DAU 04] E. DAUDÉ, "Apports de la simulation multi-agents à l'étude des processus de diffusion", *Cybergeo*, no. 255, 12 p., 2004.

[DUB 03] E. DUBOS-PAILLARD, Y. GUERMOND, P. LANGLOIS, "Analyse de l'évolution urbaine par automate cellulaire. Le modèle SpaCelle", *L'espace géographique*, pp. 357-380, vol. 32, no. 4, 2003.

[FAT 01] N. FATES, *Les automates cellulaires : vers une nouvelle épistémologie?*, mémoire de DEA, Paris I- Sorbonne, 2001.

[HÄG 67] T. HÄGERSTRAND, *Innovation Diffusion as a Spatial Process*, University of Chicago Press, 1967, Chicago and London.

[HEB 02] J. HEBENSTREIT, Principe de la cybernétique, in *Encyclopaedia Universalis* version 8, 2002.

[LAN 97] A. LANGLOIS, M. PHIPPS, *Automates cellulaires, application à la simulation urbaine*, 197 p., Paris, Hermes, 1997.

[LAN 02] P. LANGLOIS, D. DELAHAYE, "RuiCells, automate cellulaire pour la simulation du ruissellement de surface", *Revue Internationale de Géomatique*, pp. 461-487, vol. 12, no. 4, 2002.

[MAR 01] B. MARTIN, II, Automates cellulaires, information et chaos, PhD thesis, École Normale Supérieure de Lyon, 2001.

[OLL 02] N. OLLINGER, Automates cellulaires: structures, thesis, Lyon, 2002.

[POU 85] W. POUNDSTONE, *The Recursive Universe*, Oxford University Press, 1985.

[SHE 80] T. SCHELLING, "Micromotives and macrobehavior", Toronto, Norton, 1978. French edition in *La tyrannie des petites décisions*, Paris, PUF, 1980.

[VNM 44] J. VON NEUMANN, O. MORGENSTERN, *Game Theory and Economic Behavior*, Princeton University Press, 1944.

[WOL 83] S. WOLFRAM, "Statistical Mechanics of Cellular Automata", *Review of Modern Physics* 55, pp. 601-644, 1983.

[WOL 86] S. WOLFRAM, *Theory and Applications of Cellular Automata,* World Scientific, 1986.

[WOL 02] S. WOLFRAM, *A New Kind of Science*, Wolfram Media, 2002.

12.7. Websites

GDR Libergéo, Groupe modélisation, base de modèles: http://www.spatial-modelling.info.

CASA: Centre For Advanced Spatial Analysis (UCL): http://www.casa.ucl.ac.uk.

University of Utah (Paul Torrens): http://www.geosimulation.org/geosim.

Santa Fe institute: http://www.santafe.edu/projects/evca/.

Chapter 13

Multi-Agent Systems for Simulation in Geography: Moving Towards an Artificial Geography

13.1. Introduction

Regularity and persistence expressed beyond contingency are the primary concern of a modeler. The geographer, particularly those specializing in spatial analysis and quantitative geography, might be familiar with this point of view: more than the specificity of individual behaviors, it is the general tendencies beyond the "noise" of this diversity which interests the modeler. These models often sum up and describe observations established by a given scale fairly well. They are nevertheless limited in their capacity to express the conditions for such phenomena appearing. Linking with local dynamics, complexity offers concepts and resources to link global descriptions and spatial analysis.

Thinking in terms of complexity implies, among other things, conceiving observed phenomena at the level of numerous interactions which occur between elements operating at one or more lower levels. This position is not recent; it is formulated through systems theory [BER 68], which conceives systems as a collection of objects, of sub-systems in interaction. However, system theory, at least in its non-adaptive and applicative form [FOR 80], is more concerned with the modalities of a system function and its behavior than with emergence conditions and their possible evolutions.

Chapter written by Eric DAUDE.

Sciences of complexity thus appear in continuity with system theory, but they are distinguished through an approach based on sensitivity to initial conditions, emergence, self-organization, irreversibility and bifurcation processes, as well as by the technological tools which accompany their development.

The aim of this chapter is to argue in favor of a modeling process based on lower level description of systems in order to reproduce the dynamics of the higher levels, which is now compared to spatial analysis, favoring a global description of dynamics in its mezzo-level and macro-level approaches. This chapter proposes a quick synopsis of distributed artificial intelligence provided by the way of multi-agent systems. We finally propose a typology of potential research in the domain of human geography related to complex systems. *Artificial geography* may thus be understood through three overall classes of models.

13.2. From global to local description of structures and spatial dynamics

13.2.1. *Spatial analysis in practice*

Geographers explore the manner in which people produce geographical space through their own behaviors, and study the feedback, i.e. how space, be it geographical or physical, facilitates or constrains behaviors, through its rules, laws and attributes, whether they are old or new. Quantitative approaches in geography caused an overturn during the 1960s. Rich with these new advancements, geographers produce theories and propose models of spatial organization based on measures which will contribute to the enrichment and transformation of geographical knowledge.

There have been many essential contributions to spatial analysis; if only one could be mentioned it would be the provision of a scientific geographical discourse, because it is explicit and refutable. Mathematical and statistical methods have increasingly led geographers (some of them) away from the geography of the 19th century; i.e. a one-dimensional geography. If a diversity of behaviors at the individual level indeed exists[1], spatial analysis is concerned with rules and regularities explained on a global level of description, once the noise of this local diversity is eliminated. Geographers thus develop models of interaction, gravitation and diffusion in order to design and describe structures and spatial dynamics on a macro-scale. These models involve paying particular attention to distance (contiguity, hierarchy) and to quantity (population, socio-economic factors) in order to explain the development and performance of spatial organizations. In this way,

---

1 Individual taken in the statistical sense of the term.

the explanation of "local deviance"[2] or "deviance behavior" is either neglected or evoked in an often general discourse, borrowing from sociology, economics, or history[3].

Every model has its limits. Focusing research on the description of global structures and dynamics, spatial analysis has developed a way of locating and describing phenomena, hardly observable until now, although these models are "readable" only at the scale on which they have been developed. By leaving aside micro-level phenomena and their interactions in order to describe a global behavior, some behaviors remain without explanation. It can be thought that these macroscopic structures and their dynamics are the result of processes which are not visible on this macro-scale, yet they are fundamental for the global evolution of the system. We may then conceive that the constituents of the system contribute to its global dynamic, and that this is characterized by emergent properties, which appear on a macro-level when they are not directly observable on the level of elementary system constituents. If spatial analysis allows the description of these aggregated phenomena by macroscopic processes, one of the issues for *artificial geography* is to propose an explanation for these phenomena through a local description of the processes.

13.2.2. *Artificial geography in practice*

If laws and rules uncovered by spatial analysis and methodologies are sensitive to scales, this is also true for cultural geography, behavioral geography or economic geography, which work on a deeper level of analysis, which is that of the individual, putting emphasis on individual parameters first in order to explain the organization and the structuring of space.[4] Speaking about these geographies from "two extremities", the micro and the macro, is the fundamental perspective of artificial geography.[5] Understanding why certain regularities are observed at a global level, despite the absence of planning and control at this level, is one field of research for artificial geography. Intra-urban configurations or systems of cities are examples of such phenomena: beyond their descriptions, models which are sufficiently adequate to show how a system of central spaces emerges from numerous actions to

---

2 Residuals in regression analysis for example.

3 This attitude is widely present in the social science community. Reducing phenomena to one or two elementary processes contributes to the overstatement of the real importance of these processes to the detriment of others, of which the explanation is outside of the disciplinary sphere: space for the sociologist, psychology for the economist, etc.

4 These research domains are obviously distinct in their degrees of formalization.

5 This name emerged at the time of a discussion of the research group *Simulation and Artificial Territory*, J-.L. Bonnefoy, E. Daudé, P. Ellerkamp, M. Redjimi of the UMR 6012 ESPACE.

interacting agents do not exist.[6] The objective then is to discover how these regularities can emerge from autonomous agents in interaction, and governed by their own local perceptions of the environment and through their interests, whether they are rational or not. What are the effects of certain spatial configurations on people, individuals and societies? Why do certain shapes not emerge? Artificial geography must be positioned onto these research projects and those presented in section 13.1. Artificial geography is based on the principles of artificial sciences [SIM 69] and distributed artificial intelligence (DAI) [GAS 89] which puts the focus on collective and distributed resolution of complex problems, largely using individual-based models to simulate living phenomena, in the same way as artificial life [LAN 88], genetic algorithms [HOL 92] or cellular automata [WOL 94]. The name artificial geography is imposed then as an idiom, recognizable and readable by the community of researchers in DAI, of which the common denominator is complexity.

The sciences of complexity provide new research hypotheses for studies of production, organization and transformations of geographical space, as well as for the dynamics which are displayed therein. With complexity it is tempting to show how the transition between an organization (or the objects of a given level) and the elements which constitute the "bricks" of its construction are created. It may be postulated that structures, shapes and global behaviors result from a large number of interactions which take place among elements operating on one or more lower levels. In this context, all the components are involved in the global dynamic of the system characterized by emerging properties, which appear at a macro-level but are not directly observable on the level of the basic units of the system. A complex system is characterized by creations, inhibitions, fluctuations and through the existence of non-linear dynamics. Due to the non-linearity of such relationships, the behavior of such systems cannot be reduced to the sum of component behaviors. Artificial geography thus lays down a methodology to formalize open and dynamic systems.

Complex systems are often distributed, and the way in which they function is also distributed among the elements of the system. Such systems are also open to their environment, having the capacity to change and learn from its evolution. These systems are thus able to maintain a certain persistence in time and space despite the repeated reformations which affect the elements that compose them and the numerous entities which produce them. A city and its evolution through time thus reveal certain regularities in their functions, despite the changes which occur among the elements of which it is composed: population and business flow, political change, and modification of economic structure. Cities present regularities in shape and internal structure despite the presence of numerous actors which operate in these

---

6 At least they did not exist until 2003.

structures. It must be taken into account that complex systems open up into their environment and that their nature is spread in an effort to describe and explain their evolutions.

Complex systems are also dynamic, able to achieve and maintain a solution under different conditions despite the numerous disruptions which affect their constitutive elements: local disruptions are spread very little in organized systems [WEI 89]. This is far from chaos theory [MAN 04], pertinent in the universe of physics when non-linear interactions occur under a small number of equations, but where metaphors such as the "butterfly effect" [LOR 79] are certainly limited in their capacity to help us understand complex systems in human geography, which are more robust than chaotic. Self-organization theory seems to better suit the interpretation of such evolutions, characterized by amplification and self-learning processes. It also proposes a conceptual framework, with the critical self-organization [BAK 91; DAU 03], underlying the importance of phase transitions between different stages of the system. Thus, a structural regularity in the system of French cities exists, characterized by a strong urban hierarchy, despite the numerous reassignments and differentiated evolutions of the cities in this hierarchy [GUE 93]. It is a matter then of apprehending this type of system, not only by researching a single solution as in Christaller's model [CHR 33], but by trying to propose models able to recover and maintain this solution under certain conditions by accepting different combinations inside this solution.

Complex systems in geography are structured systems, which show structures with variations and evolutions. The more the number of levels is important and the interactions between many different components are numerous, the more the system will be complex, organized systems. The acceleration of history and evolution of complexity remain in an irreversible time [PRI 88].

13.3. Multi-agent systems

As they are distributed and dynamic, multi-agent systems (MASs) are particularly adapted for artificial geography. MASs are both concerned with modeling basic entities of a system, the agents, as well as their mutual and environmental interactions. According to the types of multi-agent patterns, the environment, the agent and communication can be of different natures. Some of these characteristics are presented in the subsequent sections before specifying what may constitute a geographical MAS.

13.3.1. *Environment*

In most multi-agent applications, environment provides the spatial context for the agents. The environment of a MAS is essentially characterized by five elements which allow for the definition of simulated space.

It is first a set of observed *ordered cells*, $E\{e_1, e_2, ..., e_N\}$. The order of these cells allows for the positioning of one unit e_i in group E in relation to another unit e_j. Set E possesses *a geometry* which describes the shape of the units and the topological structure of the domain. If the majority of MAS platforms are implemented according to a hexagonal or square domain, vector and reticular topologies are conceivable. Environment is also characterized by *metrics* which makes it possible to calculate the distance separating two cells possible, for example. The main distances used are Euclidian distance and Manhattan distance. Each unit of the domain possesses an *information vector (v)* which can be constructed with fixed or variable elements. In a minimal version of the environment, this vector represents the possible states of the unit; it is thus composed of a variable state {1 or 0 for present or absent}. Closer to a geographical reality, the units or cells would be characterized by static elements (altitude, presence of a river, of a road, etc.). Finally environment is characterized by *laws* which define the universe of possibilities in the structure and evolution of group E. The choice between a toroidal or finite space in a 2D cellular space constitutes a law which structures the simulated world. The laws of the environment can be: global, they thus impose themselves on the entire group of units E; and local, they are in this case specific to one category of units and have priority in reference to general rules [CAN 98]. To illustrate this, we will take a cellular world composed of units $n{\times}n$, each having an information vector v of type:

$$v = \begin{bmatrix} elevation \\ road \\ river \\ state \end{bmatrix} \quad \text{with for example one unit i: } v_i = \begin{bmatrix} 20.5 \\ 0 \\ 1 \\ vacancy \end{bmatrix}$$

The first three elements are fixed and describe the configuration of the unit at the beginning. The element *state* describes the state of the cell at a given moment (for example, "vacancy", "industry", "commerce" or "inhabited"). The general rule of this universe is that The local rule is any cell that takes the value 1 for "river" cannot change its state.

A sixth component may finally be added to characterize the environment: the *transition rules*. This modifies the nature of the environment of the MAS model because it brings it closer to the conception of cellular automata (CA). This implies that the environment is able to evolve partly independently of the actions of the agents that are situated there; it can be seen here in a configuration where a CA (Chapter 12) is coupled with a MAS. This approach is particularly well adapted for exploring systems where the relationship between environmental dynamics and human dynamics are strong, such as in natural resource management [BOU 01].

The environment of a MAS may thus be: a support, i.e. a simple space of evolution for the agents [PAG 98]; a resource, the environment is characterized by attributes which are at the origin of the agents' actions [BUR 96; EPS 96; BON 01]; a field of communication between agents [DRO 94]; an entity having its own dynamics [BOU 99; BOX 02]. The environment is thus an essential element of MAS since it is appropriate for an explicit formalization of space.

13.3.2. *Agents in the environment*

A set of agents A, is defined in environment E. This set may be finite or infinite, which implies a logical separation between elements of set E and those of set A, $A \neq E$. An agent a_i can be described as an autonomous entity capable of carrying out its own actions. The agent can communicate with others and move, but does not do so inevitably. The agent remains stationary if its relative position in the environment is important for the dynamic of the system or not, a city for example, and only its actions are thus taken into consideration. Its behavior (Figure 13.1) is the consequence of its capacities, abilities, resources, observations, knowledge and its interactions with other agents and the environment [FER 95]. This definition takes into account the difference between the *reactive* and *cognitive* agent approach.

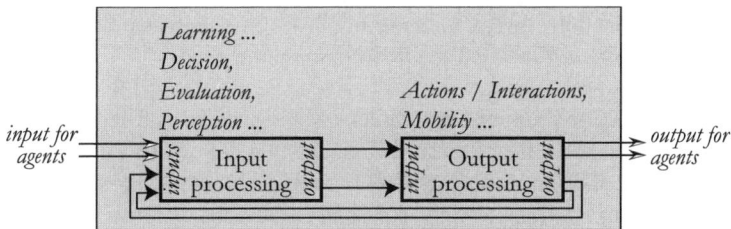

Figure 13.1. *Capacities of an agent*

Systems with reactive agents are composed of a large number of agents with simple behavior, without capacity for learning and having a limited representation of their environment. The agent is conceived, in its most radical aspect, as an entity which obeys external stimuli when taking action, without the possibility of self-control, evaluation of results, or behavioral evolution [BRO 86]. On the other hand, cognitive agent systems are composed of agents which have representations and knowledge (of themselves, of others, of the environment, etc.).The cognitive agent is capable of making decisions based on this and able to benefit from past experiences in order to evaluate the impact of its own actions in the future [VAR 89]. Reactive and cognitive approaches are the result of current research which have different concerns, the *cognitive* approach greatly impacting the work of sociologists, notably those coming from the fields of methodological individualism [BOU 92] or psychology [LEV 65], the *reactive* approach close to the universe of ethology [DEN 87], robotics [BRO 83] or artificial life [REY 87; LAN 91; BED 00].

The distinction between the two approaches actually masks a diversity of situations. The type of application for which the agent is conceived determines its characteristics: action capacities (what are its competencies?), communication capacities (can it communicate? with whom? how?), storage capacities, perception capacities (what does the agent perceive in its environment?), adaptation capacities (can it adapt itself to new situations or does it act according to strict rules?), organization dimension (with whom does it interact?), resource capacities, etc. [TRE 01]. Each of these elements set off specific studies: communication and interaction models [KON 01], coordination models [FAL 01], etc.

13.3.3. *Method of communication between agents*

The simple coexistence in an environment of autonomous agents has but little value in geography if we admit that interaction is the common denominator of manipulated objects. These interactions may be motivated by objectives, capacities and resources belonging to each of the agents. MASs are thus implemented to promote the interactions between agents, or at least between agents and their environment, and use different protocols for communication.

Interactions may be produced in a direct way by sending and receiving messages between agents, or indirectly by means of signals via the environment. Most modes of communication that pass messages refer to speech act theory with a syntax of signals for making actions [AUS 62] and relatively extensive protocols for interaction [SIM 80]. These communication modes hold all of their meaning in the domains where interactions between agents are necessary in order to resolve global problems or dynamically share a workload, i.e. controlling air traffic, electronic commerce, etc. In these various application domains, it means coordinating different agents' actions by means of communication. Interaction modes by signals via the environment are themselves largely used in ethology with, for example, the deposit of pheromones by ants [DEN 87], these messages automatically causing a reaction in the case of reactive agent. This mode of communication is much poorer than the direct transmission of a message, a signal causing one, and only one, action.

13.3.4. *Multi-agent systems and geography*

The different elements presented in the preceding section are at the core of any consideration of geographical MASs. Contingent upon the objectives of the model, one or the other components will be further insisted upon: either behavioral geography will take an interest in the cognitive aspects of the agent, or economic geography will be concerned with its types of interactions and organizations. A multi-agent geographical system is thus a set of agents situated in an environment, which can be endowed with either an internal dynamic or one resulting from the action of agents. Several levels of organization are likely to coexist in the same system, the agents being able to belong to one or several levels, which themselves form agents. These different levels are governed by their own rules, meaning agents are autonomous and in permanent interaction, meaning organization. Interactions carried out within levels and between levels allow for the system dynamic and make it possible to create emerging phenomena. Applied to geography, MASs offer the benefit of cellular robots in order to define more complex spatial interactions, as well as define inexplicit spatial relationships, such as what is produced in the case of notably commercial [BUR 96] or social [DAU 02] networks. These modalities of interaction may be combined into one agent: spatial proximity of the Moore type and social proximity of the extended selective type may be solicited in these diffusion models [DAU 04a]. It is also possible to deal with the coexistence of different kinds of interactions in the same multi-agent model or a continuum within the agent interactions. These capabilities may be associated with a differentiation of agents according to a particular attribute: the spatial range of interactions varies with the attribute. Finally, agents may modify their interaction structure during simulation, which may be the case when wanting to model the progressive construction of a social network or a system of cities.

Figure 13.2 illustrates the different points mentioned in this section. Environment is the simulation space; here it declines with different layers of information constituting the possible integration of geographical information systems (GIS) in the multi-agent domain [RAN 02]. The dynamic aspects of the environment can be managed on the basis of cellular automata, with a set of layers hierarchically organized: a layer of cells to a layer of objects according to the number of levels chosen. These different layers are set up by associative rules with cells and can be enriched by rules governing the physical processes of the environment for example. The agents act in this environment, forming the "intelligence" of the modeled system. If most spatial MASs are constructed on the basis of only one level of organization, this representation allows n levels of organization, to which n classes of agents correspond, from level-one to a level-three. Level-one agents are for example individuals or firms, level-two and three agents represent economic agents or institutional agents (cities, territorial communities). New forms of interactions, a memory of the past and capacities for distinct actions can be established at each of these levels, from which, through interactions, the evolution of the system is derived. It must be noted here that this approach is not strictly ascending, which is the case with most self-organization models.

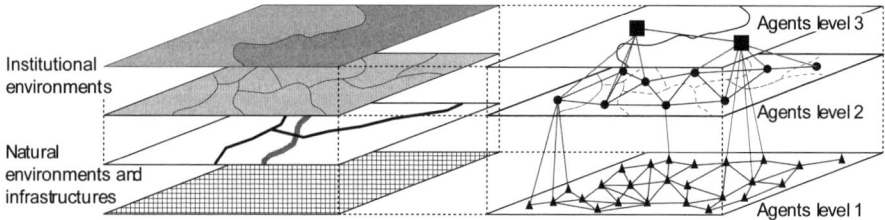

Figure 13.2. *Schematic representation of a MAS in geography*

MASs offer the possibility for geographers to associate quantitative and qualitative parameters in their models, to take into account the existence of several levels of organization and of differentiated dynamics. In this way these models become artificial laboratories, thanks to computer simulation. Individual-based simulations allow for a precise verification of the model behavior and the hypotheses associated with it. It is possible to observe the global dynamics of the phenomenon, and to follow agent behavior individually, in a group or collectively, in order to analyze data with the help of statistical techniques, thus partly avoiding the inconvenience of the "black box".

13.3.5. *A typology of MAS models*

If MASs allow us to create a high variety of models, a typology of simulation models may be proposed according to their state of performance: between parsimonious models – respectively refined – of which the performances situate themselves in their capabilities of producing results qualitatively – respectively quantitatively – close to the observed macro-structures. When using an individual-based approach, this typology should not be limited to comparisons in terms of micro-structures. Two models can in fact produce results equivalent to a macro-level based on different individual behaviors. We must differentiate between simulation models capable of producing results close to macro-structures and observed micro-behaviors[7]. According to these different elements, a typology of simulation models may be proposed such as:

_____

7 We have not considered the essential point of validation of the model in this paper. More than the verification of the simulation model which consists of making certain that the computer program is reliable and that the model is in good condition, meaning that it does not

– type 1: the model allows for the reproduction of results qualitatively close to the observed macro-structures;

– type 2: the model allows for the reproduction of results quantitatively close to the observed macro-structures;

– type 3: the model allows for the reproduction of results quantitatively close to the observed macro-structures and qualitatively close to the observed micro-behaviors.[8]

These different types are not considered in a hierarchical context; moving from one type to another does not mean that a model is more "finalized" than others. It may be necessary to look more closely at the differentiation in terms of project, which include: the state of knowledge advancement in the considered domain, the questions relative to this domain, the modeling objectives and the available data which in principle must determine the relevant state where the model will be situated and not the state that will be achieved determining the questions.

The environment, the agents and the interactions associated with MASs thus offer several research perspectives to the geographer. We might hence define an artificial geography in the words of J. Epstein and R. Axtell [EPS 96], "We view artificial societies as laboratories, where we attempt to grow certain social structures in the computer – or in silicon – the aim being to discover fundamental local or micro-mechanisms that are sufficient to generate the macroscopic social structures and collective behaviors of interest", where the social becomes spatial.

13.4. Artificial geography: simulations of structures and spatial dynamics

The theoretical developments of complex systems and technologies associated with them have been of great interest in the geographical community for the last few years. As they explicitly integrate space, these formalisms interest most disciplines which discover or rediscover the role of space in their research. This integration of space in research outside the field of geography constitutes an opportunity for

---

behave systematically in an erratic manner because of tiny changes in value, validation consisting of defining to what extent the results of simulations are in agreement with measurements in the real. Different methods of verification are possible, whether it is about original givens from simulations, for example when working on means with confidence intervals, or whether with images of simulation, spectral analyses may be carried out. Verification of this type of model continues despite an insufficiently explored domain see [COQ 97] for a proposal of guidelines of experience in ecology.

8 The typology arising from these four indicators – Macro structure; Micro behavior; Quantitative; Qualitative – may be extended. Here we limit ourselves to the three most frequent types, quantitative validation of the behaviors at a micro level often being impossible, taking into account the overall unavailability of data at this specific level.

geographers, being able to share their knowledge of space with other disciplines, the interdisciplinarity being facilitated on the basis of communal methodologies. Geographers can also profit from these new methodologies by reformulating their questions, making new hypotheses, exploring new fields at the core of the objects they study and by updating eventual gaps in their knowledge. The only choice of basic entities in the development of a model may persuade us, expanding the efforts still to be achieved in order to advance within this domain. In effect, if the current research most often concerns itself with models on two levels, a basic level of dynamics and a global level of observation, the future drives us towards multi-level modeling. However, even with a single modeled level we are sometimes driven to research information below and above this level, and it will be the same for each of the required levels in models to come. This should force us to think more precisely about the pertinence of the chosen entities in our representations of studied systems, about the relationships between entities of the same level and about inter-level relationships.

The fields of MAS applications within the array of geographical researches are numerous. Three large classes of model may be used in artificial geography for years to come. These domains of research may appear vast; they are actually held back mainly by the state of our geographic knowledge which can be formalized in terms of agents, interactions and environments. The models presented in the following sections are dedicated to each of these domains. They are abstract formalizations, which leave aside some essential aspects of the addressed themes, allowing a deeper immersion on complexity and MAS.

13.4.1. *Emergence and evolution of spatial structures*

Studying the location and distribution of objects in space is relevant to geography. Geographers construct models with this in mind, whereby the objective is to test the interpretations given from the observed regularities on the surface of the earth, as the precursory work by von Thünen [VON 26] and Christaller [CHR 33]. Von Thünen's land-rent model explains the set-up and development of a land-use concentric organization centered on the market place. On another scale, Christaller develops a model based on profit maximization and on the limited reach of economic activity in order to explain the establishment of urban distributions and hierarchies.

Modeled objects, whether they are individuals, firms or cities, have properties which are at the origin of the formation and evolution of spatial structures. The reevaluation of these properties and their formalization through different methodologies has been exposed since these original works. Among different families of models used to interpret emergence and evolution of spatial structures, multi-agent simulation stimulates new questions and offers numerous perspectives on research.

Most geographical models which explain the construction of spatial structures present hypotheses on individual behaviors of the origin of these organizations. These hypotheses depend on methodological individualism, individuals having rational and maximizing behavior, the sum of these individual behaviors bringing the system to an optimal level of organization [HOT 29; VON 26; WEB 09; LÖS 54; ALO 64; CHR 33]. It seems of interest, from a heuristic point of view at least, to reconsider the set of these models through multi-agent simulation, where individuals would no longer be homogenous and have differentiated access to information. It is in effect possible to present more realistic hypotheses on these human behaviors, such as the local circulation of information and non-systematically maximizing behaviors. Numerous models must still be explored again, such as those by Hotelling, Alonso or Weber, in order to produce a strong theoretical base, and to produce precise (computational) models based on behaviors and spatial structures.

It is possible to use MAS starting with rules found in geographical writings, such as those on urban ecology. The "concentric circle" models, [PAR 25], the "sector theory" [HOY 33] or even "multiple centers" [HAR 45], to refer only to the oldest, can taken together, provide new paths of research through simulation. Still in the prototype stage, the following model allows the global structuring of a city, of course simplified and reductive, with local rules for establishing three types of activities – dwelling, commerce and industry – and from an initial seed of population. The initial configuration is characterized by the presence of a river, a road, and a space of free land-use, which is differentiated by altitudes.

Each cell e_i of E is characterized by a set of attributes as in the vector of information described previously. This environment is the spatial support for the progression of the agents; it may in turn be either a constraint or an opportunity for the conversion of the land–cells in any given activity. Three agents are responsible for urbanization, with an agent for each activity. These agents move randomly through space and evaluate the pertinence of a given site considering implantation rules which are characteristic of their function. More precisely, if the rules are the same for the three activities, the parameters will influence the implantation of one of them more or less strongly in a place, in this case, the cell. Six localization factors are taken into account: the proximity to the river (Vf) and to the road (Vr), the presence of industry (Vi), of habitat (Vh) of commerce (Vc) and the attraction of altitude (Va). These factors are weighted by the coefficients (α) which may be negative or positive[9] according to the importance attributed to agent i in their presence in the neighborhood. The following function thus gives an indicator (PA)

---

9 The sum of these coefficients is one.

which varies from 0 to 100 and indicates the "value" of the site for the agent responsible for activity i:

$$PA_i = Vf^*\alpha_i + Vr^*\alpha_i + Vi^*\alpha_i + Vh^*\alpha_i + Vc^*\alpha_i + Va^*\alpha_i$$

The closer the value PA attributed by an agent for a site is to 100, the greater the probability of setting up its activity on this site.[10] For the simulation presented below, the coefficient values translated into symbols (Table 13.1) indicate the preferences of the three agents presented in the model. *Agent-industry* will thus have a tendency to give greater importance to sites closest to the river and a road axis, the presence of industry in the environment helping to strengthen the attraction of the place. This same place will see its potential for industry lessen if it is close to housing or if it is located on a higher altitude. The logic of these preferences which are in part in agreement with those of other agents, contribute during simulations to the development of industry on the road axis and in the direction of the river, with its rejection for altitude favoring its extension the whole corridor long.

Coefficients	*River*	*Road*	*Industry*	*Habitat*	*Commerce*	*Altitude*
Industry	+	++	++	-	-	- -
Habitat	+	+	- - -	+	+	++
Commerce		+	-	++	+	-

Table 13.1. *Weight of the parameters in the simulation*

Agent-habitat is influenced by the presence of habitats in its neighborhood and it looks for sites of medium altitude and tends to reject all places marked by the presence of industry. This is characterized by a development of residential zones which stretch towards the north of the zone (see Figure 13.3 below and also the color plate section), at altitude and in opposition to industrial zones. Finally, *agent-commerce* is principally influenced by the presence of residential sites, which favors overlapping this function in the core of residential zones, but also along the road axis and at proximity to other commerce (yellow in color figure).

---

10 If the cell is not occupied and if PA > a random number taken according to a uniform law from 0 to 100, then the cell takes state i. Otherwise if two agents are at instant t on the same cell, the priority is given to the agent with the largest PA, or a random agent if both PA are identical.

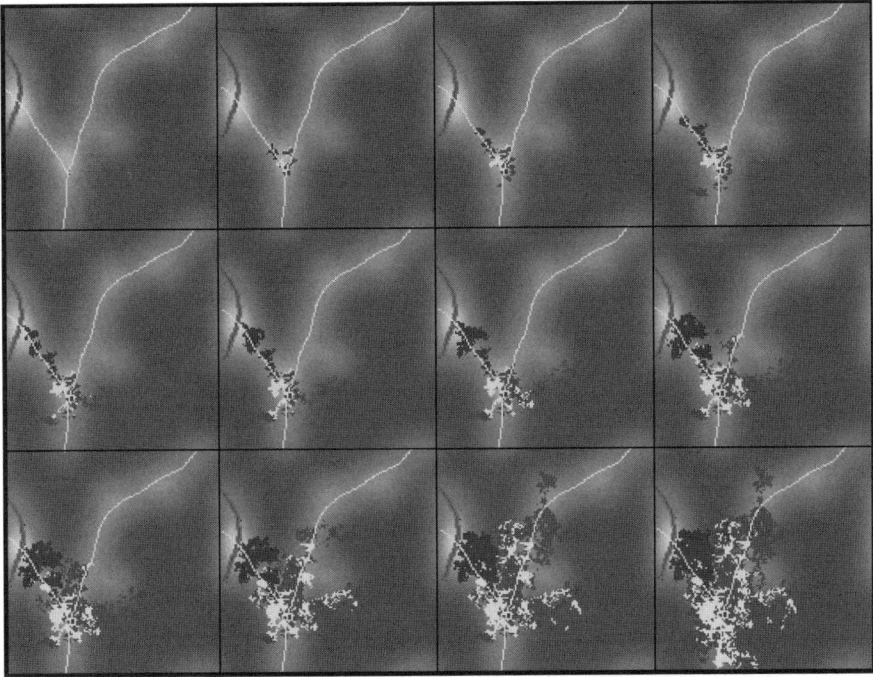

Figure 13.3. *Initial environment – land-use (variations of green according to altitude) main road (gray) and river (dark gray) – and simulations: extension of habitat commerce and industry (see also the color plate section)*

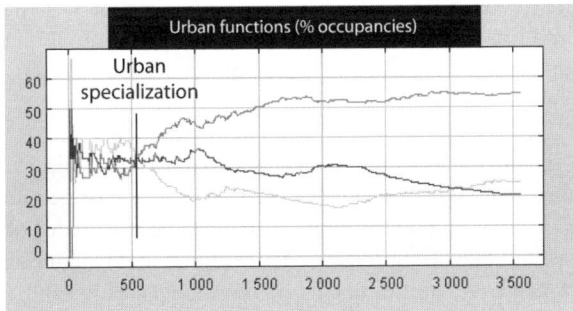

Figure 13.4. *An example of urban specialization through simulation. Ratio of urbanization: habitat, commerce, industry (see also the color plate section)*

These three urban-agent settling strategies, combined with urban growth, produce spatial effects of concurrence and blockage on certain sites. For example, the location ratio of industry declines during the simulation, while there is no limitation of the global density of this activity in the model. This phenomenon is produced because the industry is held back in its development by mountains that stretch all along the passageway, the river which presents a major barrier in this scenario, and by space already occupied by the habitat.

This model, of which the first results are presented here, allows us to "reconstruct" a city in a decentralized manner, ascending and dynamic, and tries to show that the internal urbanization and structuring of a city are the result of local processes based on an *attraction-repulsion* set. The purpose is to simulate different urban shapes based on models that only consider a few variables among the numerous dimensions that come together for the production of a city. This approach implies testing different hypotheses with location strategies, defining the coefficient parameters in order to study phase diagrams, and studying the macrostructures produced and their evaluations with the help of knowledge from spatial analysis. From an accurate knowledge of the model's behavior and different scenarios, initial observed conditions may be introduced (land-use), as well as categories of more specific actors, or even minimal global management of the density of activities in order to build a more realistic model.

13.4.2. *Exploration of dynamics in space*

If research in geography is concerned with dynamics at the origin of spatial structures and their evolutions, this research is also interested in dynamics which do not necessarily modify the space in which the phenomenon took place. This might be the case with innovation diffusion, epidemic diffusion, mobility of people, or flow of goods, for example.

Trajectories or movements are usually modeled on the basis of rules and laws belonging to the scale on which they are observed: the flow of people between cities is calculated by a gravity model, therefore it may be postulated that the relative position of places and their weight strongly determine the dynamics exhibited there, which is in part validated by the results of the models. However, this established relationship masks a real complexity, which is sometimes revealed by the size of the residuals. Beyond spatial laws, other laws may be introduced to explain and describe global dynamics in geographical space. Diffusion constitutes a good illustration of the contribution of MASs in reference to classical approaches [DAU 04a].

A simple example of diffusion in an adequate relation with the formalism of MASs is the propagation of an epidemic. In this model, the possibility of the agent becoming infected by the virus depends on the proportion of agents contaminated in its Moore type spatial neighborhood, noted V. This possibility also depends on the virulence of the epidemic, its infected neighbors being contagious for a limited period of time. The agent, i.e. human individual, is thus the basic unit in which the diffusion model is implemented. This model is constructed with a function $\tau_i(t)$ which governs the lifespan T of the virus once inside the agent and a contamination model $v_i(t)$ based on spatial proximity V. The transition of the agent from stage 0 (non-infected) to stage 1 (infected) is thus linked to the proportion of virulent agents in its spatiotemporal neighborhood $V * T$, being:

$$< v_i(t) >_{V \times T} = \frac{1}{n_R} \sum_{j \in V(i)} v_j(t)$$

where n_R represents the number of neighbors in radius R, here a first order contiguity. The probability of an agent i moving from the state 0 to the state 1 is thus expressed by the following rule:

$$P\left(\frac{s_i = 1}{s_i = 0} \right) = r \cdot < v_i(t) >_{V \times T}$$

where n_R represents the number of neighbors in radius R, here a first order contiguity. The probability of an agent i moving from the state 0 to the state 1 is thus expressed by the following rule:

$$P\left(\frac{s_i = 1}{s_i = 0} \right) = r \cdot < v_i(t) >_{V \times T}$$

This simple model shows complex shapes; it allows a non-saturation of the environment – here a simple support of propagation – and allows the observation of different global behaviors (Figure 13.5) according to the virulence and evolution of the spatial extent of interactions [DAU 04b].

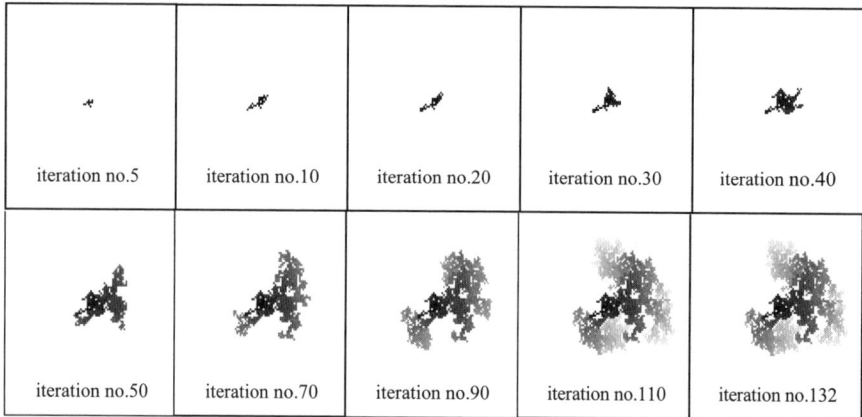

Figure 13.5. *Illustration of the epidemic diffusion, scenario with T equal to 2 and an agent infected at the center of the domain (101*101) at the initialization. The differences of gray intensity represent the date of the infection from older (darker) to more recent (lighter)*

The accuracy of such a model allows for the discovery of generic properties: at what value of T does percolation exist?, what impact does the spatial range of interactions have on the temporality of diffusion?, etc. Once used, this type of model may be improved through more realistic configurations: unequal distribution of agent populations; types of agent mobility [DAU 05]; more complicated interactions, such as through communication networks; agent heterogenity faced with contamination.

13.4.3. *Practices, representations and organization of space*

Daily spatial practices by individuals result in a series of representations of this space. These representations concern the objects situated in space just as much as space itself, and in its configuration, its laws and rules. These practices and representations refer to the individual and social sphere and participate in the development of an increasingly organized space. Simultaneously the individual producer of space is more or less constrained by the space produced through collectivity. The space observed is thus the product of numerous practices and representations which intermix and are sometimes contradictory. Thus, "individual actors, groups and institutions create space through their work and their daily actions, even if they are not directly in the space. They do this on an already defined territory, rich with 'memories' which they use or which they transform according to their means and strategies, which are guided in particular by their representations of space itself. Through its own structures, the space that they produce drives them to inflict ulterior actions, according to these representations, also modified by new practices of transformed space" [BRU 92].[11]

Geographical space is thus the result of dynamic, distributed processes where practices and representations play an essential role. Individuals, social groups and the State all represent different levels of organization in this space and play a part in the production of this space both individually and collectively. These practices and representations evolve over time, at more or less significant speeds depending on the levels on which they are situated: e.g. individuals probably have faster capacities of modifying their actions than the State would. Despite the perpetual modifications of these outlines, the dynamics of the geographical space is rarely chaotic. MASs thus constitute a precious tool for exploring dynamic links between individuals, society and space. This type of approach allows for the exploration of a domain, the individual area, perception and representations for which knowledge and research are certainly not as numerous as on a meso- or macro-geographical scale.

Similarly to the preceding domains of research in artificial geography, a simple model is presented here where the individual representations contribute to structure the space. In this segregation model by Thomas Schelling (1978), two categories of agents are presented; let us use the example of black and gray. These agents are preoccupied by the composition of their neighborhood; they are particularly attentive to the fact that a minimal fraction of their neighbors are in the same category as they are. If this condition is not filled, they move to an empty cell and the simulation keeps running until the set of individuals is satisfied by the situation. What is simulated here is the result of these representations on the global configuration of space. The remarkable thing in this model is that with relatively strong individual tolerance, and up to 66% of individuals unlike the others in the neighborhood, the simulations lead up to a relatively segregated global structure.[12] On the contrary, with a strong tendency towards regrouping, which is to say a threshold of 88% minimum alike neighbors, the model does not succeed in finding a situation of "geographical" equilibrium when the density is high, the agents being perpetually displaced in the simulation space. This phenomenon is perfectly comprehensible: when positioning agents at the initialization randomly, the probability of being in a favorable site is very low for any given agent, therefore it will have to move on the next iteration. As all agents have the same threshold of

---

11 *"Acteurs individuels, groupes et institutions créent l'espace par leur travail, par leurs actes quotidiens, même ceux qui ne portent pas directement sur l'espace. Ils le font sur un terrain déjà défini, muni de "mémoires", qu'ils utilisent ou qu'ils transforment selon leurs moyens et leurs stratégies, lesquelles sont guidées en particulier par leurs représentations de l'espace lui-même. Par ses propres structures, l'espace qu'ils produisent les amène à infléchir leurs actions ultérieures, selon ces représentations, également modifiées par les nouvelles pratiques de l'espace transformé"*.

12 In a domain of 51*51 cells, with 2,500 agents divided into two categories.

tolerance and the same behavior, the majority of them are led to move from this iteration and the following ones, which make them go back to recreating the initial conditions at each iteration.

The idea of the following model is to determine a simple rule which allows individuals to find a place that is suitable for them, and to do this in a decentralized manner. When individuals have a relatively low tolerance threshold, meaning a low tolerance for differences, we can either favor the gathering of individuals according to their category with a "gated communities" rule, which excludes all individuals from one category of the neighborhood from another category, or give the possibility for individual preferences to evolve, taking into account for example the representations that individuals have of their neighborhoods over time. Thus when an individual is in proximity to one or several agents who are satisfied with their positions, which is to say they do not "wish" to move, and whatever their type may be, the individual reduces his or her level of exigency for socio-spatial similitude. With this hypothesis it is supposed that an individual receives certain benefits from choices and from the satisfaction of his or her neighbors. It is also supposed that he or she is thus influenced by his or her socio-spatial environment. Figure 13.6 shows the results of a simulation with a relatively low difference threshold for the entire population at initialization, with 12% of the neighbors different, (similitude at 88%). However, a globally balanced situation is reached because unlike in the Schelling model under the same conditions, individuals are progressively influenced by their contacts and lower their initial "exigencies", which translates into a decline in average preferences for socio-spatial proximity at the core of the population studied.

This simplified model shows the potential of MAS in capturing the dynamic evolution of individual preferences and their effects on the structuring of space. A large research field is developing in the domain of social composition of cities and its evolution.

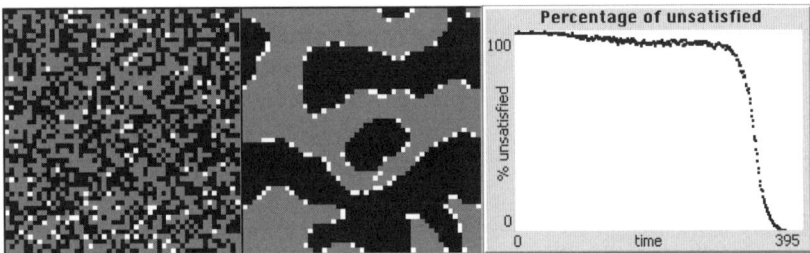

Figure 13.6. *Individual representations and structuring of a space*

The collective subconscious brings ideas which impose themselves upon each individual arriving at a given site, with its rich neighborhoods, poor neighborhoods, connecting neighborhoods, etc. These collective ideas sometimes have relatively long temporalities in regard to urban reality. The individual thus has to make a choice between his or her global representations, emerging experience of the city and personal characteristics. In the end these numerous dimensions produce a choice of location which may be different from global ideas. The evolutionary dynamics of the internal structure of the city are thus in part the product of interacting individual choices, with MAS thus contributing to a better understanding of regularities observed on the level of a city despite the absence of planning or absolute control on this level.

13.5. Conclusion

If multi-agent simulation offers numerous perspectives for geographical research, at the moment it is more or less in the prototype stage. The essential reason for this achievement is probably that these simulation methods are advanced in relation to our knowledge and force us to reformulate and rethink our questions.

The transition to simulation models able to reproduce phenomena which use a smaller scale based on entities of fine granularity implies the collaboration of numerous researchers, sometimes from domains of different disciplines. Dynamic models created in such a way require numerous parametric analyses in addition to their operative validation, so as to calibrate the model to the observations. Moreover, the necessity of collecting information not yet available is a consequence of any introduction of new tools. A supplementary argument for this type of technology will be found here, apt to be quickly made into a project with a minimum amount of computer knowledge for theoretical simulation models, or on the other hand, capable of mobilizing a team of dynamic model researchers to realize more functional ambitions.

13.6. Bibliography

[ALO 64] ALONSO W., *Location and Land Use*, Harvard University Press, Cambridge, 1964.

[ART 90] ARTHUR W. B., "Positive feedbacks in the economy", *Scientific American*, no. 2, p. 80-85, 1990.

[ASC 00] ASCHAN C., MATHIAN H., SANDERS L., MÄKILÄ K., "A spatial microsimulation of population dynamics in Southern France: a model integrating individual decisions and spatial contraints", in G. Ballot, G. Weisbuch (ed.), *Applications of Simulation to Social Science*, Hermes Paris, 2000.

[AUS 62] AUSTIN J.L., *How to Do Things with Words*, Clarendon Press, Oxford, 1962.

[AXE 97] AXELROD R., "Advancing the art of simulation in the social sciences", in R. Conte, R. Hegselmann, P. Terna (ed.), *Simulating Social Phenomena*, Springer-Verlag, Berlin, p. 21-40, 1997.

[AXT 94] AXTELL R., EPSTEIN J., "Agent-Based modeling: understanding our creations", *Bulletin of the Santa Fe Institute*, vol. 9, no. 2, p. 28-32, 1994.

[BAK 91] BAK P., CHEN K., "Self-organized criticality", *Scientific American*, p. 46-53, 1991.

[BAR 00] BARRETEAU O., BOUSQUET F., "SHADOC: a multi-agent model to tackle viability of irrigated systems", *Annals of Operations Research*, no. 94, p. 139-162, 2000.

[BED 00] BEDAU M., "Artificial Life VII", *Proceedings of the Seventh International Conference on Artificial Life*, Bradford Books, 2000.

[BEN 98] BENENSON I., "Multi-agent simulation of residential dynamics in the city", *Comput. Envir. and Urban Systems*, vol. 22, p. 25-42, 1998.

[BER 68] BERTALANFFY L. Von., *General Systems Theory, Foundation, Development, Applications*, New York: G. Braziller, 1968; French translation: *Théorie générale des systèmes*, Paris, Dunod, 1973.

[BER 82] BERLEKAMP E., CONWAY J., GUY R., "Winning way for your mathematical plays", in *Games in Particular*, vol. 2, London: Academic Press, 1982.

[BON 01] BONNEFOY J.-L., BOUSQUET F., ROUCHIER J.,, "Modélisation d'une interaction individus, espace et société par les systèmes multi-agents: pâture en forêt virtuelle", *L'Espace géographique*, no. 1, p. 13-25, 2001.

[BON 03] BONNEFOY J.-L., "Des ménages aux structures urbaines: des représentations spatiales comme moteurs de dynamique dans des simulations multi-agents", *Cybergéo: Revue Européenne de Géographie*, no. 234, 2003.

[BOU 92] BOUDON R. (ed.), *Traité de sociologie*, PUF, Paris, 1992.

[BOU 99] BOUSQUET F., GAUTIER D., "Comparaison de deux approches de modelisation des dynamiques spatiales par simulation multi-agents: Les approches spatiales et acteurs", *Cybergéo*, http//www.Cybergeo.press.fr, 1999.

[BOU 01] BOUSQUET F., LE PAGE C., "Systèmes multi-agents et écosystèmes", in J.P. Briot, Y. Demazeau (ed.), *Principes et architectures des systèmes multi-agents*, Hermes Science Publications, Paris, 2001.

[BOX 02] BOX P., "Spatial Units as Agents: Making the Landscape an Equal Player in Agent-Based Simulations", in H. Randy Gimblett (ed.), *Integrating Geographic Information Systems and Agent-Based Modeling Techniques for Simulating Social and Ecological Processes*. Santa Fe Institute, Oxford University Press, p. 59-82.

[BRO 86] BROOKS R., "A robust layered control system for a mobile robot", *IEEE Journal of Robotics and Automation*, RA-2 (1), p. 14-23, 1986.

[BRU 92] BRUNET R., FERRAS R., THERY H., *Les mots de la géographie*, RECLUS – La Documentation Française, Paris, 1992.

[BUR 96] BURA S., GUERIN-PACE F., MATHIAN H., PUMAIN D., SANDERS L., "Multi-agents systems and the dynamics of a settlement system", *Geographical Analysis*, vol. 28, no. 2, 1996.

[CHR 33] CHRISTALLER W., *Die Zentralen Orte in Südeuntschland*, Gustav Fisher, Jena, 1933.

[CIL 98] CILLIERS P., *Complexity and Postmodernism: Understanding Complex Systems*, Routledge, London, 1998.

[COQ 97] COQUILLARD P., HILL D., *Modélisation et simulation d'écosystèmes: Des modèles déterministes aux simulations à évènements discrets*, Masson, Paris, 1997.

[DAU 02] DAUDE E., Modélisation de la diffusion d'innovations par la simulation multi-agents. L'exemple d'une innovation en milieu rural, PhD thesis, Université d'Avignon et des Pays du Vaucluse, 2002.

[DAU 03] DAUPHINE A., *Les théories de la complexité chez les géographes*, Anthropos, Paris, 2003.

[DAU 04a] DAUDE E., "Apports de la simulation multi-agents à l'étude des processus de diffusion", *Cybergéo: Revue Européenne de Géographie*, no. 255, 2004.

[DAU 04b] DAUDE E., LANGLOIS P., "Les formes de la diffusion", *Colloque GéoPoint*, Avignon, 2004.

[DAU 05] DAUDE E., ELIOT E., "Effets des types de mobilités sur la diffusion des épidémies: l'exemple du Sida à Bombay", *Colloque Théo Quant*, Besançon, 2005.

[DEL 94] DELAHAYE J.-P., *Information, complexité et hasard*, Hermes, Paris, 1994.

[DEN 87] DENEUBOURG J.-L., PASTEELS J., *From Individual to Collective Behaviour*, Bâle, Birkhauser, 1987.

[DRO 93] DROGOUL A., De la simulation multi-agents à la résolution collective de problèmes. Une étude de l'émergence de structures d'organisation dans les systèmes multi-agents, PhD thesis, University of Paris VI, 1993.

[DRO 94] DROGOUL A., FERBER J. (1994), "Multi-agent simulation as a tool for studying emergent processes in societies", in N. Gilbert, J. E. Doran (ed.), *Simulating Societies: the Computer Simulation of Social Phenomena*, London, University of London College Press, p. 127-142.

[EPS 96] EPSTEIN J. M., AXTELL R., *Growing Artificial Societies: Social Science from the Bottom Up*, The MIT Press, London, 1996.

[FER 95] FERBER J., *Les systèmes multi-agents: vers une intelligence collective*, InterEditions, 1995.

[FOR 80] FORRESTER J., *Principles of System*, The MIT Press, London, 1980.

[GAS 89] GASSER L., HUHNS M., *Distributed Artificial Intelligence*, vol. 2, Morgan Kaufmann, San Mateo, 1989.

[GIL 99] GILBERT N., TROITZSCH K., *Simulation for the Social Scientist*, Open University Press, Philadelphia, 1999.

[GUE 93] GUERIN-PACE F., *Deux siècles de croissance urbaine*, Anthropos, Paris, 1993.

[HAR 45] HARRIS C., ULLMAN E., "The nature of cities", *Annals of the American Academy of Political Science*, no. 242, p. 7-17.

[HEG 98] HEGSELMANN R., FLACHE A., "Understanding complex social dynamics: a plea for cellular automata based modelling", *Journal of Artificial Societies and Social Simulation*, vol. 1, no. 3, 1998.

[HEL 85] HELPMAN E., KRUGMAN P., *Market Structure and Foreign Trade*, The MIT Press, Cambridge, 1985.

[HEU 94] HEUDIN J.-C., *La vie artificielle*, Hermes, Paris, 1994.

[HEU 98] HEUDIN J.-C., *L'évolution au bord du chaos*, Hermes, Paris, 1998.

[HOL 92] HOLLAND J. H., *Adaptation in Natural and Artificial Systems: An Introductory Analysis with Applications to Biology, Control and Artificial Intelligence*, 2nd ed., The MIT Press, 1992.

[HOL 95] HOLLAND J. H., *Hidden Order: How Adaptation Builds Complexity*, Reading, MA: Addison-Wesley, 1995.

[HOT 29] HOTELLING H., "Stability in competition", *Economic Journal*, no. 39, p. 41-57.

[HOY 33] HOYT H., *One Hundred Years of Land Values in Chicago*, University of Chicago Press, Chicago, 1933.

[KAU 95] KAUFFMAN S., *At Home in the Universe: the Search for Laws of Self-organization and Complexity*, Oxford University Press, Oxford, 1995.

[KRU 97] KRUGMAN P., "How the economy organizes itself in space: a survey of the new economic geography", in W.B. Arthur, S.N. Durlauf, D.A. Lane (ed.), *The Economy as an Evolving Complex System II. A Proceedings Volume in the Santa Fe Studies in the Sciences of Complexity*, Perseus Books, p. 239 262, 1997.

[LAN 88] LANGTON C.G., *Artificial Life*, Addison-Wesley, Boston, 1988.

[LAN 91] LANGTON C.G., "Life at the edge of chaos", in C.G. Langton, C. Taylor, J.D. Farmer, S. Rasmussen (ed.) *Artificial Life II*, Boston (SFI Studies in the Sciences of complexity), vol. X, Addison-Wesley, p. 41-91, 1991.

[LAN 97] LANGLOIS A., PHIPPS M., *Automates cellulaires: application à la simulation urbaine*, Hermes, Paris, 1997.

[LEV 65] LEVY A., *Psychologie sociale, textes fondamentaux*, Dunod Paris, 1965.

[LOR 79] LORENTZ E., "Predictability: does the flap of a butterfly"s wings in Brazil set up a tornado in Texas?", in *The Essence of Chaos,* The Jessie and John Danz Lecture Series, University of Washington Press, 1993.

[LÖS 54] LÖSCH A., *The Economics of Location*, Yale University Press, 1954.

[MAN 04] MANDELBROT B., *Fractals and Chaos: The Mandelbrot Set and Beyond*, Springer, New York, 2004.

[PAG 98] PAGE S. E., "On the emergence of cities", Santa Fe Institute, Working Paper 98-08-075, 1998.

[PAR 25] PARK R., BURGESS E., McKENZY R., *The City*, University of Chicago Press, Chicago, 1925.

[PRI 79] PRIGOGINE I., STENGERS I., *La nouvelle alliance, Métamorphose de la science,* Gallimard, Paris, 1979.

[PRI 88] PRIGOGINE I., STENGERS I., *Entre le temps et l'éternité*, Fayard, Paris, 1988

[PUM 97] PUMAIN D., "Pour une théorie évolutive des villes", *L'espace Géographique*, no. 4, p. 119-134, 1997.

[RAN 02] RANDY H., *Integrating Geographic Information Systems and Agent-Based Modeling Techniques for Simulating Social and Ecological Processes*, Santa Fe Institute, Oxford University Press, 2002.

[RES 94] RESNICK M., "Beyond the centralized mindset", *Journal of the Learning Sciences*, vol. 5, no. 1, p. 1-22, 1994.

[REY 87] REYNOLDS C., "Flocks, herds and schools: a distributed behavioral model", *Proceeding of SIGGRAPH'87*, 1987.

[SAN 92] SANDERS L., *Système de villes et synergétique*, Anthropos, Paris, 1992.

[SAS 03] SASAKI Y., BOX P., "Agent-based verification of von Thünen's location theory", *Journal of Artificial Societies and Social Simulation*, vol. 6, no. 2, 2003.

[SCH 78] SCHELLING T., *Micromotives and Macrobehavior*, Norton, New York, 1978.

[SIM 69] SIMON H., *Les sciences de l'artificiel*, Folio essais, Paris, 2004 (1st Ed. 1969).

[TES 02] TESFATSION L. (2002b), "Agent-based computational economics: growing economies from the bottom up", *Artificial Life*, vol. 8, no. 1, p. 55-82, 2002.

[TRE 01] TREUIL J.-P., MULLON C., PERRIER E., PIRON M., "Simulations multi-agents de dynamiques spatialisées", in L. Sanders (ed.), *Modèles en analyse spatiale*, Hermes, Lavoisier, Information géographique et Aménagement du territoire, Paris, p. 219-252, 2001.

[VAN 00] VANBERGUE D., "Modélisation de phénomènes urbains: Simulation des migrations intra-urbaines", in Pesty S., Sayettat-Fau C. (ed.), *Systèmes multi-agents, JFIADSMA'00: méthodologie, technologie et expériences*, Hermes, Lavoisier, Paris.

[VAR 89] VARELA F., *Connaître. Les sciences cognitives, tendances et perspectives*, Seuil, Paris, 1989.

[VON 26] VON THÜNEN J. H., *Der Isolierte Staat in Beziehung auf Landwirtschaft und Nationalökonomie*, (1826), complete edition by Scumacher-Zarchlin H. (1875), Hempel und Parey, Wiegandt.

[WEB 09] WEBER A., *Über den Standart des Industrien*, Tübigen (1909); English translation: *Alfred Weber's Theory of the Location of Industries*, University of Chicago Press, Chicago, 1929.

[WEI 89] WEISBUCH G., *Dynamique des systèmes complexes: Une introduction aux réseaux d'automates*, InterEditions CNRS, Paris, 1989.

[WOL 84] WOLFRAM S., "Universality and complexity in cellular automata", *Physica D*. 10, p. 1-35, 1984.

[WOL 94] WOLFRAM S., *Cellular Automata and Complexity*, Reading, MA: Addison-Wesley.

Conclusion

The "confrontation" with practical problems, which was the purpose of Chapters 3 to 8 of this book, has been supported by the progress achieved since the 1970s by the geographic school of spatial analysis. "Analysis" may be defined as the splitting of a problem to separate it into its constituent elements and the links unifying them. Spatial analysis has been constantly based on a firm theoretical framework, which made it possible to guide the research goals and the choice of variables and processing methods. The now traditional models, which are at the disposal of geographers, are able to give an answer to social demands and are relatively easy to manage for various problems.

This theoretical and methodological background of spatial analysis has been frequently limited to the single macro-geographic scale, thus orienting the choice of variables to this single scale. It can be said that significant progress in computing processes in the early 21st century has enlarged the methodological approach, opening new fields of research, which are fundamental at this initial stage. The new tools make it possible, starting from a work on the parcels, pixels or individual behavior, to develop research on the emergence of structures and global dynamics, starting from interactions produced at a local level. They should allow a deepening in the study of retroactive effects from the highest levels down to agents and local spatial units. Geography is thus again placed at the heart of the human and social sciences, offering a bridge between micro, meso and macro-levels. Spatial analysis and simulation are hence replying to each other, allowing a qualitative progression of research into the complexity inherent in society.

On a theoretical level, instead of starting from observations we must then analyze and understand, simulations make it possible now to proceed to experiments in the same conditions as in the laboratory, starting from theories expressed at the

---

Written by Yves GUERMOND.

outset, to setting them up, to materializing them by numerical experiments, to testing different possible configurations. We may then observe the evolution of the simulated phenomenon, both to infer from it what could happen in the real world under similar conditions, and to explain the phenomena observed afterwards in a geographic context. Simulation therefore offers an alternative to the direct experiments which are almost unthinkable in human and social sciences. Simulations may then become the means to realize, in geography as in other sciences, an experimental verification of the proposed hypotheses.

The new methods explored here are part of a theoretical corpus which is changing our vision of reality. They induce us to think about our research objectives, and set a new perspective on social phenomena. If our geographic knowledge, expressed about macro and meso-geographic scales, is fundamental for validation and orientation of our research in this field, it is not fully sufficient to be exploited on sharper scales. This work must be done, and will inevitably have to go through thorough research orientated towards the construction and intelligibility of abstract models, which make it possible to build theories as strong as those developed on larger scales. It is at this very moment that a real bridge could be built between the different organizational levels present in geographic reality. This could restore the mediation links between the individual "constructor" and the "regulatory structure" formed by society, and more generally by the complex systems formed by the Earth's space, both social and natural.

List of Authors

Michel BUSSI
University of Rouen
France

Eric DAUDE
University of Rouen
France

Daniel DELAHAYE
University of Caen
France

Bernard ELISSALDE
University of Rouen
France

Yves GUERMOND
University of Rouen
France

Gilles LAJOIE
University of La Réunion
France

Patrice LANGLOIS
University of Rouen
France

Françoise LUCCHINI
University of Rouen
France

Jean-François MARY
Urban Planning Agency of Le Havre
France

Daniel REGUER
University of Le Havre
France

Thierry SAINT-GERAND
University of Caen
France

Jean-Manuel TOUSSAINT
International Health Market Trends
Rouen
France

Alain VAGUET
University of Rouen
France

Index

cultural services, 44, 47, 50-57, 60-63, 66-68

D

determinist, 8-11, 27-30
diffusion, 24, 30, 40, 45, 96-97, 103, 153, 174, 200, 217, 231, 271-272, 285, 293, 295-297, 303, 310, 317, 324-326
digital terrain model (DTM), 203, 293
distance(s), 16-21, 33, 49, 61-64, 76, 92, 102, 104-107, 109, 111, 115, 121, 126, 130-131, 138, 140, 150-151, 157, 163, 184-185, 192, 197-198, 201, 209-211, 217-219, 289, 295-296, 299, 310, 314

E

ecology
    methodological ecology, 143-146
emergence, 16, 26-27, 74, 76, 260, 262, 266, 268, 299-301, 309-310, 320
entropy, 18-19, 257
environment, 8-9, 16, 31, 34
epistemology/epistemological, 101-103, 106, 142, 243, 264
ethics, xxi, 10
European Union, 45, 50-53, 68, 82, 85-88, 102, 151, 161-163, 166-188, 268, 303
experiment/experimental, 10, 22, 30-35, 40, 74, 130, 139, 142, 168, 219, 226, 230, 260-262, 277, 302, 335-336

F

field of contact, 295
forecasting, 28, 71, 78-90, 96-97, 163, 278
Forrester, 17, 22, 76, 164, 255

fractal/fractality, 15, 30, 198-200, 211, 259
France, 23, 27, 41-45, 47, 50, 53-54, 60, 63, 65, 93, 106, 109, 115, 137, 149, 153, 167, 170, 172, 178, 183, 210, 218
function
    output function, 280-281
    spatial interaction function, 298
    transition function, 29, 279-283, 292, 296
    width function, 197-200

G

game
    cooperative games, 156-157
    game of life, 76, 279, 284, 289, 291-292, 299
Garden of Eden, 289
geometry, 15, 198, 207, 241, 269, 274, 286, 290, 297, 305, 314
geosystem, 266, 267
GIS, 3, 115, 117, 120-123, 130-131, 155, 208, 219-226, 231, 235, 239, 243, 245, 252, 268-269, 290, 304, 317
graph
    neighborhood graph, 263, 286
    outflow graph, 205-206
    transition graph, 292
    topological graph, 204-206, 241, 250, 252

H

Hägerstrand, 20, 285, 293, 295
hazard, 27, 82, 246-250
hierarchy, 31, 45, 47, 53, 60, 74, 138, 150, 163, 169, 175-176, 186-187, 195, 199, 207, 209, 224-230, 235, 238, 248, 263, 270-274, 303, 310, 313, 317-320
highlights, 43
human geography, 28, 31, 310, 313,